AF334206

EXTRAGALACTIC ASTRONOMY

EXTRAGALACTIC ASTRONOMY

GEOPHYSICS AND ASTROPHYSICS MONOGRAPHS

AN INTERNATIONAL SERIES OF FUNDAMENTAL TEXTBOOKS

Editor

B.M. McCORMAC, *Lockheed Palo Alto Research Laboratory, Palo Alto, Calif., U.S.A.*

Editorial Board

R. GRANT ATHAY, *High Altitude Observatory, Boulder, Colo., U.S.A.*
W.S. BROECKER, *Lamont-Doherty Geological Observatory, Palisades, New York, U.S.A.*
P.J. COLEMAN, JR., *University of California, Los Angeles, Calif., U.S.A.*
G.T. CSANADY, *Woods Hole Oceanographic Institution, Woods Hole, Mass., U.S.A.*
D.M. HUNTEN, *University of Arizona, Tucson, Ariz., U.S.A.*
C. DE JAGER, *The Astronomical Institute, Utrecht, The Netherlands*
J. KLECZEK, *Czechoslovak Academy of Science, Ondřejov, Czechoslovakia*
R. LÜST, *President Max-Planck Gesellschaft für Förderung der Wissenschaften, München, F.R.G.*
R.E. MUNN, *University of Toronto, Toronto, Ont., Canada*
Z. ŠVESTKA, *The Astronomical Institute, Utrecht, The Netherlands*
G. WEILL, *Service d'Aéronomie, Verrières-le-Buisson, France*

VOLUME 20

J. L. SÉRSIC

Observatorio Astronómico, Universidad Nacional, Córdoba,
CONICET, Buenos Aires

EXTRAGALACTIC ASTRONOMY

Lecture Notes from Córdoba

D. REIDEL PUBLISHING COMPANY

DORDRECHT : HOLLAND / BOSTON : U.S.A.
LONDON : ENGLAND

Library of Congress Cataloging in Publication Data 

Sérsic, J. L. (José Luis)
    Extragalactic astronomy.

    (Geophysics and astrophysics monographs ; v. 20)
    Includes index.
    1. Astronomy. I. Title. II. Series.
QB43.2.S47   520   81–23419
ISBN 90–277–1321–9   AACR2

Published by D. Reidel Publishing Company,
P.O. Box 17, 3300 AA Dordrecht, Holland.

Sold and distributed in the U.S.A. and Canada
by Kluwer Boston Inc.,
190 Old Derby Street, Hingham, MA 02043, U.S.A.

In all other countries, sold and distributed
by Kluwer Academic Publishers Group,
P.O. Box 322, 3300 AH Dordrecht, Holland.

D. Reidel Publishing Company is a member of the Kluwer Group.

Printed in The Netherlands

PREFACE

This book is an outgrowth of the notes made for the semester lectures on 'Problems of Extragalactic Astronomy' given almost annually during two decades at the Observatorio Astronómico of the Universidad de Córdoba. Shorter versions were also given at La Plata, Santiago de Chile, São Paulo, Rio de Janeiro and Paraiba. E. Scalise made a Portuguese language version of the notes and encouraged me to publish them; although my friend J. Kleczek is to be blamed for the idea of this book.

Not every subject on Extragalactic Astronomy has been touched in this book: instead I have followed those which interested me during 25 years of professional practice in this part of the world.

I acknowledge helpful suggestions from M. Pastoriza and G. Carranza, the comprehension of Director L. Milone, and the collaboration of the staff of the Observatory in Cordoba. R. Tschamler's humor and wit made light the task of producing the English version and M. Pizarro's devotedness produced the edited MS. To both of them I am in deep gratitude.

"A book is published out of necessity, otherwise the author would spend his entire life polishing the originals" was the answer given by J. L. Borges to an inquisitive journalist. These words explain why this book is so different from the lecture notes, and also from the book I was hoping for. I thank B. McCormac and the D. Reidel Publ. Co. for my salvation from Borges' inferno.

San Antonio de Arredondo,   J. L. SÉRSIC
Córdoba
December 1980

C. D. PERRINE

(1867–1951)

To the memory
of C. D. Perrine

TABLE OF CONTENTS

CHAPTER III: ACTIVE GALAXIES

CHAPTER IV: GALAXIES AND THEIR ENVIRONMENT

CHAPTER V: MEASURING THE UNIVERSE

CHAPTER VI: COSMOLOGY

CHAPTER I

FORMS AND STRUCTURES

After a few eons, biological evolution on this planet has led to a special adaptation of the sense of sight of 'homo sapiens' to the frequency range in which solar radiation is emitted most efficiently. It is not by chance that the sensibility curve to the color of human sight reaches its maximum where the solar spectrum-filtered by terrestrial atmosphere – attains its peak. During centuries the variety of 'homo sapiens' whom somebody (Strughold, 1958) – not without a smile – risked qualifying as 'erectus nocturnus montanus' has scanned the skies with or without instruments but basically using his eyes: this determined a particular image of the universe, a 'stellar' conception of it. As a consequence of the conditioning of our sight by a star, we are qualified to observe many others.

We can now understand why astronomers call a galaxy 'normal' when it is composed mainly of stars and whose radiation is almost entirely the sum of the optical radiation of individual stars plus, in some cases, a minute component coming from gaseous nebulae. The use of modern technological facilities has allowed astronomers to expand their capacity to detect radiation in frequencies different from the optical ones. This has resulted in the discovery of galaxies which emit other types of radiation in quantities comparable with and even larger than those we would expect if they were 'normal'. Some nuclei of galaxies emit much radiation in infrared and ultraviolet bands and even in X-rays. Radio galaxies emit much of their energy in radio frequencies and in most cases this emission has its sources in regions where there are neither stars nor other visible objects. Moreover objects have been discovered of characteristics significantly different from those of galaxies, such as quasars, which emit quantities of energy presumably larger than galaxies do in some or all frequencies of the spectrum.

The aim of a scheme of classification of galaxies is to put in systematic order all the variety of objects observed in the universe. Such a scheme must fulfill two conditions: it must be objective and practical. Objectivity is achieved by basing oneself exclusively on the observable features of the objects, independent of the interpretations which can be given, however reliable they may seem. In order to be practical, a system of classification must be a brief means of easy identification of the prominent features of the objects.

As we shall see next, several systems have been advanced which satisfy these conditions to a different degree.

I.1. Normal and Peculiar Galaxies

According to Payne (1925), when classifying a certain number of objects, the criteria should be selected so that the material is distributed in the most natural groups. Morgan (1951) refined this concept and defined a Natural Group as a collection of

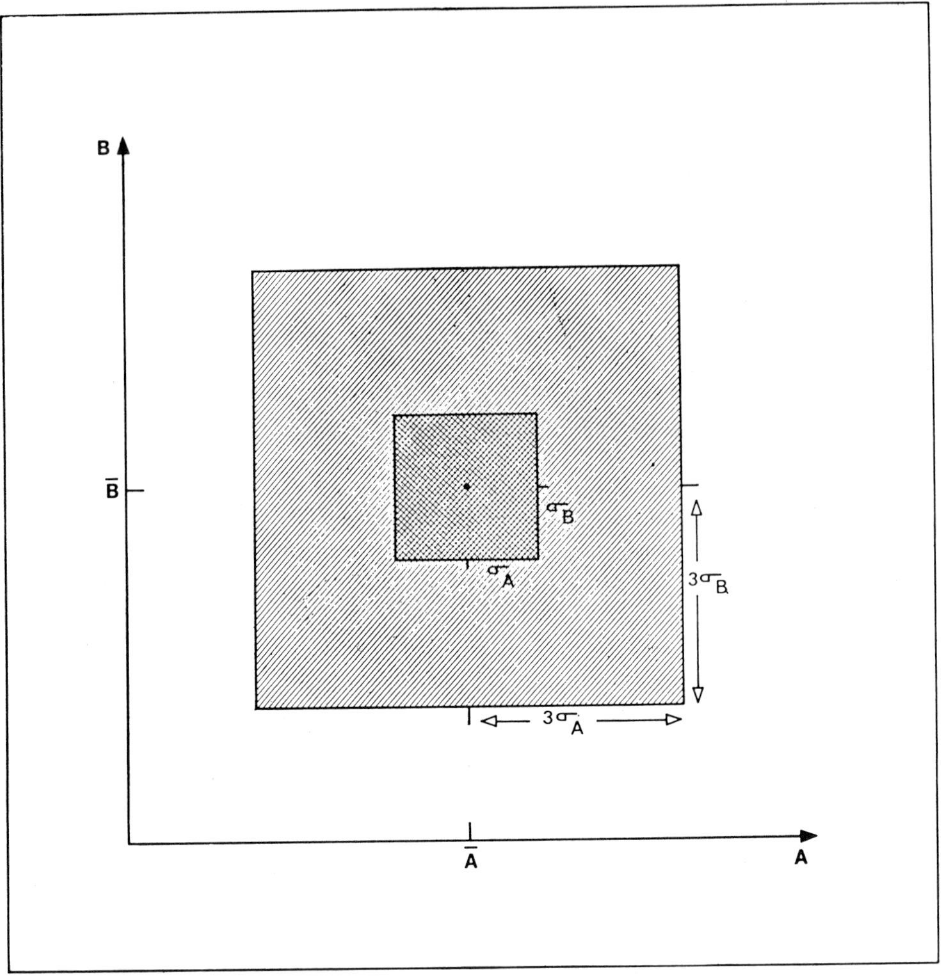

Fig. I.1.   Natural groups are defined as the dark gray area within σ_A, σ_B in the space of the parameters
   A, B. In the penumbral area (light gray) the points lie between σ and 3σ in A, B, respectively.

objects which have in common some obvious characteristics and which, at the same
time, occupy a limited area in the space of such characteristics.

   A Natural Group would thus consist of two parts: a nucleus and a zone of penumbra (Figure I.1.). The mean values $\bar{A}$, $\bar{B}$, etc. of the parameters A, B, etc. correspond
to the baricenter of the nucleus. The radius of the nucleus is of the order of the
dispersion σ_A, σ_B etc. associated with these mean values. The penumbra zone
extends from a radius σ as far as 3σ, where the probability of association of a particular case with the nucleus decreases to less than 1 %.

   Although a 'Normal Galaxy' has obvious characteristics in common with a great
number of similar ones within a certain range of variation, we are still far from being
able to put these characteristics in quantitative terms. This is why the classification

of galaxies has been founded on purely morphological criteria based on the examination of 'blue' photographic plates obtained with telescopes of a certain type. This produces in fact a selection of the sample which has repercussions on the definition of what we consider 'normal'.

In this context we may say that the elements making up a galaxy which follow a certain type of symmetry in their spatial distribution form a sub-system. This symmetry can be central, rotational, etc. Examples of sub-systems are: the halo, the disc, a bar, the nuclear region, the spiral arms taken by pairs, etc.

The subsystems can be ordered according to their importance, measured by their contribution to the surface brightness of the galaxy. This is a strictly morphological but not physical criterion since the surface brightness is the determining factor for the detectability of the subsystem on the photographic plate.

Let us say then that the 'normal galaxies' form the nucleus of a Natural Group, defined by the condition that their subsystems be concentric and coplanar. That is to say, that they have a common center of symmetry (nucleus of the galaxy) and that, if there exist planes of symmetry, these should be coincidental among themselves according to their respective order of importance. The penumbra zone of the Natural Group thus defined is populated by 'peculiar galaxies'. According to the aforesaid, a galaxy constituted by one single subsystem must be normal. Such is the case with elliptical galaxies. Irregular galaxies may also be considered normal within this scheme, as morphologically they show only the plane subsystem*. Because of their low surface brightness, its fluctuations break up the expected symmetry. Anyway, these objects are located in the periphery of the nucleus of the Natural Group under consideration.

The ultimate source of radiated energy in a normal galaxy is in the nuclear processes taking place in the interior of the stars. If a small part of the radiated energy is of non-stellar origin, this is mainly due to the fact that interstellar gas and dust redistribute the radiation towards the ultraviolet and infrared ends of the spectrum. Other parts of the galaxies also make small contributions to their spectra in X-ray and radio frequencies but they do not represent any significant contribution to the total radiation. If we add this condition to the morphological definition stated above, we deduce that a 'normal galaxy' as a physical system has to satisfy the conservation of mass and the constancy of flux of the emitted radiation. It must also be dynamically stationary along many characteristic times $(G\rho)^{-1/2}$.

I.2. Classification of Normal Galaxies

The study of the fine structure of the nucleus of the Natural Group defined in Section I.1 with the same parameters together with other additional ones is equivalent to establishing a classification system of normal galaxies.

If we use as additional criteria the number of subsystems (one or more) and their importance (high or low surface brightness), the following natural sub-groups can be distinguished at once (Figure I.2):

Elliptical Galaxies (E): Objects consisting of a single subsystem with ellipsoidal symmetry and usually high surface brightness.

---

* Under observing conditions mentioned above.

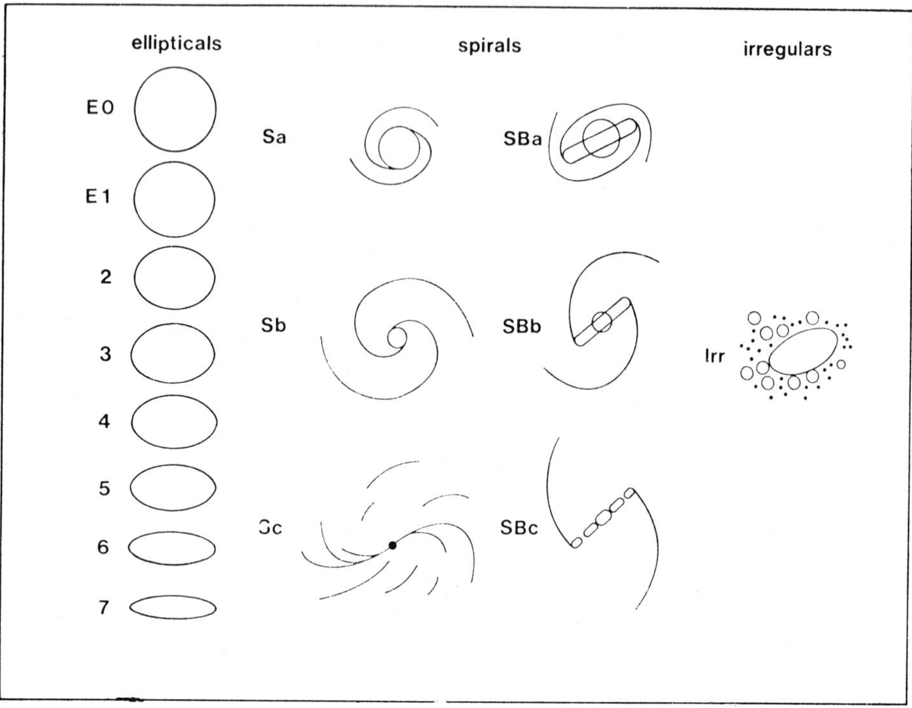

Fig. I.2.   Morphological types of galaxies.

Spiral Galaxies (S): Formed by juxtaposition of more than one subsystem. If there is rotational symmetry we say that it is pure spiral (S). If, on the contrary, there is a bar subsystem, the galaxy will be called barred (SB). Both classes of spirals show an ample range of surface brightnesses. Irregular Galaxies (I): According with what was said in Section I.1, these objects show a single plane subsystem of low surface brightness with marked fluctuations.

The succession of morphological types E–S–I is called Sequence of Galaxies or Hubble Sequence. The objects located at the left are called 'early' and those at the right 'late'.

The different systems of classification discriminate to a higher or lesser degree the three aforementioned categories. In the following we shall describe the three most used systems of morphological classification in order of complexity.

The Holmberg Classification

This system was proposed by Holmberg (1958) as the result of a comprehensive photometric study of 300 galaxies (cf. Chapter VIII). The separation of spirals in (S) and (SB) is considered unnecessary, since its author points out that all possible intermediate states between pure and barred spirals can be found. This simplistic criterion is also present in the elliptical galaxies, which are only designated as (E), as the apparent flattening can be deduced from the dimensions listed in the catalogues.

The (S) spirals are ordered according to the relative importance of the spheroidal
subsystem with respect to the disk. As to the Irregular galaxies, two types are distinguished: Irr I, coincident with subgroup (I) and Irr II, the so-called type II irregulars which, as we shall see later on, belong to the category of peculiar galaxies.

Fig. I.3.   NGC 4486 = M87. A giant elliptical galaxy in the Virgo Cluster. It has more than 2000
associated globular clusters and a peculiar feature in the nuclear region. (Section III.6)
(*Courtesy of Hale Observatories.*)

Description of the Holmberg Types

E: Elliptic galaxies irrespective of their apparent flattening (Figures I.3 and
            I.8).
            Examples: NGC 205, 4472, M32.

Fig. I.4.   NGC 4736. The Sa type in clearly shown here, with a dominant nuclear bulge and
tight spirals. (*Courtesy of the Hale Observatory*.)

SO: Object of type S, but without the spiral-arm subsystem. The spheroidal subsystem is always dominant (Figures II.10 and I.8).
Examples: NGC 1291, 3115, 5102.

Fig. I.5.   NGC 300. This galaxy is the southern counterpart of M33 = NGC 598 (Figure I.15). Notice the broad broken arms and the inconspicuous nucleus, which makes it a SC$^+$ object.
(Courtesy of CTIO.)

Sa: Coincides with Hubble's definition: galaxies with incipient spiral structure, with a dominant nuclear region.

Examples: NGC 1433, 4594 = m 104, 6753.

Sb^-:   Extensive nuclear region responsible for a considerable fraction of the total luminosity. Systems with fairly closed arms (many revolutions) without any pronounced contrast between arms and nuclear region. There is no appreciable resolution in most cases.

Examples: NGC 224 = M31, 1515, 1617, 3031 = M81.

Sb^+:   A comparatively small nuclear region. Systems with more open arms and good contrast with nuclear region. Noticeable resolution in structural details.

Examples: NGC 1097, 3953, 3992, 6744.

Sc^-:   Small nucleus, sometimes semi-stellar. Slight loss of symmetry in the open and well pronounced arms. Advanced degree of resolution (Figure I.4).

Examples: NGC 1566, 2442, 2997, 5194 = M51, 5457 = M101, 7424.

Fig. I.6.   The Small Magellanic Cloud exemplifies the late-type in the Hubble sequence of normal galaxies. (*Courtesy of Córdoba Observatory.*)

Sc⁺:   Lack of a sensible nuclear region, sometimes semi-stellar in appearance. The spiral arms appear fragmented, ill defined, and short. High resolution comparable to that of Irr I galaxies (Figure I.5).
Examples: NGC 55, 300, 598 = M33, 1313, 2403, 7793.

Irr I: There is no nuclear region. Neither is there any defined system of spiral arms. Morphologically a flat subsystem dominates with the highest degree of resolution and fragmentation. Low surface brightness (Figure I.6).
Examples: LMC, SMC, NGC 6822, IC 1613.

The simplicity of the Holmberg system allows its immediate use. It is especially adequate for the study of the intrinsic and integrated properties along the E–S–I sequence, as it is essentially a unidimensional classification.

The Hubble Classification

Historically this is the first successful classification of galaxies. Dating from 1926, it was modified and enlarged in 1936 and reached its maximum development in the Hubble Atlas of Galaxies (published by Sandage, 1961) which practically defines it.

According to Baade (1958), the simplicity and completeness of Hubble's scheme permits the inclusion of 95% of the galaxies. This argument can be reversed, saying that the nucleus of the Natural Group of galaxies, the normal galaxies, encloses 95% of the sample, leaving only 5% in the penumbra zone as peculiar galaxies.

The elliptical galaxies are designed with the letter E followed by a digit from 0 to 7, which represent the integer part of $10\,(1\text{-}a/b)$, where b, a, are the minor and major axes of the image of the galaxy.

As refers to the spirals, the Hubble classification distinguishes the pure (S) and barred (SB) ones. The classification criteria for spirals are:

(a) The relative importance of the spheroidal subsystem over the disk subsystem; and

(b) the appearance of the spiral arms, which goes from those very closed, structureless ones with more than one turn in the Sa galaxies, and through the more open and structured ones in the Sb's to the short, fragmented and well resolved ones in the Sc's.

The arms in the barred spirals (SB) start perpendicularly to the bar, while in the S galaxies they run tangentially from the nucleus, or, sometimes, from a small annular system (Figure I.7).

The Irr I galaxies show a high degree of fragmentation and low surface brightness. There are neither traces of spiral arms nor evidence of a nuclear region.

Hubble's system as well as Holmberg's also include type Irr II, to which we will refer in Section II.3.

From a morphological point of view, the SO are a transition type between the E and S galaxies. They are easily distinguishable from the ellipticals because they have a well defined disk system (Figure I.8) although they are not spirals as they lack spiral arms.

Fig. I.7.   NGC 2442. The barred structure is underlined by the dark-lanes in this low galactic latitude southern galaxy. (*Courtesy of CTIO.*)

Fig. I.8.   NGC 1549–1553, The difference of structure between Ellipticals and SO galaxies is clearly shown in this picture of the pair NGC 1549 (E) and NGC 1553(SO). (*Courtesy of Córdoba Observatory.*)

The de Vaucouleurs Classification

With this elaborated system de Vaucouleurs (1959) has succeeded in improving and enlarging the possibilities of the Hubble Classification, at the price of certain lack of practicality as a whole. de Vaucouleurs describes the details of this system which is also given in the introduction to the Second Reference Catalogue of Galaxies (Cf., Chapter VIII.1).

The main innovation de Vaucouleurs introduces in his classification takes into account omissions or minor structural details in the spirals. His most interesting contribution is the consideration of three *families* of spiral galaxies: pure spirals (SA), pure barred spirals (SB) and intermediate spirals (SAB) which partake in various degrees the characteristics of the preceding ones. With the introduction of the SAB family de Vaucouleurs overcomes the ambiguity implicit in the Hubble

Fig. I.9.   NGC 1566 is a southern Sy 1 galaxy which has a spectacular double arm system. The de Vaucouleurs type is SAB(s)bc Cf. Figure III.6.

System which admits only a bi-modal situation (S and SB). Examples of SAB galaxies: NGC 1566, 5236 = M83 (Figure I.9).

When describing the Hubble classification we have pointed out that the spiral arms start from the ends of a bar or tangentially to the nucleus or to an annular structure. de Vaucouleurs takes these situations into consideration by introducing three varieties (r, rs, s) according to whether the arms start from bars and nuclei (s) or tangentially to an annular structure (r). He also considers an intermediate situation (rs).

Examples:

$$SA(s) \quad : NGC \ 3031 = M81, \ NGC \ 224 = M31.$$
$$SB(s) \quad : NGC \ 1365, \ NGC \ 1300.$$
$$SA(r) \quad : NGC \ 4736 = M94, \ NGC \ 6744.$$
$$SB(r) \quad : NGC \ 1433, \ NGC \ 1512.$$
$$SAB(rs): NGC \ 4303, \ NGC \ 5457 = M101.$$

Finally, de Vaucouleurs (1959) also considered as an explicit element of his classification the existence of external annular structures (R). Although this feature seems not to be very frequent because its weak surface brightness, modern telescopes and cameras of small f-ratio have put in evidence the abundance of structures of this type, especially associated with lenticular galaxies and early spirals. For example: NGC 1291, (R)SBO (Figure II.10).

The sequence of galaxies (E–S–I) has also been revised and enlarged. Table I.1 shows this sequence and introduces the stages t that de Vaucouleurs uses to codify it.

As we shall see in Section II.1, the t stage correlates with various fundamental

TABLE I.1

Revised Hubble sequence

Stage t	Type T	q_0	Notes
−6	E⁻	0.33	Compact E
−5	E		Plus dE
−4	E⁺	0.29	Morgan cD's
−3	L⁻	0.26	Lenticulars
−2	L⁰	0.23	Lenticulars
−1	L⁺	0.21	Lenticulars
0	SO/a	0.19	also IO's
1	Sa	0.17	
2	Sab	0.15	
3	Sb	0.13	
4	Sbc	0.12	
5	Sc	0.11	
6	Scd	0.09	
7	Sd	0.08	
8	Sdm	0.12	
9	Sm	0.16	
10	Im	0.20	plus dIm
11	Im⁺	. .	Compact Im

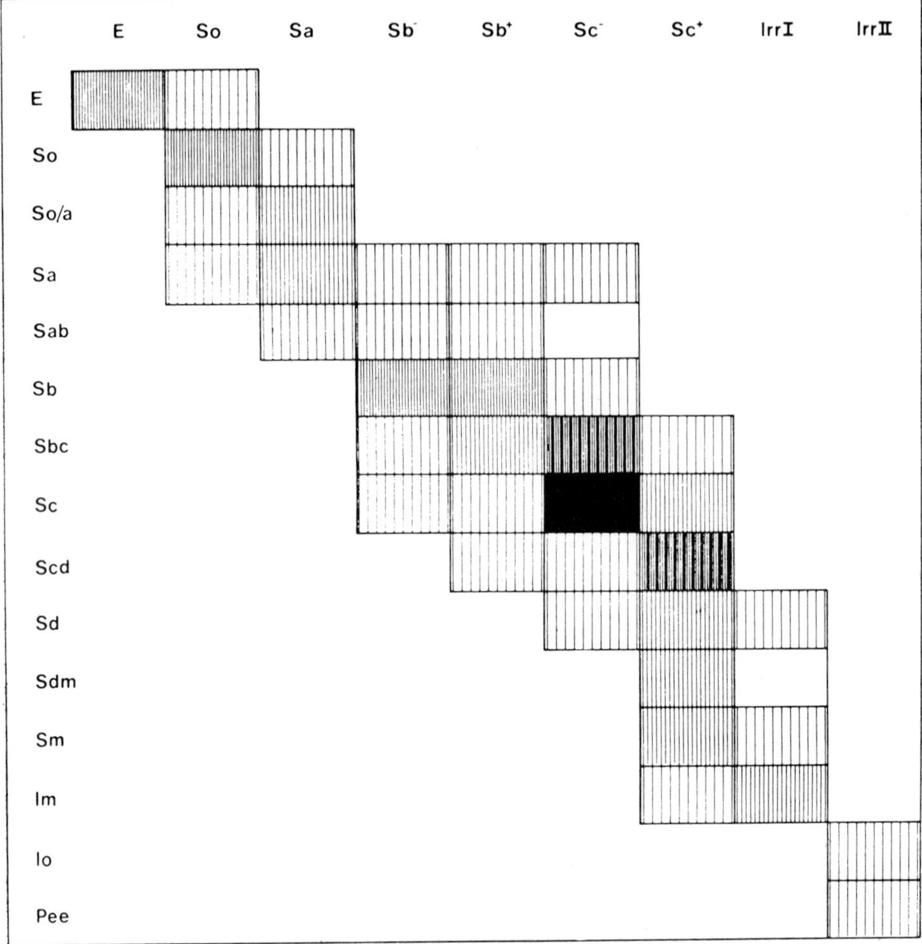

Fig. I.10.   Correlation between Holmberg's and de Vaucouleurs' classification systems. The greyness
is proportional to the number of galaxies in each cell.

physical parameters. According to the author of this classification system, the precision with which a trained observer can estimate t on plates of adequate resolution,
is measured by the standard error $\sigma = 1$, resulting from the comparison of independent classifications done by different observers (Figure I.10).

I.3. Apparent and True Flattening of Galaxies

Observations provide the ratio $q = b/a$ between the apparent diameters a, b on the
projected image of a galaxy. If we assume an ellipsoid of revolution as a first approximation to its tridimensional figure, the inclination i of the principal plane with
respect to the plane tangent to the celestial sphere is given by the formula

$$\cos^2 i = (q^2 - q_0^2)/(1 - q_0^2) \tag{1}$$

Fig. I.11.   The flattening of a late type galaxy is clearly noticeable in this plate of NGC 4244. (*Courtesy of Hale Observatories.*)

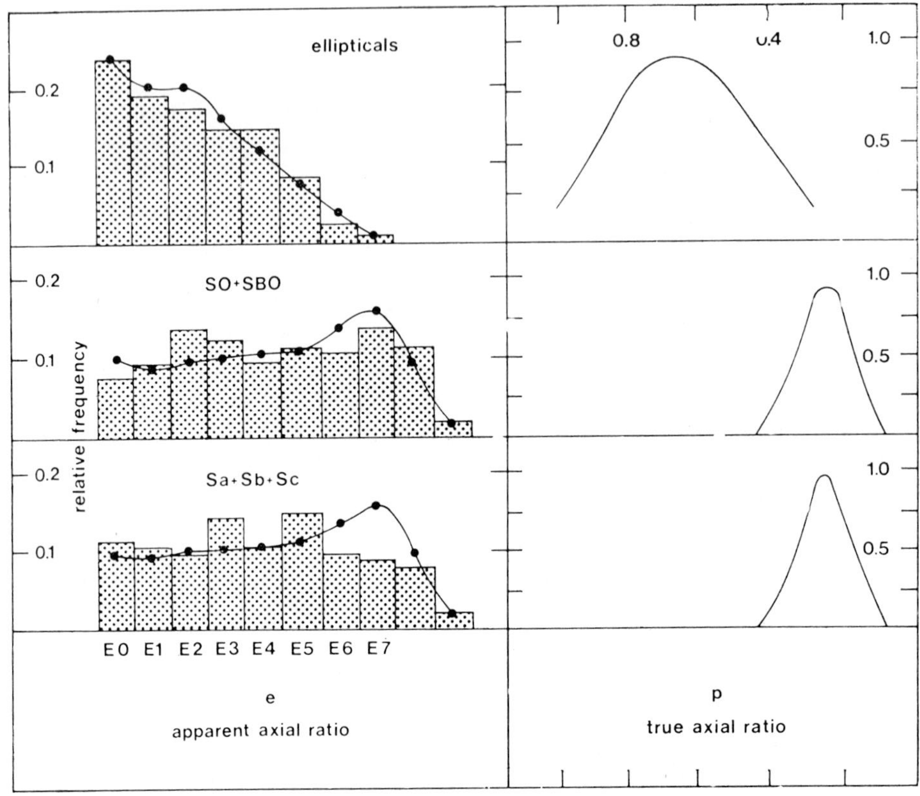

Fig. I.12.   Freguency distribution of axial ratios in galaxies. (*Adapted from Sandage et al.*, 1970.)

due to Holmberg. Here $q_0 = c/a$ is the intrinsic flattening of the ellipsoid which is assumed to represent the galaxy. Holmberg found $q_0 \approx 0.2$, and recently Heidmann et al. (1971) have estimated q_0 as a function of the morphological type along the sequence, on the basis of statistical material derived from the RCG (Table I.1). The intrinsic flattening q_0 decreases from $t = -5(E)$ to $t = 7(Sd)$ and then increases up to $t = 10$ (Im). In order to compute the inclination i with (1), Heidmann et al. (1971) recommend the use of the value q_0 specific for the morphological type of the given galaxy (Figure I.11).

The appearance of elliptical galaxies with the same luminosity differs only because of the apparent flattening q. There are no elliptical galaxies with $q < 0.3$. Those objects satisfying these conditions do not show isophotes of elliptical shape, their forms come from the superposition of a disk subsystem concentric to the spheroidal subsystem: They are the socalled lenticular or SO galaxies (Figure I.4)

The frequency curve of q for field galaxies has been discussed by several authors. Sandage et al. (1970) conclude that the intrinsic flattening of E galaxies follow a gaussian distribution with mean $\bar{q}_0 = 0.65$ and dispersion $\sigma = 0.18$, while the distribution for the SO's is very similar to that of the spirals, with $q_0 = 0.25$ and $\sigma = 0.06$ (Figure I.12).

A recent discussion of the same material used by Sandage *et al.* (1970) but including the possibility of the existence of prolate galaxies was carried out by Noerdlinger (1979). The parameters defining the preceding distributions are slightly modified but no convincing statistical evidence was obtained in favor of their existence.

According to a careful research by Rood and Sastry (1967), there is no correlation whatsoever between q and the linear dimensions of E, SO, and cD galaxies respectively, in opposition to previous conjectures.

I.4. Other Classifications Systems

There exist other systems of classification of galaxies, directed towards specific aims, such as relating the morphological characteristics with certain parameters obtained by non-morphological means. Two of these systems are renown both by their conceptual simpleness and by the success obtained in their application. They are the Yerkes classification, developed originally by Morgan and Mayall (1957) and then improved by Morgan over several years, and the system of luminosity classes (L_c) introduced by van den Bergh (1960a, b).

The Yerkes Classification

This classification shows the correlation between the appearance of galaxies and the type of population dominant in them as shown by their integrated spectra.

Morgan (1958) is only interested in the stellar population that contributes most to the integrated light which does not necessarily imply that the same should happen with its contribution to the total mass. This leads to the development of a system of spectral classification for large stellar groups starting from their integrated light. Certain classes of stars can be spectroscopically detected in an univocal way in this fashion, and thus avoid the complications of absorption and reddening produced by the interstellar medium.

The bright nuclear region of a galaxy is the most sensitive feature for discriminating the characteristics of the stellar population. The transition of the spectral characteristics from B, A, and F, up to those of type K, is accompanied by a systematic change of the appearance of that region.

Morgan introduces two parameters of classification: the 'population group' and the 'morphological family'. The first one is defined by one of the criteria of the Hubble Classification: the relative importance of the spheroidal subsystem over the disk. A sequence of seven concentration stages, namely a, af, f, fg, gk, and k is then introduced, with concentration increasing towards k. The objects classified a have a high percentage of hot stars and gas, whereas in the k group, yellow giants predominate and there is little or no trace of gas.

The morphological families are similar to the usual Hubble types: E, S, B (instead of SB), and I. Other six additional families belong to the peculiar galaxies and will be discussed in Chapter III.

Finally, for all the families a digit indicates the apparent ellipticity of the projected image of the galaxy.

Examples:

Object	Spectrum (MK)	Yerkes
NGC 224 = M31	K	gkS5
NGC 4214	OB	aI
NGC 4321	FG	fgS1
NGC 4472	K	kE2

The application of this classification technique has led to interesting conclusions. Morgan has shown that within the limitations of the spectral range used in the classification, the range of integrated spectral types and of spectral characteristics among stellar clusters closely agrees with the observed range among galaxies. When this kind of analysis is extended to systems of galaxies (Chapter IV), such as groups and clusters, Morgan finds that there are no clusters of galaxies populated with a majority of objects belonging to early population groups (a, af).

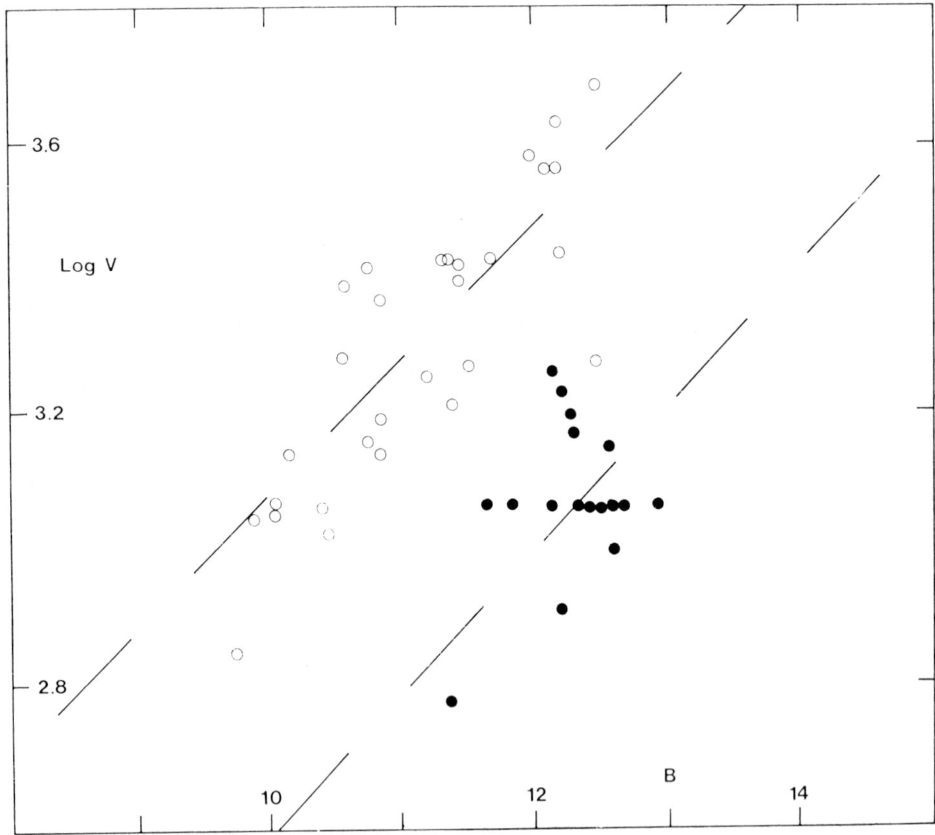

Fig. I.13.   The Hubble diagram permits to discriminate luminosity classes. In the figure (due to S. van den Bergh, 1960a, b) we see the different relationships defined by the Sb I and Sb III galaxies.

Fig. I.14.   A Sc I spiral which has played a critical role in the assessment of the scale of distances (cf., Section V.1.4). (*Courtesy of Hale Observatories.*)

Fig. I.15.   M33 = NGC 598. A normal spiral Sc II–III. Compare with Figure I.14. (*Courtesy of Hale Observatories.*)

Luminosity Classes

van den Bergh (1960a, b) has studied systematically all galaxies of the Sb, Sc, and Irr I types on prints of the Palomar Sky Survey for which apparent magnitudes m and redshifts cz were available. If $m_c = m - A$ is the apparent magnitude corrected for galactic absorption, M the corresponding absolute magnitude, and $cz = H_0 \times D$,

Hubble's law, (Chapter V) we have

$$m_c = (25 + 5 \log H_0) + M + 5 \log (cz) \tag{2}$$

if D is expressed in Mpc and H_0 in km s^{-1} Mpc^{-1}.

For a sample of galaxies with different m_0 and cz, Hubble's law leads us to a linear relation with slope 5 in the log (m_c) and m_c plane whose zero point depends on the mean absolute magnitude M of the sample. van den Bergh (1960a, b) showed that if we select the spirals (S) according to the degree of development of their spiral structure or the Irr I according to their surface brightness, the dispersion of relation (2) for this sample is reduced considerably.

In other words, the morphological characteristics of the spiral structure and the surface brightness in the irregulars are not distributed at random in the Hubble diagram but behave as luminosity indicators (Figure I.13). According to van den Bergh (1972), the luminosity class L_c of a spiral can be estimated within a factor of 2 by comparing the degree of development of its spiral structure to that of standard galaxies of known luminosity (Figures I.14 and I.15).

Through the consideration of morphological variables such as length, regularity and degree of resolution of spiral arms or surface brightness and degree of fragmentation in the case of irregulars, van den Bergh was able to distinguish up to five luminosity classes which, following the stellar nomenclature, he designed I, II, ..., V. The preliminary calibration of his sample was done assuming a value of $H_0 = 100$ km s^{-1} Mpc^{-1} (Table I.2). Then he estimated the luminosity classes L_c for some galaxies of the Local Group with known distances and simultaneously determined a corrected value of H_0 (Chapter V), and the final calibration of his system which has been published as a catalog (Chapter VIII).

TABLE I.2

Calibration of luminosity classes

Type	LC =	I	II	III	IV	V	VI
Sb	$M_p =$	−20.4	−19.4	−18.0			
Sc, IrrI	$M_p =$	−20.0	−19.4	−18.3	−17.3	−16.1	−15

The absolute magnitudes M_p are photographic and are given for $H_0 = 100$ km s^{-1}Mpc^{-1}.

de Vaucouleurs (1964) carried out a statistical analysis of the material of this catalog looking for possible systematic effects depending on declination and galactic latitude, without finding significant correlations excepting a slight excess of high luminosity systems in low galactic latitudes which he attributed to selection effects in the Shapley–Ames catalog.

The luminosity classes have been used chiefly in connection with the distance scale (Cf., Chapter V) or for studying the metagalactic velocity field (de Vaucouleurs, 1964). Various distance indicators correlate with L_c, as we shall see in Chapter V, so that this classification system has become indispensable in dealing with the problem of the scale of distances (Figure I.3).

I.5. Classification of Peculiar Galaxies

The peculiar nature of a galaxy reflects a transient non-stationary state in its evolution. This is a conclusion *a posteriori* which should not be used as a criterion in a morphological classification.

Any dynamically isolated system relaxes towards an equilibrium configuration (Chapter IV). Non-isolated systems disturb each other through tidal interactions. If this interaction is very strong or even if two galaxies collide, the structural symmetry is broken up to such a point that they are not recognizable any longer as 'normals'.

But it is also possible to alter the dynamics of an isolated system by means of the liberation of great amounts of internal energy in lapses which are short compared with its characteristic time scale $(G\rho)^{-1/2}$. This type of phenomenon is typical of 'active' galaxies (Sections III.2) and is made evident or confirmed by the study of its spectrum. In this way it is found that the liberation of internal energy not only is transferred to the mass movements, thus changing the visible structures, but also that a large part of it escapes outside as radiation (and even in some cases as matter).

We can thus divide the peculiar galaxies into two groups:

(1) Interacting galaxies (IG).

(2) Active galaxies (AG).

Now the criterion followed is no longer morphological.

An excellent collection of photographs of peculiar galaxies can be inspected in the *Atlas of Peculiar Galaxies* published by Arp (1966) (Chapter VIII).

The Irr II galaxies introduced by Holmberg (1958) belong to the AG group (Section III.2) whereas many classified originally by Hubble as Irr I are today recognized as IG. These do not need to be classified morphologically because, as de Vaucouleurs said "interacting or colliding systems form an interesting subject of study, but are not new types of galaxies". We will come back to the subject of IG's in Section IV.1.

Reciprocally some members of the AG's do not have evident morphological peculiarities and can be classified as normal galaxies.

Five percent of the galaxies are peculiar and approximately 2% show signs of activity.

References

Arp, A.: 1966, *Atlas of Peculiar Galaxies*, California Institute of Technology.

Baade, W.: 1958, *Lecture Notes*, (from tape), Harvard, College Obs.

de Vaucouleurs, G.: 1959, *Handbuch der Physik* **LIII**, 275.

de Vaucouleurs, G.: 1964, *Astron. J.* **69**, 737.

de Vaucouleurs, G.: 1974, in J.R. Shakeshaft (ed.), 'The formation and Dynamics of Galaxies', *IAU Symp.* **58**, 1.

Heidmann, J., Heidmann, N., and de Vaucouleurs, G.: 1971, *Mem. Roy. Astron. Soc.* **75**, 85.

Holmberg. E.: 1958, *Lund. Medd.* **136**.

Morgan, W.W.: 1951, *Publ. Obs. Michigan* **X**, 33.

Morgan, W.W. and Mayall. N.U.: 1957, *Publ. Astron. Soc. Pacific* **69**, 291.

Morgan, W.W.: 1958, *Publ. Astron. Soc. Pacific* **70**, 364.

Noerdlinger, P.D.: 1979, *Astrophys. J.* **234**, 802.

Payne, C.: 1925, *Stellar Atmospheres*, Harvard Coll. Obs., p. 190.

Rood, H. and Sastry, G.: 1967, *Astron. J.* **72**, 223.

Sandage, A.R.: 1961, *The Hubble Atlas of Galaxies*, Carnegie Institution, Washington, D.C.

Sandage, A.R., Freeman, K.C., and Stokes, N.R.: 1970, *Astrophys. J.* **160**, 831.
Strughold, H.: 1958, *Publ. Astron Soc. Pacific* **70**, 64.
van den Bergh, S.: 1960a, *Astrophys. J.* **131**, 215.
van den Bergh, S.: 1960b, *Astrophys. J.* **135**, 558.
van den Bergh, S.: 1972, *J. Roy. Astron. Soc. Can.* **66**, 237.

Bibliography

Curtiss, H. D.: 1918, *Lic. Obs. Bull.* **13**, 12.
de Vaucouleurs, G.: 1972, *Mem. Roy. Astron. Soc.* **77**, 1.
Gouguenheim. L., Bottinelli, L., Chamaraux, P., and Heidmann, J.: 1973, *Astron. Astrophys.* **29**, 217.
Hodge, P.: 1974, *Astron. J.* **79**, No. 11, 242.
Holmberg, E : 1946, *Medd. Lunds. Astron. Obs.* **II**, No. 117.
Holmberg, E.: 1964, *Uppsala Astron Obs. Medd.* **148**.
Hubble, E.: 1926, *Astrophys. J.* **64**, 321.
Hubble, E.: 1936, *The Realm of Nebulae*, Yale University Press.
Hubble, E. and Mayall, N.U.: 1934, *Publ. Astron. Soc. Pacific* **46**, 139.
Huchra, J. and Sargent, W.: 1973, *Astrophys. J.* **186**, 433.
Lundmark, K.: 1926, *Uppsala Medd.* **19b**, No. 8.
Morgan, W.W.: 1959, *Publ. Astron. Soc. Pacific* **71**, 394.
Rood. H. and Baum, W.: 1967, *Astron. J.* **72**, 398.
Sérsic, J.L. and Pastoriza, M.G.: 1966, *Publ. Astron. Soc. Pacific* **77**, 287.
Sérsic. J.L. and Pastoriza, M.G.: 1967, *Publ. Astron. Soc. Pacific* **79**, 152.
Sérsic, J.L., Pastoriza, M.G., and Carranza, G.J.: 1972, *Astrophys. Space Sci.* **19**, 469.
Sérsic, J.L., Bajaja, E., and Colomb, F.R.: 1977, *Astron. Astrophys.* **59**, 19.
Séarle, L. and Sargent, W.: 1972, *Astrophys. J.* **173**, 25.
van den Bergh, S.: 1959, *Publ. Dominion Observ. Ottawa* **2**, 147.
van den Bergh, S.: 1975, *J. Roy. Astron. Soc. Can.* **69**, 57.

CHAPTER II

NORMAL GALAXIES

II.1. Contents

The purpose of any classification system is to provide for a deeper understanding of the intrinsic properties of the classified objects.

The sequence of galaxies was morphologically defined (cf. Section I.2) on the basis of the forms and structures of the objects which, by definition, constitute the so called normal galaxies.

Observations other than morphologic provide quantitative and qualitative data on luminosities, color, mases, composition, etc. of both the stellar and gas components in galaxies. It is then found that most of these parameters smoothly correlate with the morphological properties used to define the sequence of galaxies, providing in this way a wealth of information of the understanding of its physical meaning.

II.1.1. PHOTOMETRIC PROPERTIES

The 10th century Persian astronomer Abd-al-Rahman-al-Sufi tells in his writings about a "small celestial cloud already observed by the Arab astronomers, not appearing however in the ancient catalogs, although their authors should had seen it as well as ourselves". Sufi was refering to M31, the galaxy in Andromeda. The first registered sight of M31 in Europe was due to Simon Marius of Franconia, who said it was seen the first time "with the help of an eye-glass" on December 15th, 1612 A.D. "The intensity of its light increases from the periphery to the centre. It looks like a candle light seen across a corny plate ..."

The first written record about the Magellanic Clouds was due to Pietro Martyr d'Anghiera, an Italian historian at the service of the Catholic Kings of Spain, circa 1512. Almost a decade before Pigaffeta's well known Diary, d'Anghiera tells "... here an there appeared between the stars whittish and bright clouds, similar to those visible in that part of the skies known as the Milky Way".

Although in these writings there is no speculation regarding the nature of the described objects, we find in them the primitive elements which later had to be used to define the photometric properties of galaxies: the surface brightness distribution described by S. Marius and d'Anghiera's estimate of it by comparison with the Milky Way.

A comprehensive 'mise a point' on the same subject four centuries later may be found in (Evans, 1979).

The photometry of extended objects (Section VIII.2) supplies information about the distribution of the specific intensity $I(m) = \text{Dex}(-0.4m)$ on the projected image of a galaxy in a given photometric system (Figure II.1).

It is useful to define a function $S(m)$ proportional to the solid angle (or area in the plate, measured in arc sec^{-2} or arc min^{-2}) subtended by all those points of the object for which the specific intensity $I(m')$ satisfies the condition $I(m') \geqslant I(m)$. The function

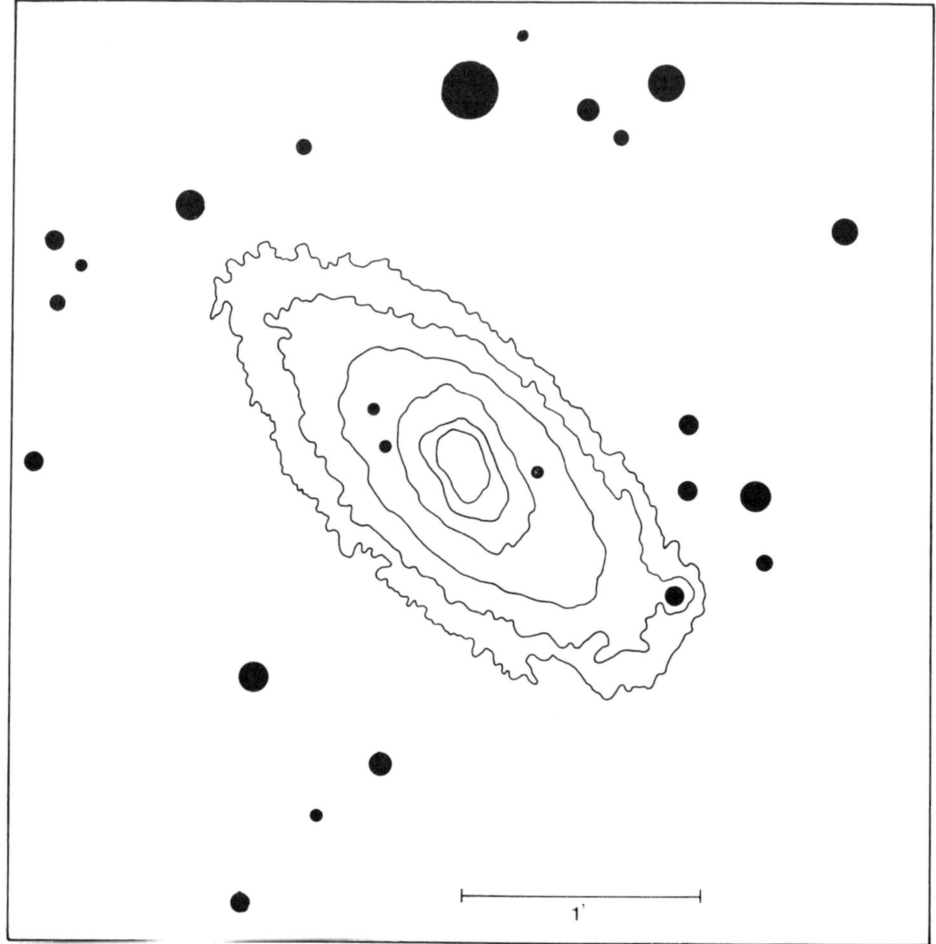

Fig. II.1. Isodensities of NGC 5253 traced on basis of a Córdoba IIIa–J plate.

$S(m)$ thus defined will then be a definite positive and monotonous increasing function of m for all $m \leqslant m_0$ and zero for $m = m_0$. The specific intensity $I(m_0)$ is the peak intensity in the galaxy. For normal galaxies m_0 varies between 15 and 21 mag arc sec^{-2}. The mean background sky brightness is $B_c = 21.5$ mag arc sec^{-2}, so that the photometry must be carried to levels quite lower than B_c to cover a sensible arc of the function $S(m)$.

Since the function $S(m)$ has been defined operationally, it generally takes the form of a numerical table for discrete values of the argument m or $I(m)$. It has been found, however, that the expression

$$S(m) = K(m - m_0)^N \tag{II.1}$$

reasonably represents the numerical function $S(m)$ for most galaxies (Sérsic, 1968) (Figure II.2).

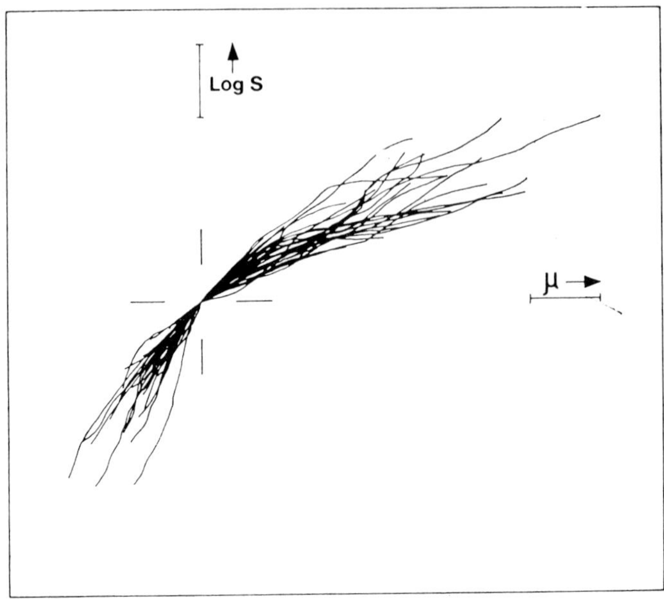

Fig. II.2.   Luminosity profiles (log$S$ vs m) for several galaxies of different morphological types.
Sérsic (1968).

Total Luminosities

The total luminosity L_T is obtained directly from observational data by means of

$$L_T = \int_0^{S'} I \, dS. \tag{II.2}$$

But if we recall that $S(m_0) = 0$ and require S' to be finite for a very large m, the partial
integration of Equation (II.2) leads to

$$L_T = 0.921 \int_{m_0}^{\infty} IS \, dm. \tag{II.3}$$

It is easily seen that the luminosity of the object up to the limit magnitude m is
given by

$$L(m) = I(m)S(m) + 0.921 \int_{m_0}^{m} IS \, dm. \tag{II.3b}$$

If a sufficiently large arc of the numerical function $S(m)$ is known, the integrals in
Equation (II.3) and (II.3b) can be calculated. The values of IS as a function of m
define a curve in the plane IS vs m (Figure II.3) which increases from zero at m_0 and
decays again to zero for large m, but whose intermediate structure contains important
information on the photometric structure of the galaxy.

The substitution of Equation (II.1) into (II.3) permits its analytical integration. If

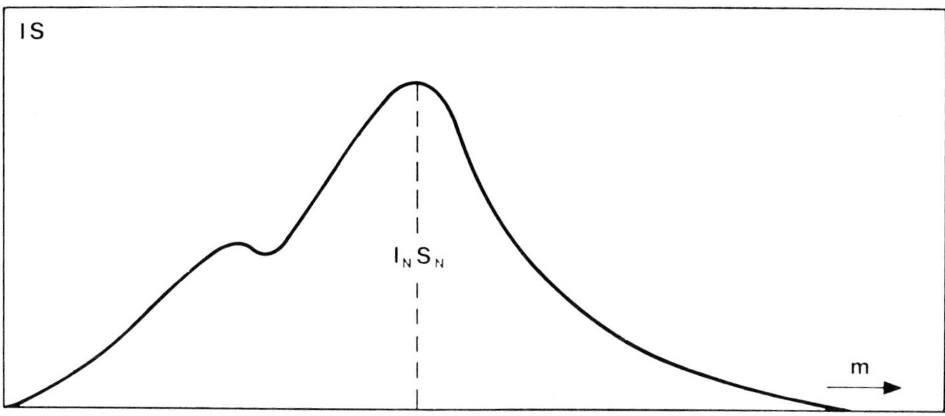

Fig. II.3.   Diagram IS vs m.

we observe now that the integrand takes the form

$$K(m - m_0)^N \, \mathrm{Dex}(-0.4m)$$

and that it has a maximum in

$$m_N = m_0 + 1.086N$$

we can use the values I_N, S_N of I and S in $m = m_N$ as parameters in substitution of m_0 and K, so that we easily obtain

$$L_T = Q_N I_N S_N. \tag{II.4}$$

The numerical coefficient Q_N is expressed by $Q_N = (e/N)^N \Gamma(N+1)$ can be tabulated (Table II.1).

If we introduce now the variable

$$\mu = N + 0.921(m - m_N)$$

and define the function $E_N(x) = 1 + (x/1!) + (x^2/2!) + \cdots + (x^N/N!)$ Equation (II.2b) can now be written

$$L(m) = L_T[1 - e^{-\mu}E_{N-1}(\mu)]. \tag{II.4b}$$

When we have integrated with Equation (II.3b) an incomplete brightness distribution, Equation (II.4b) allows us to compute L_T provided we have some idea of N.

TABLE II.1

Values of Q_N

No.	Q_N	N	Q_N
1	2.72	6	6.22
2	3.69	7	6.71
3	4.46	8	7.17
4	5.12	9	7.60
5	5.70	10	7.99

This can be done by an iterative process whenever the observed arc of $S(m)$ includes the value S_N.

The first homogeneous two-color set of magnitudes for galaxies is due to Holmberg (1958). The apparent luminosities are not total, they were computed up to the limiting isophote of 26.5 mag arc sec^{-2} in the photographic and 26.0 mag arc sec^{-2} in the photovisual ranges.

The most extensive and homogeneous set of total magnitudes in the U, B, V system is presently collected in the Reference Catalogue of Bright Galaxies (I and II) by de Vaucouleurs and de Vaucouleurs (1964), de Vaucouleurs and Corwin (1976), and de Vaucouleurs *et al.* (1976) for some 1500 galaxies.

It is possible to know N (or Q_N) for any galaxy through numerical integration of Equation (II.3) and $Q_N = L_T/I_N S_N$. The values of Q_N obtained in this way correlate with the morphological type t (Sérsic, 1968). However, the scatter in the values of Q_N is too large to be used as an acceptable quantitative substitute for t.

Since L_T is obtained through a definite photometric window, Q_N will also be dependent on it. In general $Q_N(V) > Q_N(B)$, which means the brightness distributions looks more concentrated (earlier) in the V-window than in the B one. This results from color differences between the subsystems contributing to the total light distribution on the projected image of a galaxy.

Brightness Distributions

The earliest quantitative attempts to represent the brightness distribution in galaxies is due to Reynolds (1913), who measured intensity distribution in the Andromeda galaxy. Later Hubble used Reynolds' formula for elliptical galaxies and the distribution law $I/I_0 = (a + r)^{-2}$ became known as Hubble's luminosity distribution law.

de Vaucouleurs (1948) proposed a different expression (Equation (II.7) below) which is the best known for spherical subsystems. A standard brightness distribution was also found for disk subsystems (Equation (II.8) below) which has been particularly useful in the study of this kind of objects and also in the photometric separation of subsystems.

The luminosity distributions are particular cases of the general one implied by Equation (II.1) as we will see next.

By means of the reduced variables μ and $s = S/S_N$, Equation (II.1) takes the simple form

$$S = \left(\frac{\mu}{N}\right)^N. \tag{II.5}$$

Let now $r = (S/\pi)^{1/2}$ be the equivalent radius of the isophote of the area S. The reduced equivalent radius will then be written $\rho = (S/S_N)^{1/2} = s^{1/2}$ while the reduced specific intensity $J = I/I_N$ is connected with μ by the expression $\log J = (N - \mu) \log e$. By means of Equation (II.5) we have at last

$$\log J = - N \log e \, (\rho^{2/N} - 1) \tag{II.6}$$

the reduced form for the law of luminosity distribution in the equivalent profile associated with Equation (II.1). Note that there are no free parameters, so that once N is fixed, the form of the profile is always the same.

If we make $N = 8$ in Equation (II.6) we obtain

$$\log J = - 3.474 \, (\rho^{1/4} - 1) \tag{II.7}$$

the law found by de Vaucouleurs (1948) for spheroidal subsystems, in particular for elliptic galaxies.

The $\rho^{1/4}$ law, as Equation (II.7) is also called, has been tested recently by de Vaucouleurs and Capaccioli (1979) with very precise and extensive observations of NGC 3379, an E1 galaxy adopted as photometric standard by the International Astronomical Union. It was found that Equation (II.7) represents the profile of NGC 3379 in a range of 19 magnitudes with residuals smaller than 0.08 mag. The application of Equation (II.6) to these data leads to $N = 7.936$ and equal residuals in the same interval of magnitudes.

The $\rho^{1/4}$ law has a very sharp peak in its central region. If we take into account the dispersive effect of seeing we find that the profile is still represented by Equation (II.7). In certain cases the observed luminosity peak is definitely stellar and this is interpreted as having originated in a subsystem of high compacticity (Section III.6).

If we make $N = 2$ in Equation (II.6), we obtain

$$\log \; J = - 0.868(\rho - 1) \tag{II.8}$$

the exponential luminosity law discovered by Patterson already in 1940, valid for the disk subsystems in late spirals and irregular magellanics.

Equation Equation (II.8) may also be written as an exponential law

$$I(r) = I_0 \exp(- r/r_2),$$

where $I_0 = I_2 e^2$ is the central brightness and r_N the scale of length. The central surface brightness is, in consequence, $m_0 = m_N - 2.18 = m_e - 1.82$ (see below). Freeman (1970) has shown that r_2 has a range between 0.5 and 5 kpc for types earlier than Sc but the maximum value of r_2 decreases from 5 kpc at Sc ($t = 5$) to about 1 kpc at the Im ($t = 10$) galaxies. With some exceptions, in most of the disk galaxies the values of m_0 duly corrected by galactic absorption and inclination, are concentrated around $(m_0)_c = 21.65$ mag arc sec^{-2} which is independent of the morphological type. A constant m_0 implies a correlation between total disk luminosity $L_D = Q_2 I_2 S_2 = Q_2 e^{-2} r_2^2$ and the scale of length r_2, which is in agreement with earlier photometric work (Holmberg, 1958; de Vaucouleurs, 1959).

Further work is still needed in this direction because the sample used by Freeman was small and also selection effects can be present (de Vaucouleurs, 1974).

The difference between the numerical coefficients in Equation (II.7) and (II.8) from the usual expression (de Vaucouleurs, 1974) comes from the different normalization used: de Vaucouleurs takes as units of J and ρ the 'effective' values of the specific intensity I_e and the equivalent radius r_e of the isophote enclosing one half of the total luminosity of the object. The conversion of normalization systems can be done observing that, according to Equation (II.4b) μ_e is the root of the equation

$$2E_{N-1}(\mu_e) = e^{\mu_e}$$

and whose value, within the 0.5% in the $2 \leqslant N \leqslant 8$ interval can be approximated with $\mu_e = N - 0.33$. This tells us that the effective isophote is -0.36 mag more luminous than m_N.

Other Luminosity Distributions

The *dE* galaxies associated with giant systems seem to be gravitationally truncated (Section IV.1) i.e. the presence of the intense gravitational field of the more massive system favors the escape of stars of greater velocity which is translated into an alteration of the luminosity distribution. A situation of this kind arises also in the globular clusters, for example in our Galaxy. For them King (1962) has demonstrated that the projected stellar density (and therefore the luminosity) can be presented by the expression

$$I(r) = K[(1 + r^2/r_c^2)^{-1/2} - (1 + r_t^2/r^2)^{-1/2}]^2,$$

where r_c is the core radius, r_t the tidal radius and K a scale factor. With these three parameters we can obtain a good representation of a great variety of luminosity distribution curves. Hodge (1961) has used Equation (II.9) with much success in these studies of dwarf galaxies, particularly the Sculptor and Fornax systems which are tidally limited by our Galaxy.

The application of King's formula to E galaxies has been discouraged by its author, as the fast luminosity increase in the central region cannot be represented by it. Besides, de Vaucouleurs (1979) has demonstrated that for these objects the tidal radius r_t lacks any physical sense and depends only on the limit reached by the surface photometry.

Apparent Dimensions

An often used system of apparent dimensions of galaxies is that defined by Holmberg's photometric work (1958). The catalogue gives the major (a) and minor (b) apparent diameters corresponding to the part of the galaxy contained within the standard isophote of 26.5 mag arc sec^{-2}.

A whole set of homogeneous diameters for galaxies is given in de Vaucouleurs and de Vaucouleurs (1964), Corwin (1976), and de Vaucouleurs *et al.* (1976). They were standardized to a surface brightness of 25 mag arc sec^{-2} (*B*-magnitudes). This level corresponds closely to the maximum diameters detectable by visual inspection of the plate in blue prints of the Palomar Sky Survey. From that primary set of apparent diameters, de Vaucouleurs *et al.* derived isophotal 'face-on' diameters where inclination effect is taken into account.

Finally, through correction by the galactic extinction effect the so-called 'corrected face-on' diameters are obtained.

Colors

When the photometric information regarding the brightness distribution on the projected image of a galaxy has been obtained through more than one photometric window, the local color distribution is given simply by $C(k, d') = m_k - m_{k'}$, whilst the integrate color $C_T(k, k')$ may be found through

$$C_T(k, k') = -2.5 \log(L_T(k)/L_T(k)),$$

where the indices k, k' denote two different photometric windows. $L_T(k)$ and $L_T(k')$ are obtained with Equation (II.3) and all the preceding consideration regarding the brightness distribution are valid for each case.

There are, however, few galaxies for which the surface photometry has been produced in more than one color; the larger contribution in this sense is due to Holmberg (1958). The surface photometry (unpublished) was made for 300 northern galaxies in the photographic (P) and photovisual (V) ranges. The sample was divided into the nine morphological types which constitute Holmberg's classification scheme (Section I.2). After integrating the photographic and photovisual magnitudes (limited up to the isophote of $m = 26.5$ mag arc sec^{-2}), an observed color $C = P - V$ was obtained for all objects. However, a more complete and extended information about colors of galaxies is presently based on photoelectric photometry, principally with the broad-window 3 color U, B, V-system which yields the two color indices $U - B$, $B - V$. These data are duly corrected by the aperture effect and the absorption effects respectively produced by the obscuration in our own Galaxy, and the absorbing material in the object itself. Other corrections take into account the reddening caused by redshift and the anomalous colors due to emission lines in the spectrum. The total range of color in galaxies is from 1.0 to 0.3 in $(B - V)$ and -0.2 to $+0.6$ in $(U - B)$, which is less than one magnitude in both cases. Since the measuring errors are rather large, color indices are only determined to within 10% of their range. These data are now available for some 1500 galaxies (de Vaucouleurs and Cappaccioli, 1979, and references therein).

The general properties of the corrected total colors of normal galaxies can be summarized as follows:

(i) Both $(U - B)_T^0$, $(B - V)_T^0$ corrected total color indices correlate smoothly with stage t along the Hubble sequence (Figure II.4) but the intrinsic scatter is large enough to preclude its use as a quantitative classification parameter.

(ii) A well defined two-color relationship for galaxies (Figure II.5) represents the Hubble sequence in the $(U - B)$, $(B - V)$ plane. Emission lines in the spectrum of a galaxy displace the representative point towards bluer colors along a line similar to the interstellar reddening (de Vaucouleurs, 1951).

The sequence of normal giant galaxies goes from blue for late spirals to red in the giant ellipticals. The change in color is mostly due to the increasing deficiency of the young-population stars when we move along the sequence in that sense (Section II.3). The lower half sequence in the two-color diagram is populated by the central bulges of spirals and also by ellipticals and SO galaxies, with increasing luminosities toward the red end. This is due to a luminosity effect discussed below (iv) and also in Section II.1.2.

(iii) The color distribution on the face of a galaxy show the ellipticals are only slightly redder in the center. In case of the spirals, this effect increases along the sequence reaching its maximum at the early spirals ($t = 4$) and then decreasing until $t = 9$, 10 which are bluer in the center, as one would expect from the distribution of spherical and disk subsystems in galaxies.

(iv) The correlation between color and intrinsic luminosity in early type (E–SO) galaxies was first established by Baum (1959) and later confirmed by several observers (Figure II.6) (e.g. de Vaucouleurs, 1974, and references therein). This effect seems to be absent in spirals (de Vaucouleurs, 1977). Visvanathan and Sandage (1977) have applied this luminosity effect to use E and SO galaxies as distance indicators (Section V.1.2).

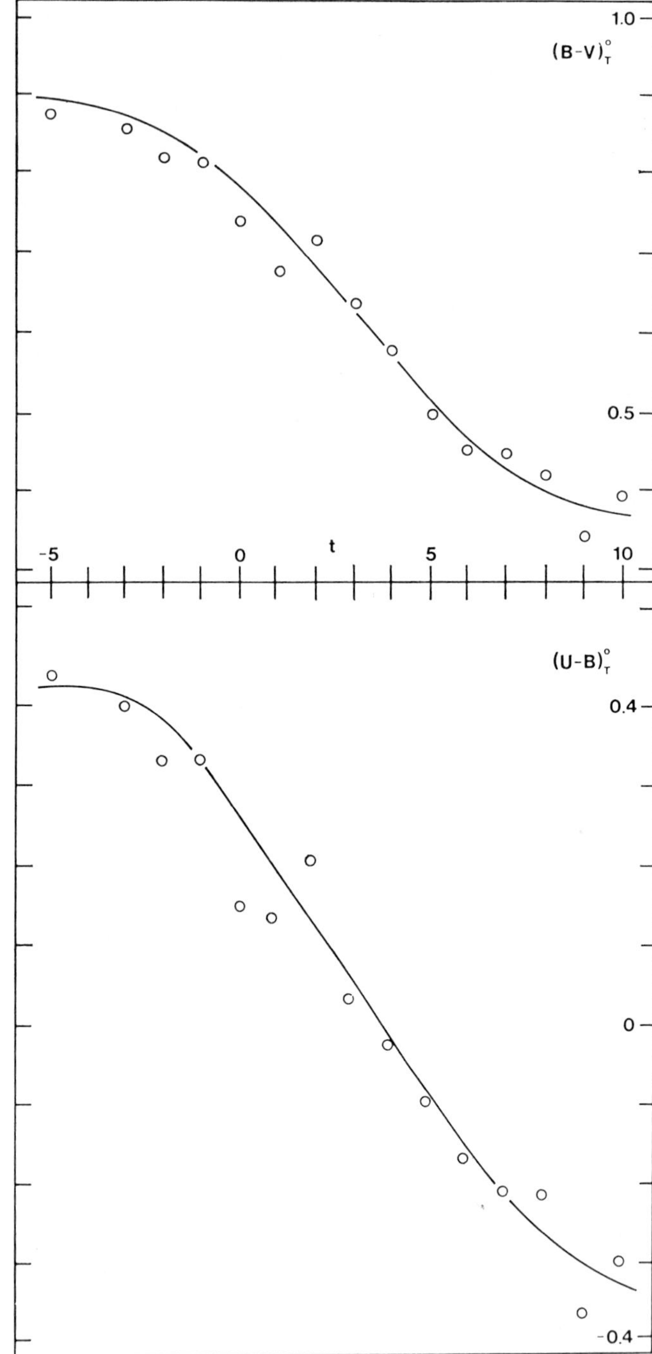

Fig. II.4.   Color Indices of galaxies as a function of stage *t*. (*After de Vaucouleurs* (1977).)

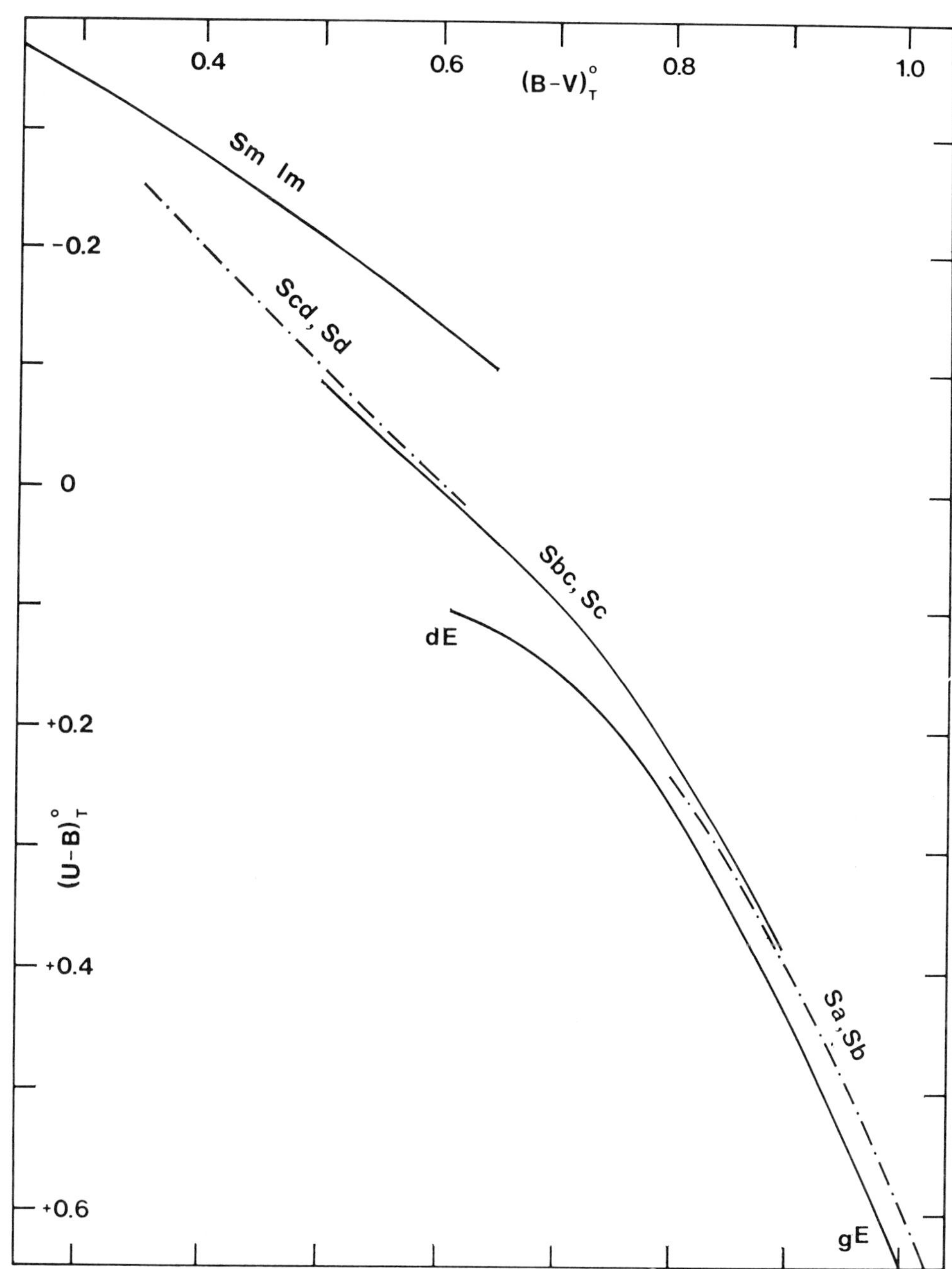

Fig. II.5.   Two-color diagram for galaxies.

The color of a particular galaxy or region within a galaxy depends on both of the types of stars contributing to the light and on their metal abundances (Section II.1.2). It is not possible, in general, to separate both effects on basis to *UBV* data alone but the use of spectoscopic information help to solve the ambiguities (Chapter VIII).

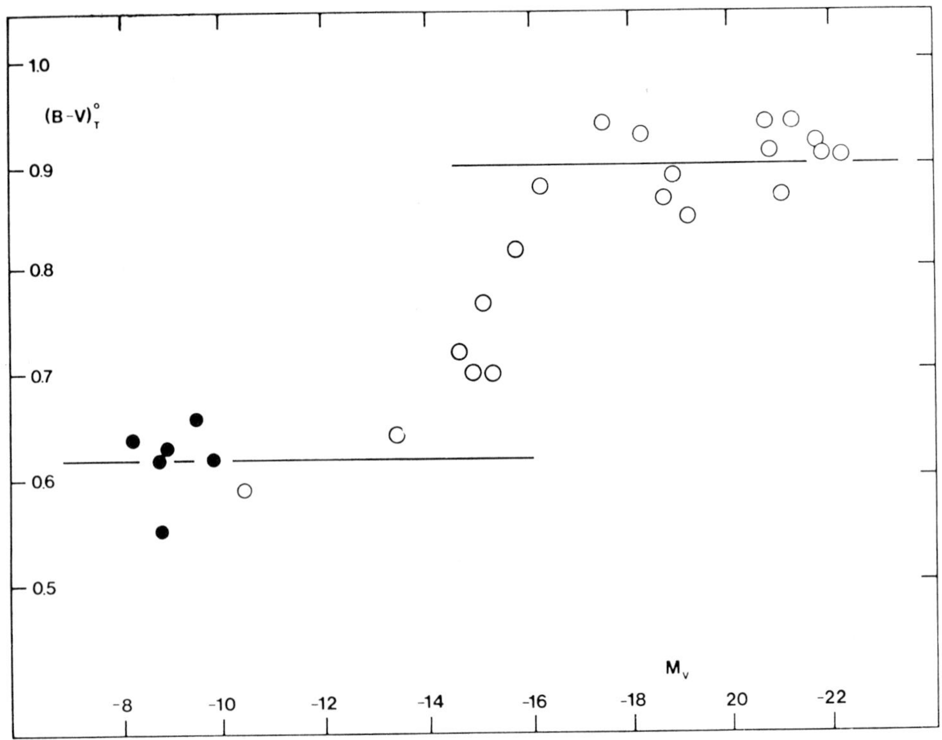

Fig. II.6.   Color-luminosity effects for early-type galaxies. (*After Baum* (1959).)

II.1.2. STELLAR CONTENT

The stellar content of a galaxy can be investigated, in principle, through the study of
its H–R diagram. The topology of the bivariate distribution of stars in the color (C)
or spectrum (S) versus luminosity (L)-plane defines the so-called stellar population
type. It is expected that the topology varies from one place to another within a galaxy:
in that way it would be possible to trace the space distribution of the population
types.

This technique has been succesfully applied to open and globular clusters, as well
as to stellar associations in our Galaxy. Since the theory of stellar evolution provides
an interpretation of the topology of the H–R diagram in terms of age and composition
of the stars belonging to a given population group, the procedure seems straightforward.

The procedure, however, suffers from severe limitations when applied to galaxies.
Although in the case of the nearest members of the Local Group, particularly the
dwarf satellites of our Galaxy, it has been succesfully applied, it can not be extended
farther away because most of the individual stars cannot be observed, either due to
their superposition or because they are too faint.

To circumvent these limitations, astronomers rely on the integrated properties of the composite radiation coming from the whole or a part of the galaxy. Both multi-color photometry or, preferably, direct continuum spectrophotometry can give, in principle, some indication on the relative contribution of stars of different temperatures to the total flux (cf. Chapter VIII). On the other hand, the absorption lines which define the spectral types of stars behave as a sensitive indicator for the elementary components in the composite radiation of a stellar system. In this argumentation lie, precisely, the foundations of the Yerkes classification (Section I.2).

The concept of stellar populations in galaxies was introduced by Baade (1944) on basis to the gross properties of the H–R diagrams he could deduce from the colors and magnitudes of their brightest resolved stars. Population I was then defined as constituted by the bright blue supergiants observed in the spiral arms, whilst the brightest Population II stars were assimilated to the red giants in globular clusters. The Yerkes classification by Morgan and Mayall lead to a revision of the original scheme by Baade and now it is accepted that the extreme populations in galaxies are a young-star rich equivalent to Baade's extreme Population I and a young-star deficient population similar to Population II but usually metal enriched. Most galaxies seem result from a mixture of these extreme populations in different proportions.

Composite Spectra

The qualitative study of composite spectra have led to important results regarding stellar populations in galaxies. Most of the galaxies so studied have been ellipticals or early-type spirals belonging to Morgan's population group k (Section I.2), which means the K stars make the dominant contribution to the spectrum in the visual region (Spinrad, 1962). There is, however, a wide variation in the intensity of the sodium D-lines within the same group; the same occurs with the Mg I blend. Because the D-lines exhibit a negative luminosity effect in late-type stars, it was concluded that the late giant stars dominate the composite visual spectrum in galaxies with weak D-lines, whilst the late dwarf stars are the main contributors in the same range for galaxies with strong D-lines. In the latter case it can be estimated that late dwarfs are over-abundant by a factor of one hundred in relation to the proportion prevailing in the solar neighborhood.

The stellar content of the nuclear and main body regions in latter type galaxies (Sb–Sc) belongs to population groups f and fg of Yerkes classification which mean a substantial contribution from stars of types earlier than k. Through the use of stellar population synthesis Turnrose (1976) was able to show that substantial intrinsic reddening exists in most of the nuclear regions studied, the existence of a significant upper-main sequence population in most of the cases, and a lack or a very heavily populated lower main sequence in all cases.

Luminosity Effects

Spinrad (1961) and Deutsch (1964) have shown that the luminous early-type galaxies have stronger lines than objects with lower luminosities. These observations were confirmed by McClure and van den Bergh (1968). A very comprehensive study of the luminosity effect in ellipticals is due to Faber (1973). According to that, elliptical galaxies form a single-parameter family with correlated colors (see Section II.1), the strength of the absorption features such as the CN bands and the Mg I blend steadily increase with luminosity.

This is in concordance with the studies by McClure *et al.* (1968) on the correlation of the reddening-free parameter $Q = (U - B) - 0.72 (B - V)$ vs luminosity of the E galaxies in the Virgo Cluster. The most luminous E galaxies are much redder than the faint ellipticals. When these studies are extended to E galaxies belonging to rich and poor clusters, similar average values for Q where found. To interpret these results, van den Bergh (1975a) proposes that either the evolutionary history of ellipticals is independent of the surrounding medium from were they formed or field ellipticals have escaped from clusters.

Composition Differences

It is easy to study the differences in composition among elliptical galaxies because their integrated colors and spectra come from old stars only. Since age here is not a significant parameter, abundance differences stand out. In spiral and irregulars, on the other hand, the properties of the integrated light depends on the contribution of younger populations and its detailed age distribution, so that the manifestations of the chemical composition are hidden behind the age-effect.

van den Bergh and McClure have shown that the stellar population typical of a metal-poor globular cluster cannot reproduce the high luminosities observed in the giant ellipticals. A fit was only possible if the dwarf stars which populate the giant ellipticals had a larger abundance of metals, by at least a factor of two with respect to the abundance in the dwarf ellipticals.

On the other hand, the stars of the dwarf ellipticals which are tidally limited are very metal-poor, as can be directly inferred from their H–R diagrams (e.g., Draco, Fornax, Sculptor). The color of these systems are similar to those of halo globular clusters, which points to the same sense (Section II.1.1).

The relation between metalicity and intrinsic luminosity (or mass) in E galaxies may be understood as a consequence of the ability the first-generation OB stars have to sweep clean the remaining gas in less massive systems: The resulting galaxy is left with the first-generation metal-poor stars. In the more massive systems the energy input from the OB stars cannot overcome self-gravitation, the gas remains in the galaxy, becomes contaminated with heavier elements through the usual process of stellar evolution (mass loss, supernovae), and further stellar formation continues. The new generations of stars now become metal-enriched and the observed correlation arises.

The preceeding argumentation does not exclude a similar scenario for spiral and irregular galaxies, since most of their mass belongs to the spherical and disk subsystems. A correlation in the same sense is also to be expected which is strongly suggested by the observation that the sequence: Galaxy, M33, LMC, SMC, and NGC 6822 is one of decreasing metalicity and also luminosity (or mass).

From a discussion of all available information on metalicity in galaxies, particularly those nearby members of the Local Group, van den Bergh (1974, 1975b) have concluded that the galaxies with high mean density exhibits a higher metal abundance than galaxies with a low mean density. This effect is also apparent within individual galaxies, where gradients of composition on show that the highest heavy element abundance also appear to be in the region with the highest density.

II. 1.3. The Interstellar Medium in Galaxies

The interstellar gas and dust represent a significant constituent in most types of galaxies. In ours it accounts for a fraction of the order of 5% of the total mass. In the irregular galaxies this fraction could amount to one third.

Several physical processes permit gathering information on the density and composition of the interstellar medium through observations in the optical and radio domains: emission of radiation in the neutral and ionized gas, absorption produced by the gas and the electrons, dispersion caused by electrons, and scattering by electrons and also by dust and gas. Other ways to explore the interstellar medium are also X-, γ-, and cosmic-rays.

Neutral Hydrogen Observations

The radio frequency observations of the 21 cm (1420 MHz) line originated in the hyperfine structure of the ground state of H I (cf. Chapter VIII) provides a wealth of information regarding the neutral gaseous component of the interestellar medium in galaxies, namely
- The total H I content.
- The H I distribution.
- The systemic radial velocity.
- The radial velocity field of the gaseous component (Section II.2.1).
- The total mass of the galaxy (Section II.2.2.)

The surface brightness of flux B is usually expressed in terms of the brightness temperature T_b given by the Rayleigh's formula $T_b = B\lambda^2/2k$, where k is Boltzman's constant, λ the wavelength of the radiation being measured, and B is expressed in erg sec^{-1} cm^2 Hz^{-1} sterad^{-1}. To relate T_b to the surface density (projected on the plane of the sky) it is generally assumed that the H I is optically thin and also that there is no interaction with the continuum radiation. In such circumstances the number of H I atoms cm^{-2} along the line of sight, at a point of coordinates θ, ϕ in the sky is given by

$$N(\theta, \phi) = 1.823 \times 10^{18} \int_{-\infty}^{+\infty} T_b(\theta, \phi, V_R) \, dV_R,$$

where the integration is extended to all radial velocities (V_R) expressed in km s^{-1}. The typical column densities N are of the order of 10^{20} to 10^{21} atoms cm^{-2}.

The total mass of H I, $\mathfrak{M}_{HI}$ is now obtained integrating $N(\theta, \phi)$ over the solid angle subtended by the H I-distribution in a galaxy,

$$\mathfrak{M}_{HI} = 1.236 \times 10^3 D^2 \iiint T_b(\theta, \phi; V_R) \, d\theta \, d\phi \, dV_R$$

solar units. D is the distance in Mpc and θ, ϕ are expressed in arc minutes. The typical values found are of the order of 10^9 solar masses.

There are three sources of error in the derivation of H I masses:

(i) Observational errors such as a deficient signal – to – noise ratio, incomplete coverage of the source, and errors in the calibration. From the comparison of measurements by different observers, a typical dispersion of 20 to 30% is found in the masses.

(ii) The distance error, arising from the uncertainties in the knowledge of D.

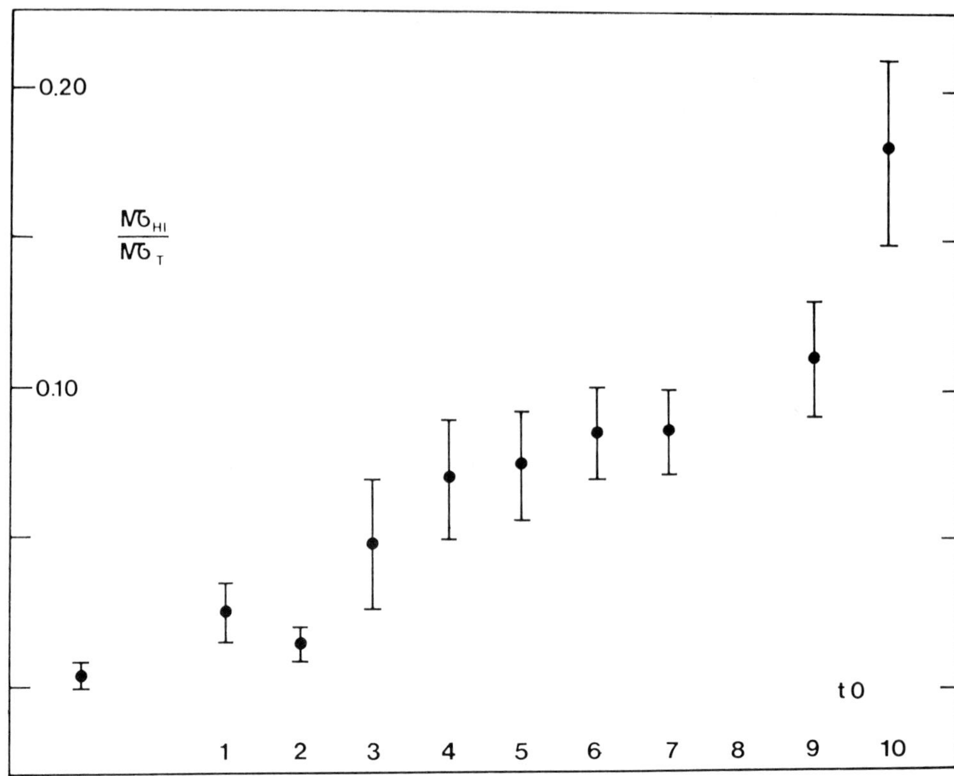

Fig. II.7.   H I to total mass ratio as function of stage *t*. (*Adapted from Roberts* (1972).)

(iii) The assumption of a small optical thickness, which is improved. In the case of our Galaxy this assumption seems adequate, although there are many regions where the optical thickness is high. Because of this, the H I masses can be underestimated. The inclination of the plane of the observed galaxy increases the optical thickness for edge-on objects in relation to those which are observed face-on, although the larger velocity spread the first case works against this effect. Heidmann *et al.* (1971) have provided tables to correct for those effects.

The Neutral Hydrogen Content along the Sequence

The ratio $\mathfrak{M}_{\mathrm{H\,I}}/\mathfrak{M}$ steadily increases from $\leqslant 0.1\%$ in ellipticals to about 25% in irregular galaxies (Figure II.7). The upper limit for ellipticals is somewhat surprising, because stars return gas to the interstellar medium in amounts which should be detectable in these objects, well above the observed limit. (Faber and Gallagher, 1976). Mathews and Baker (1971), and Coleman and Worden (1977) suggest any mass lost by the stars is blown out of the galaxy but not necessarily lost by the system as a whole: a massive non-luminous halo could retain it. From the point of view of the H I content, the SO galaxies are a transition between ellipticals and spirals, although

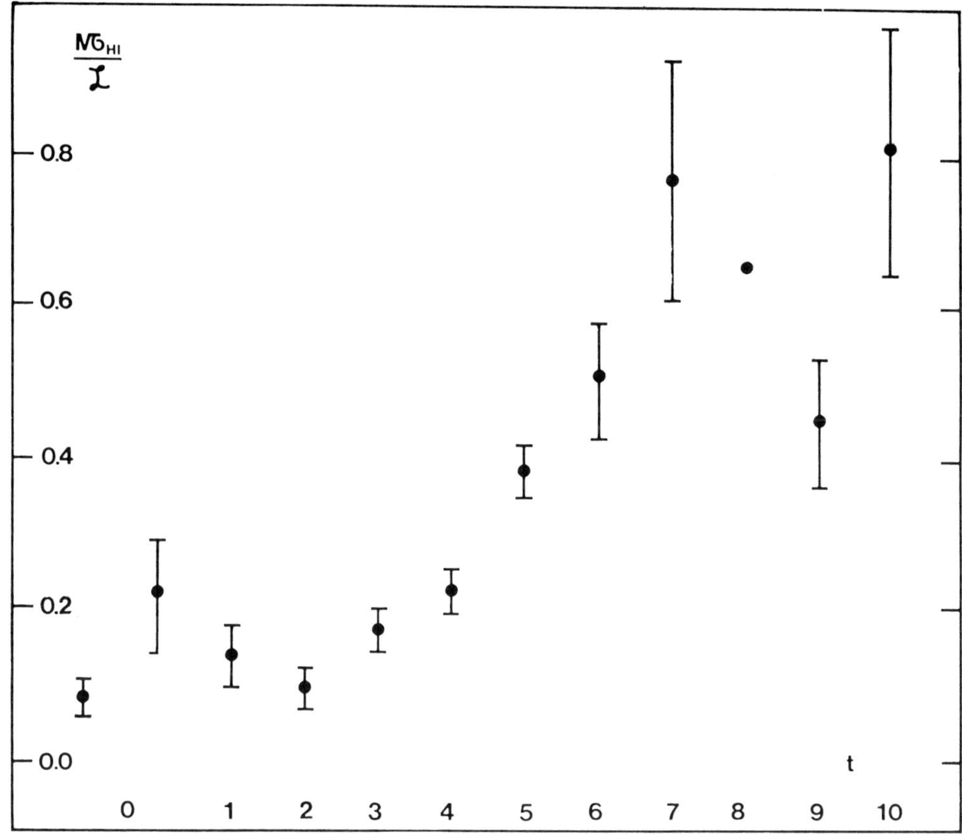

Fig. II.8. H I mass to luminosity ratio as a function of state. (*Adapted from Roberts* (1972).)

there are some individual cases whose H I abundance can be compared to that in the Sb-type galaxies.

A small number of spirals which can be considered as otherwise normal, have a noticeable deficiency in H I, i.e. NGC 613, and NGC 4321 in the Virgo Cluster. These galaxies which are underabundant in H I have been termed 'anemic' by van den Bergh (1977). It has been suggested that they are normal objects whose gaseous content has been partially removed as a consequence of the environment (Section IV.1.2).

The hydrogen mass-to-luminosity ratio $\mathfrak{M}_{HI}/\mathfrak{L}$ also is a function of the morphological type along the sequence of galaxies (Figure II.8). A constant increase of $\mathfrak{M}_{HI}/\mathfrak{L}$ towards the late-types is clearly noticed.

The $\mathfrak{M}_{HI}/\mathfrak{L}$ ratio is independent of distance (both are proportional to D^2) so that this parameter is a well determined intrinsic property. The correlation between $\mathfrak{M}_{HI}/\mathfrak{L}$ and t implies a characteristic time scale during which a galaxy keeps its present luminosity $\mathfrak{L}_0$ which converts its H into stars. According to Roberts (1963) it can be estimated that all spirals can maintain its present rate of star formation during some units of 10^{10} yr. By reversing this argument it is easy to realize that the duplication of the interstellar mass would have kept the present rate of stellar formation during a time which is of the order of the age of the galaxies. This means that the large H I abundance in late-type galaxies does not necessarily imply them to be younger, but simply that they have a lower efficiency in the conversion of gas into stars.

Neutral Hydrogen Distribution

The H I surveys of nearby galaxies show that the large-scale distribution of H I is often found displaced with respect to the optical center. These asymmetries are found in a 30% of all Sb and Sc galaxies, and are also reflected in the rotation curves (e.g. M81). The overall extension of the H I distribution is wider than the optical dimensions of the galaxy.

In spiral galaxies the main concentration of H I lies in the regions with well developed spiral arms, within the optical dimensions defined by Holmberg (1958). Early-type spirals show a minimum in the projected surface density in the central regions. Shu (1974) has suggested that this central depression could be a result of the increased stellar formation due to the more frequent passages of the gas through the density wave in the central regions.

The surface density reaches a maximum at some distance from the center which is 10 kpc in the case of M31 and 12 kpc for our Galaxy. From this maximum the density decreases with increasing distance from the center.

High-resolution studies of several nearby spirals (i.e. M31, M33, M51, M101) have detected the strong H I concentration along the spiral arms (Figure II.9). A density contrast between arm-interarm gas of 3 to 5 has been found for M81 and M51. In the M51 galaxy the H I arms are coincident with the absorption features, slightly offset from the arms delineated by the H II regions. This phenomenon is also present in M81 (Allen, 1975; Shane, 1975; Rots, 1975). The displacement between H I and H II arms has been interpreted in terms of the difussion of newly formed stars away from the shock region in the density-wave model (Roberts, 1969).

Appreciable quantities of extended low density clouds of H I at large distances (2 to 5 Holmberg radii) from the center of galaxies have been detected in nearby galaxies (Davies, 1974). In certain cases, such as M31, M33, NGC 300, and M51, the H I clouds seems to be a normal feature of galaxies, remnants perhaps, of the process of stellar formation which cannot continue there because of their low densities. The associated masses are of the order of $\leqslant 10^9$ solar masses.

The Magellanic Stream is an outstanding example of H I obviously associated by position and dynamics to the Magellanic Clouds. In this case, as well as in the M81 complex, tidal interaction between galaxies is responsible for the H I features, although material left over from earlier times cannot be excluded (cf. Section IV.1.2).

The origin of this outlying H is not yet clear, and at least four causes may be considered: gas from the galactic disk where the density is too low to allow star formation (i.e. NGC 6822), tidal interactions (cf. Section IV.1.1), activity in galactic nuclei (cf. Section III.5) and accretion from the intergalactic medium (cf. Section IV.1.2).

The Ionized Medium in Galaxies

In an early contribution Mayall (1958) discussed the frequency distribution of the [O II] λ 3727 emission, which is characteristic of H II regions the same as [N II] λ 6583,

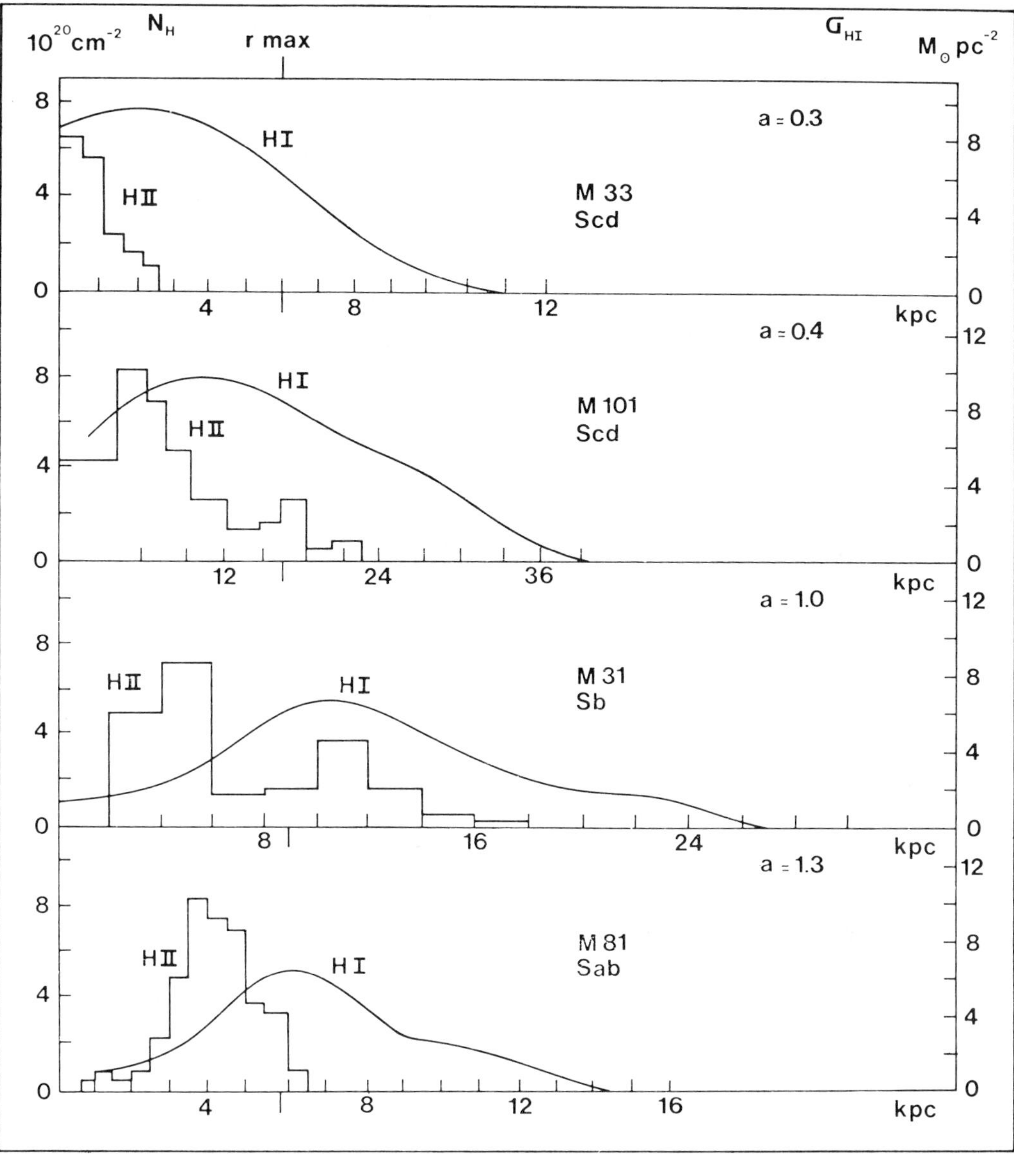

Fig. II.9.   Distribution of H I and H II in the disk of spiral galaxies. (Adapted from Trimble (1975).)

since the first ionization potential of O and that of H are closely equal, whereas that of N is only slightly larger. The ionized gas is found in all types of galaxies, including ellipticals, with frequency increasing from early – to late – types. No significant differences were found between normal and barred spirals. Page (1948, 1952) noticed, moreover, that all galaxies with emission in λ 3727 showed a stronger emission in the Hα line.

A systematic survey in the red region of the spectrum was made by Burbidge and Burbidge (1962, 1965), whose results can be summarized as follows:

– The Irr I galaxies have the largest proportion of ionized gas decreasing steadily toward earlier types of the sequence.

– On Irr I galaxies and in the nuclear as well as in spiral arm regions of the SBb and SBc and of some Sbc and Sc galaxies there must be a considerable population of massive exciting stars. The nuclei of other Sc and Sbc galaxies and those of the majority of Sb galaxies contain less ionized gas than the spiral-arm regions.

– The E, SO galaxies only show the [N II] λ 6583 line.

– Among galaxies with emission lines in the red region, the intensity ratio Hα/[N II] is ~ 3 in spiral arms, and < 1 in all ellipticals, in 81% of the S I and in 55% of the spirals. The same ratio is $\geqslant 3$ in all brighter regions of the irregular galaxies.

The intensity ratio Hα/[N II] depends on the excitation mechanism operating on the gas. For radiative processes such as excitation by the ultraviolet radiation from OB stars, Hα/[N II] ~ 3. This is the mechanism prevailing in spiral arms, irregular galaxies and also in certain nuclear complexes (cf. Section III.3). Detailed studies of H II regions in our Galaxy confirm this interpretation.

The intensity ratio Hα/[N II] $\leqslant 1$ observed in E, SO and nuclear regions of some spirals can not be explained in that way. Minkowski and Osterbrock (1959), Osterbrock (1960) have proposed that collisional excitation is responsible for it. The source of energy comes from the conversion of stellar kinetic energy into heat through degradation of turbulence in the gas located in the deep potential well of the galaxy. The late giant stars which populate these regions (Section II.1.2) eject mass which collides with the gas already settled in the central regions. The velocity of collision is comparable to the velocity dispersion of stars, several hundred km s^{-1}, and enough energy becomes available to keep the gas radiation in a very long time scale.

The H II regions are typical representatives of the interstellar medium in general. The electron density of the H II medium covers a vast range of values, from 10^{-2} cm^{-3} in the diffuse inter-arm medium to 10^4 cm^{-3} in the denses H II complexes.

They are good sources for studying the abundances of N, O, S, and He and provide information regarding to differences of chemical composition between galaxies. The observations of Peimbert and Spinrad (1970) show that the emission regions in the dwarf Irr I galaxy NGC 6822 are deffisient in N and O by factors of the order of 6 and 2 respectively, compared with the H II regions near the Sun. In its most extreme form, this phenomenon appears in the low-luminosity galaxies studied by Searle and Sargent (1972) in which some heavy elements are about 10 times less abundant than in our own Galaxy.

The He–H ratio has been measured in several galaxies, usually from emission line spectra of H II regions. Aller and Faulkner (1962) obtained $N(\text{He})/N(\text{H}) \approx 0.11$ for NGC 346 in the SMC. Johnson (1959), Faulkner and Allen (1965), and Mathis (1965) obtained $N(\text{He})/N(\text{H}) \simeq 0.13$, 0.08, 0.14 respectively for the 30 Doradus nebula in the LMC. Schmidt (1962) obtained 0.10 for the ratio in NGC 6822 which is an Irr I belonging to the Local Group. For other Irr I galaxies like NGC 4214 and IC 1569, Mathis found $N(\text{He})/N(\text{H}) = 0.1$–$0.2$ and 0.11 for NGC 604 in M33. Peimbert and Spinrad (1970) have derived the abundance ratio for He in H II regions of several galaxies, obtaining $N(\text{He})/N(\text{H}) = 0.1$ a figure similar to that found in

H II regions of our Galaxy. These values agree with the predictions of cosmological explosive models (Section VI.4).

The distribution of H II regions in galaxies has been discussed by Hodge (1974): on intermediate – and late-type spirals the surface number-density has a minimum in the central part of the galaxy and decreases to small values at about the boundary of the optical main body. The maximum density is then found in an annular region: On scales of the order of a kpc this distribution correlates with that of H I, a circumstance which has its bearing on the rate of formation of exciting stars (Section II.3.2). On scales smaller than 1 kpc, however, most H II regions are found near the edge of H I condensations (Boulesteix *et al.*, 1974). The best studied cases are those of M31 and M33 (Emerson, 1974; Israël and van der Kruit, 1974). Random asymmetry is, on the other hand, the characteristic of the H II distribution in irregular galaxies.

There is evidence for the existence of diffuse, extended H II regions between radii of 3 to 9 kpc in our Galaxy (Metzger, 1975) in correspondence with the region where most of the H II complexes are found. A similar situation has been found by Monnet (1971) in several Sc and some Sb galaxies, where the diffuse H II medium pervades the whole interarm region.

The possible use of H II diameters as distance indicators was considered by Sandage (1958). Sérsic (1959, 1960) has done an extensive study of H II regions in galaxies, and finds that their sizes and absolute magnitudes vary with the morphological type (Table II.2), from 60 pc in the Sa to a maximum of 200 pc in the Sc$^-$ galaxies. The trend is in the sense that the dimensions increase with increasing H I content (relative to the total mass) until the irregulars and dSc galaxies are reached where the falling off is connected with a luminosity effect (Chapter V).

TABLE II.2

Mean diameters and absolute magnitudes of the average
three largest H II regions in galaxies*

Galaxy type	D(pc)	M_B
Sa	60	(10.6):
Sb$^-$	90	11.1
Sb$^+$	150	11.8
Sc$^-$	200	12.1
Sc$^+$	155	11.6
Irr I	120	10.4

* Adapted from Sérsic (1959. 1960).

Sérsic (1964) has tried to relate these sizes with the rate of stellar formation of the exciting stars (Section II.3.2). He did this by relating the fraction of space filled with ionized gas to the rate of stellar formation, applying then the relation to interpret the frequency function of the H II diameters in galaxies.

Gas in molecular form, mostly H_2, probably makes a significant contribution to the total gas content of galaxies. Molecular hydrogen may be estimated directly from the ultraviolet absorption most of the data having been obtained with the Copernicus satellite. The H_2 molecule is present in clouds which are dense enough to protect themselves from the dissociative effect of the ultraviolet radiation from stars. Hydrogen molecules form when neutral atoms (H I) collide with the surfaces if interstellar dust particles.

Major contributions of CO have been identified in our Galaxy which implies gas in molecular form near the galactic center and also in a ring of 4.6 kpc from it. The total mass has been estimated as comparable with that of H I.

No galaxies are known with an H I mass larger than 10^{10} solar masses. This may be a consequence of a molecular formation in regions of high optical thickness, besides of the limiting factor imposed by stellar formation.

II.1.4. SUBSYSTEMS

A more comprehensive discussion of the properties of the subsystems which integrate the structure of a galaxy (Section I.1) can be given now through the use of observational techniques other than morphology. We describe next the properties of subsystems and their interrelations in normal galaxies.

Spheroidal Subsystems

Some elliptical galaxies have been found generally to have very large low-surface-brightness halos, whereas spirals show smaller extensions (Arp and Bertola, 1969; de Vaucouleurs, 1969). This has been confirmed by further investigations by Kormendy and Bahcall (1974) although for E and SO galaxies in poor clusters the same author (Kormendy, 1977) found that the haloes surrounding ellipticals with massive companions are not present in more isolated objects, being therefore of tidal origin.

A rather tight relationship which probably represents an intrinsic property of spheroidal systems was found between the brightness of the effective isophote, B_e and the effective radius r_e, namely $B_e = 3.02 \log r_e + 19.74$. Here r_e is in kpc and B_e in B-mag arc sec^{-2}. This means that the more luminous spheroidal systems are also more extended. Because the dispersion in B_e is only 0.28 B-mag arc sec^{-2}, this expression can be used for distance estimates.

Globular Clusters

The subsystem of globular clusters, because of its overall symmetry, belongs to the spheroidal subsystem. The projected radial density of globular clusters in galaxies (Young, 1976) follows closely the $\rho^{1/4}$ law given by Equation (II.7), which is characteristic of the spheroidal subsystems. The 'effective' radius R_c of these distributions correlates with the morphological type, from 3 kpc for our Galaxy, 7 kpc for M31, and to 100 kpc for M87, the giant E galaxy in the Virgo Cluster (Harris and Racine, 1979).

The number N_c of globular clusters increases with the luminosity of the parent galaxy (Harris and Racine, 1979)

$$\log N_c = -0.4(M_V^0 - 12.9).$$

Earlier work in this sense was done by Jaschek (1957) and Wakamatsu (1977). There is a suggestion that spirals are too bright for their number of clusters, which is in agreement with de Vaucouleurs (1958) that N_c actually correlates with the luminosity of the spheroidal component. Anyway, the relation between N_c and M_V^0 has to be taken as what it is, a statistical one: an outstanding case is the high luminosity, globular cluster under-polulated, nearby galaxy NGC 5128.

The luminosity function for globular clusters in a galaxy has been represented reasonably by a gaussian (de Vaucouleurs, 1977; Hanes, 1977) which for the Galaxy is defined through the mean $\langle M_V \rangle_{GC} = -7{.}^m3 \pm 0.1$ and a dispersion $\sigma_V = 1{.}^m2 \pm 0{.}^m1$. There are suggestions (Hanes, 1977) about the universality of this function, which would be thus of great importance for distance determinations (Section V.1.2).

The similitude of properties in the globular cluster systems associated to a wide variety of galaxies suggests their intrinsic simplicity. This is an important fact which points again to its association with the spheroidal systems.

The Nuclear Subsystems

Until recent times the observations of the nucleus of a galaxy were restricted to optical wavelengths and seeing was the limiting factor. Structures smaller than 0.5 arc sec were indiscernible, which means 1.5 pc in M31 and some 45 pc in the Virgo Cluster. The VLBI technique has made possible resolving powers of 10^{-3} arc sec or even smaller ones in radiofrequencies (Section III.6.4) but information regarding the stellar distribution and ionized gas content still remains seeing-limited. In this sense distance is the major obstacle and the reason why our knowledge of the nuclei of normal galaxies rests on the observations of nearby galaxies.

In the earlier literature (before world war II) what was referred to as nucleus is now called 'nuclear region' of a spiral galaxy, that is, the old stellar population bulge emerging over the central part of the disk and following the brightness distribution law which characterizes the spheroidal subsystem. The nucleus itself is a spheroidal-shaped well-defined centrally-placed light concentration with dimensions of the order of a few parsecs.

TABLE II.3

Radial motions in nuclear regions of normal galaxies*

Object	Type	Radial motions
NGC 224 = M31	Sb	outward
253	Sc	
2903	Sc	inward
3351	SBb	
3672	Sc	
5383	SBb	

* After Pismish (1979).

From a photometric comparison between the nucleus of M31 and the globular cluster M33, Sandage (1971) derived a density of 2×10^5 stars pc^{-3} in the few central parsecs. More detailed calculations by van den Bergh (1965) for M32 give an average density of 3×10^6 pc^{-3} within a radius of 1.5 pc around the nucleus of this galaxy.

The stellar and gaseous content has been already discussed in (Sections II.1.2 and II.1.3). As a rule, the gas motions in the nucleus partake of the rotation of the galaxy but large deviations from circularity are also observed superposed to the rotational motions. These deviations are compatible with radial motions on the plane of symmetry and also with vertical motions (Table II.3).

Disk Subsystems

The photometric decomposition of a brightness distribution into a spheroidal Equation

(II.7) and a disk Equation (II.8) component has been performed by several authors through both subjetive and objetive criteria (see e.g. de Vaucouleurs, 1977). The ratio L_D/L_B between the total luminosities of the disk (L_D) and the spheroidal component or 'bulge' (L_B) was found by Yohizawa and Wakamatsu (1975) to correlate with stage t along the sequence and being smaller for early-type galaxies. This result, although plausible, has to be considered preliminary because of the smallness of the sample used.

Burstein (1979 a, b, c) has used the photometric estimates of L_D/L_B to test the nature of SO galaxies. These objects can be interpreted either as former spirals which are now gas-free or they are true transition objects between ellipticals and spirals. For the first alternative several mechanisms have been advanced to explain the gas-stripping usually resorting to the intracluster medium or in galaxy collision. In the second case the SO galaxies are thought able to keep themselves clean of gas.

If the SO galaxies resulted from gas stripping, their basic subsystems (spheroid, disk) should be comparable to those of normal spirals. Burstein has determined L_D/L_B for several SO galaxies and found that those figures contradict the gas-stripping model. In fact, few L_D/L_B ratios in SO are as large as in M31 and no SO galaxy in the sample had L_D/L_B comparable to that of the Sc galaxies (Section IV.2.2).

Arm Subsystems

The disk- and arm-subsystems have been separated by Schweitzer (1976) in several spiral galaxies, The disks follows the exponential (Equation (II.8) law with an extrapolated central surface brightness $B_0 = 21.7 \pm 0.4$ mag. arc sec^{-2}. The disk colors were found to be very uniform, within a narrow range of values between $B - V = 0.7$ and 0.9, and their spectra dominated by late-type stars. The arm subsystem is significantly bluer than the disk, with a highly composite spectrum, as follows from the color indices. The intensity ratio between arm and disk increases outwards, giving origin to the well known bluing in the outer regions of spirals (Section II.1.1).

A broad spiral pattern was also detected in the underlying disk of spiral galaxies, which represents density variations in the surface mass density of the disk population.

Kormendy and Norman (1979) have shown, however, that many galaxies do not show the global spiral pattern required by the density-wave theory. These galaxies have instead filamentary arms (like NGC 2841) or are simply irregulars (Section II.2.1).

The Bar and its Associated Subsystems

In a detailed study Kormendy and Norman (1979) has shown that lenses and inner rings are important subsystems associated to barred spirals.

The concept of a lens subsystem is due to Sandage (1961) and is defined as an intermediate ellipsoidal feature placed between the spherical and disk subsystems. It is neatly differenciated from them by a flat gradient in the surface brightness before a sharp edge after which the gradient becomes steeper again. The stellar content of lenses is similar to that of the bars. Their intrinsic shapes seem to be triaxial ellipsoids with the major axis along the bar, whose longitude exactly fills the lens.

The presence of inner and outer rings in galaxies was explicitly aknowledged by de Vaucouleurs in his Classification System (Section I.2), but Kormendy has been the

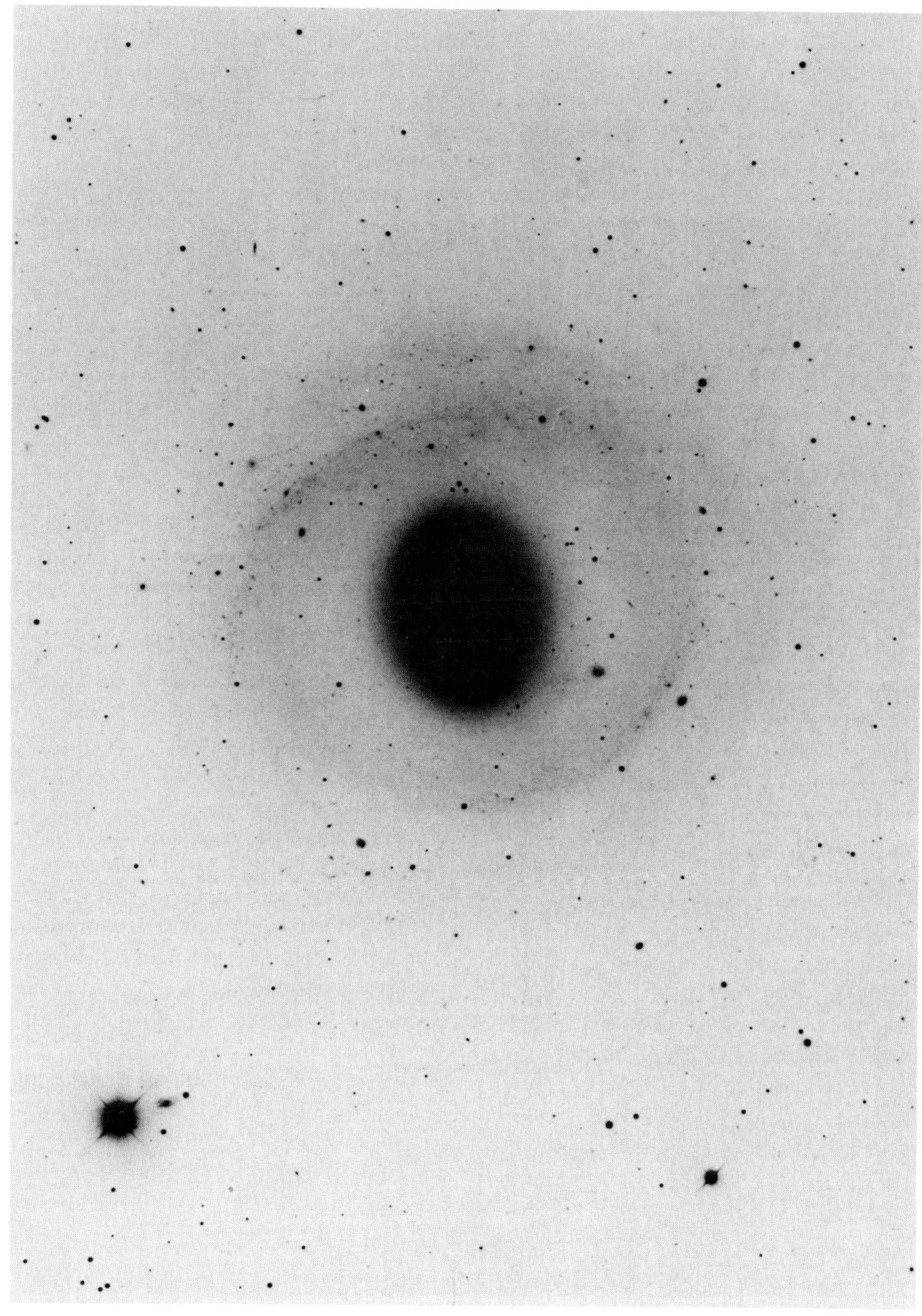

Fig. II.10.   NGC 1291 a giant SO ringed galaxy. (*Courtesy of CTIO*.)

first to throughly analyse the interrelations between these and other better known subsystems.

Lenses and inner rings occur in 54% of the SOB–SBa and 76% of SBab-SBc galaxies. The rings are scarce between early-types and almost no late-type galaxies have lenses. The inner rings and lenses have the same diameters for galaxies of equal luminosity which means the bar length is not the decisive factor to determine whether a galaxy has a ring or a lens (Figure IV.3).

Duus and Freeman (1975) have proposed that inner rings are originated from the disk material rearranged by the bar. Sérsic and Calderon (1979) have suggested the inner rings are formed by material being accreted by the galaxy, The interplay of attractive forces and Coriolis acceleration makes the ring tilted with respect to the bar (Section IV.1.2).

The outer rings on the average are noncircular and have a diameter 2.2 times the length of the bar. At this distance for a rotating bar, a potential minimum should be found and Kormendy conjectures the blue color, abundant H I content and observed structural details (like knots and filaments) are an index of stellar formation. (Figure II.10).

II.2 Motions and Masses

The internal motions in late-type galaxies are usually detected in the emission line spectrum originated in diffuse or H II regions of the ionized gas component, or the 21 cm line of H I. In the case of elliptical galaxies they have to be derived from the absorption lines of the composite stellar spectrum. This circumstance marks the essential difficulty in obtaining rotation curves for early type galaxies, because the absorption lines result from the contributions of individual stars across the volume of the galaxy whilst the emission lines in late-type systems arise from a thin gaseous disk in circular motion.

Slit spectrograms provide the velocity data on a single line, usually selected to be along the minor axis of the projected image of the galaxy. Fabry–Perot interferometer data give such information in two dimentions, covering in a single observation a large part or all the galaxy image (Figure II.11). H I observations require beamwidths at last one order of magnitude smaller than the Holmberg diameter to have enough resolving power to explore the velocity field. When the beamwidth is larger than the Holmberg radius, an integrated profile of the 21-cm line is obtained. From its half-width, masses in order of magnitude can be derived.

The knowledge of total masses, mass distributions and mass-luminosity ratios for a fair sample of galaxies is of the utmost importance to our understanding of the origin and evolution of these objects.

II.2.1. KINEMATICS

Rotation is dominant in the velocity field of most galaxies. This property seems to be valid for all galaxy types, including the irregular objects in spite of their disordered appearance (Roberts, 1975). Departures from circular motion are, however, frequent and may be large in some cases. They usually are related with conditions external to the galaxy (Section IV.1).

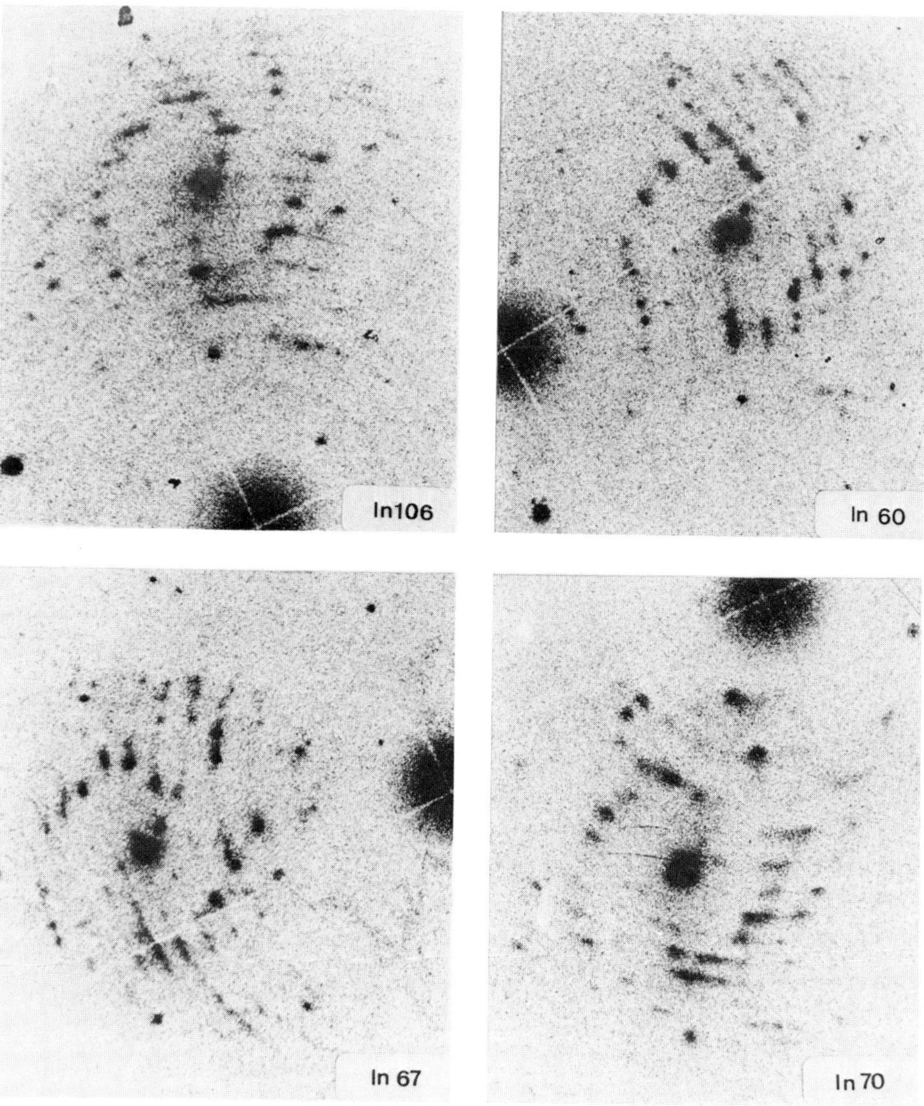

Fig. II.11. Fabry–Perot interferograms of NGC 5236 = M83. (*Courtesy of Carranza* (1972).)

The emission-line rotation curves may be seriously affected by non-rotational motions in the inner part of the galaxy, as is seen in the large asymmetries observed in M31 and M81 which reaches 100 km s^{-1}. The rotation curves in these regions might not be good indicators of the mass distribution (Faber and Gallagher, 1979). At larger distances the interpretation is not free of difficulties either, because of the smoothing produced by the side lobes of the radio telescope. The agreement between radio and optical approaches however, permits some confidence in the measurements.

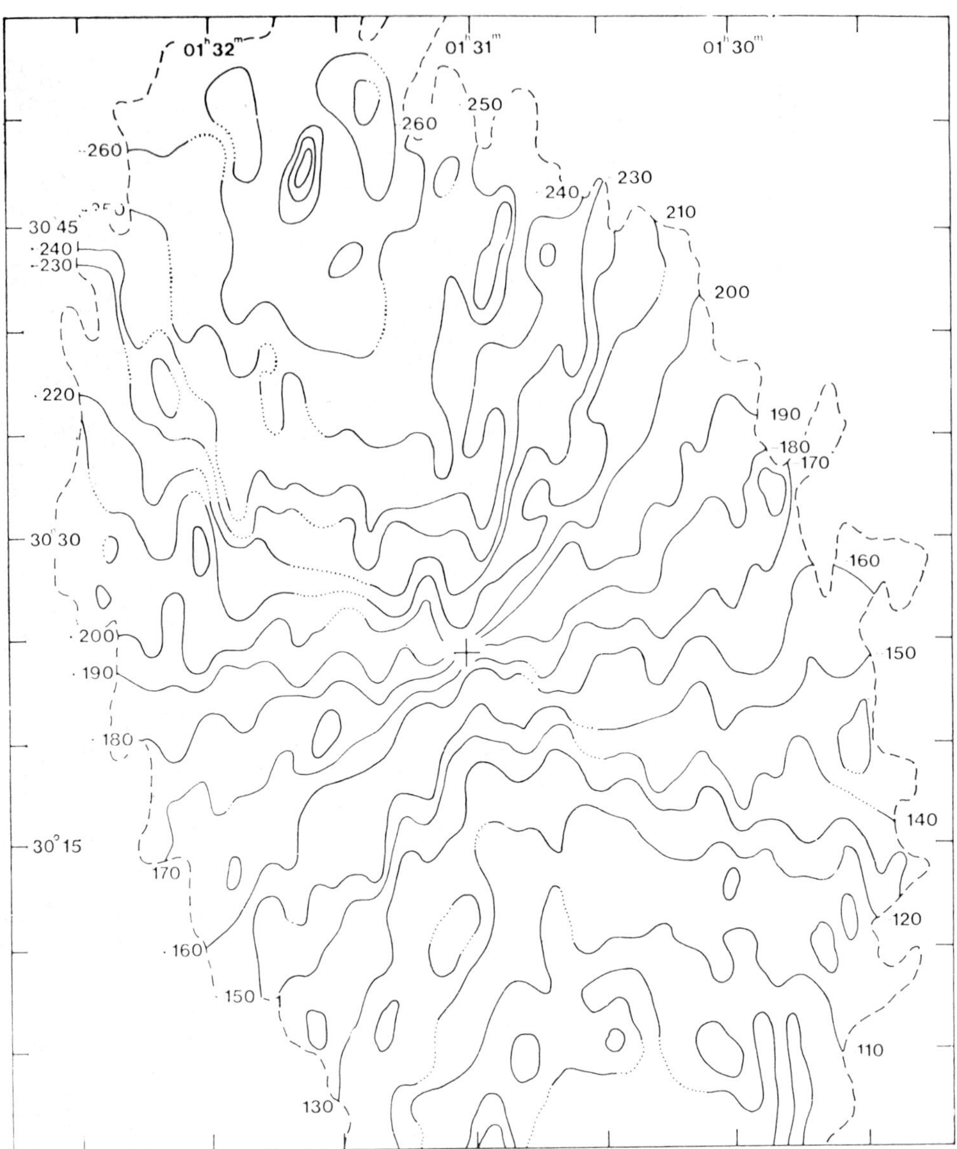

Fig. II.12.   H I velocity field of M33. (*Adapted from Warner et al.* (1973).)

Rotation Curves in Late-type Galaxies

Observations provide the radial component of the velocity field averaged along the line of sight across the galaxy. Instead of three functions depending on the three space coordinates (the velocity components), we only obtain one function (the radial component V_R) of the two variables which specify the position on the projected image of the galaxy. To make the problem tractable, some symmetry conditions must be

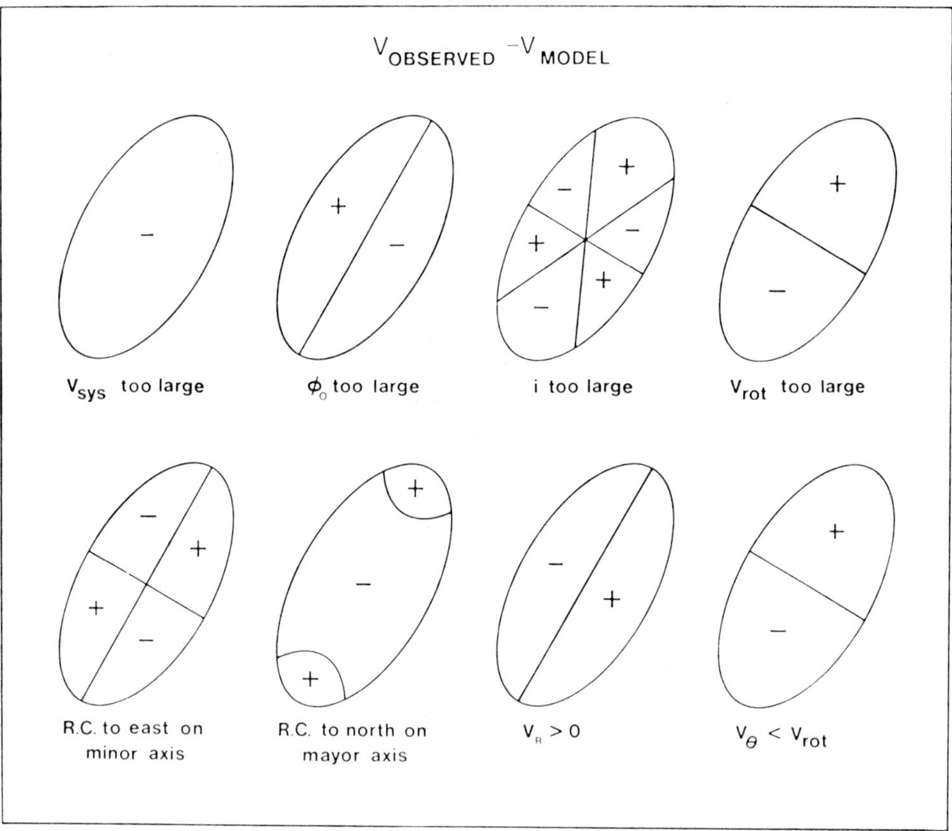

Fig. II.13. Residuals produced by different errors in the parameters defining the curve of rotation in a galaxy. (*After van der Kruit and Allen* (1976).)

imposed, namely: (i) the measured velocity refers to a thin symmetry plane where the gas component of the galaxy is found, and (ii) rotation is the principal regime of motion, non-circular deviations being assumed non-systematic and small. The only objects which approximate these conditions are spiral and irregular galaxies.

The orientation of the plane is fixed by the angle i between the symmetry axis and the line of sight, and the position angle of the line of nodes. Both can be estimated by photometric or spectroscopic means. If the isophotes of a galaxy are of elliptical shape, the position angle of the major axis should agree with that of the line of nodes and the inclination is given by the axial ratio $i = \cos^{-1}(b/a)$. In practice, this is not a simple affair: the mass distribution may be reasonably axisymmetric, but the brightness distribution is modified by the spiral arms and chaotic distribution of OB stars and H II regions in the irregulars. The isophotes are then far from elliptical, their center does not agree with the rotation center and the position angle of the major axis changes from one isophote to another.

The use of the radial velocity field all over the galaxy image is the best way to solve the problem (Figure II.12). This has been done by several authors (Warner *et al.*, 1973; van der Kruit and Allen, 1976) particularly to deal with high resolution H I measurements, but it also can be applied to Fabry–Perot data. Let R, θ be a set of polar coordinates on the plane of the galaxy, the origin is in the center of rotation and the angle θ is measured from the line of nodes. The observed radial velocity is then given by

$$V_R = V_G + V(R)\cos\theta \sin i$$

for pure rotation. V_G is the systematic radial velocity (i.e. the radial velocity of the center of rotation) and $V(R)$ is the circular velocity at a distance R from the origin.

Starting from assumed first approximation values of V_G, i, the position angle of the line of nodes, and the position of the origin, a least-square solution can be performed to determine $V(R)$ (Section VIII.4). The smooth curve drawn through the computed discrete values of $V(R)$ is then adopted as the rotation curve of the galaxy. A study of the residuals in the observed velocity field (Figure II.13) can be made to improve the solution. In this way both the orientation of the plane of symmetry, and the rotation curve are obtained.

The barred spirals have different symmetry from pure spirals. There is a symmetry plan but instead of axial symmetry we have $\rho(r) = \rho(-r)$ and $V(r) = -V(-r)$. Three angles must be known then to fix the orientation of the galaxy: besides the inclination i and the position angle of the line of nodes we need a third which fixes the position of the bar with respect to the line of nodes. From morphology and isophotes or spectroscopy no information whatsoever can be gathered with respect to these angles. The postulated stationary state applies only to the solid rotation of the bar and to the velocities with respect to a system rotating with it. Since the mass distribution in barred spirals has no axial symmetry, the same happens with the velocity field and it becomes necessary to know the radial and rotational velocity components (Section II.2.3).

Observed Rotation Curves in Late-type Galaxies

A considerable number of spiral galaxies was spectroscopically observed in the sixties by several authors (Burbidge and Burbidge, 1975) in order to measure its rotation curves and to made mass estimates. Because of instrumental limitations the mean extent of the rotation curves obtained through optical means prior to 1975 was only one third of the Holmberg radius of the galaxy (Roberts, 1975). For these curves a characteristic steep rise in rotational velocity around the nucleus in found. Mayall (1960) has given extensive data on rotation periods on basis of information gathered from this part of the rotation curve (Section VIII.5) for several galaxies. A weak dependence of periods on the morphological type was found, in the sense that the latter the galaxy, the longer the period.

Beyond the nuclear region the optical rotation curves follow a short horizontal branch after which no more data could be obtained because of the rapid decrease in the surface brightness of the galaxy. For most of the optically derived masses in the past a further hypothetical keplerian decrease of the rotational velocity was assumed (Section II.2.3). 21-cm H I observations however showed that the H extended further out (Section II.2.1) and the radial velocities led to rotation curves remaining flat to

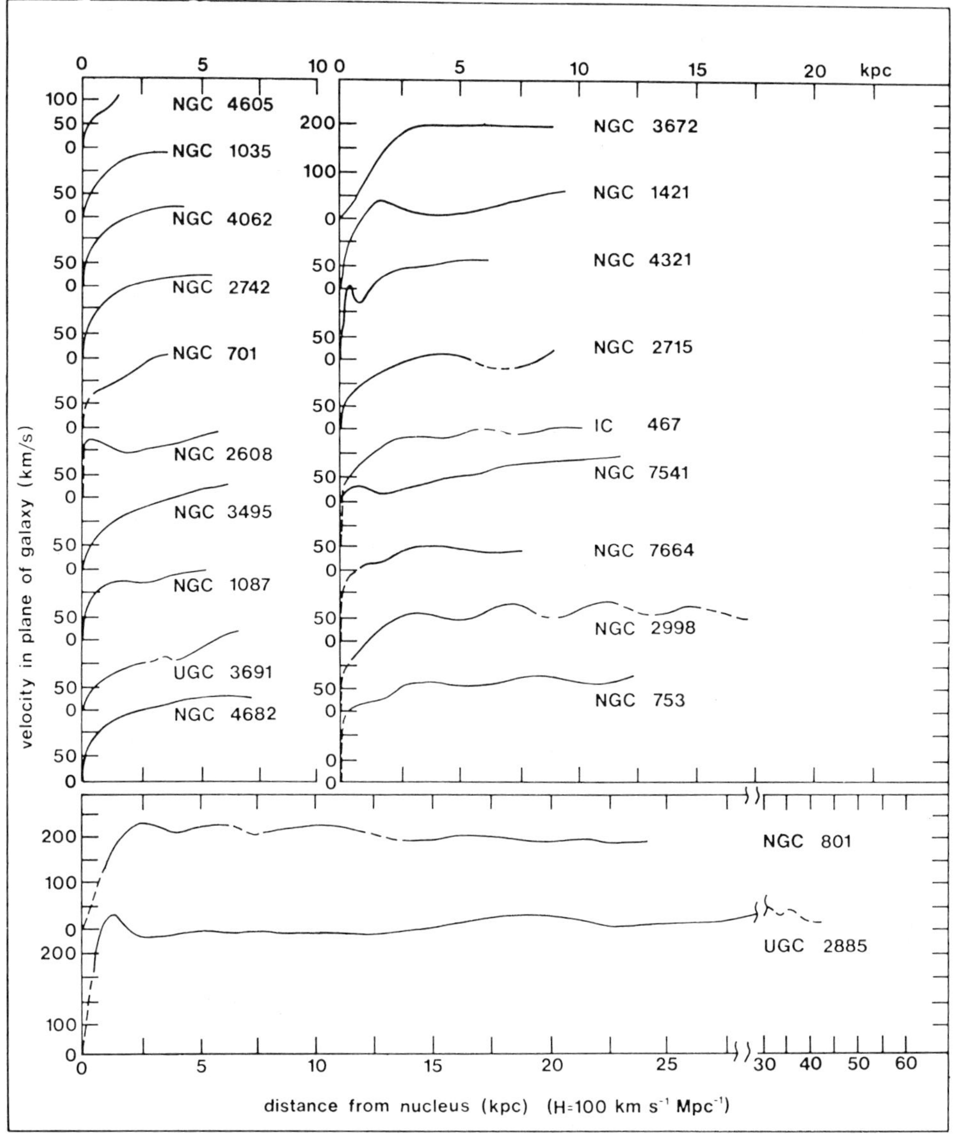

Fig. II.14.   Extended rotation curves in spiral galaxies. (*After Rubin et al.* (1979).)

the limit of the observations, which in some cases went out to 50 kpc (Roberts and Rots, 1973). In the medium-central part of the curves the agreement between the H I and H II measurement is good, with exception in the nuclear region where non-circular motions are usual. Both components of the gaseous medium behave, in consequence, in a kinematic equivalent fashion.

On the optical side, Rubin *et al.* (1979, 1980) have extended the range of the former optical observations through the use of improved spectrographs and image tubes, finding essential agreement with the 21-cm observations, The circular velocity is near constant (the rotation curve is flat) to large distances from the nucleus across the optical image of the galaxies. This means significant mass exists outside of the optical image.

Along the sequence of galaxies the maximum rotational velocity V_M decreases from 320 km s^{-1} for the Sa to about 200 km s^{-1} for the Sc galaxies (Figure II.14), which implies a decrease in density at a given R along the sequence.

The properties of the rotation curve near the nuclear region exhibit great variation between galaxies: those being small have lower velocity gradients than the larger systems. Some of them have such steep gradients that V_M is reached almost immediatly outside the nuclear region. Beyond this point, velocities are very similar for each galaxy at a given R, independently of the galaxy size. The rotation curves of all Sc galaxies investigated by Rubin *et al.* (1979, 1980) can be superposed without scaling to form a common rotation curve.

This latter property leads to some understanding of the luminosity effect found by van den Bergh which is the basis for his luminosity classification (Section I.2). Since small galaxies cover only the inner part of the standard rotation curve, the quantity $d(V(R)/R)/dR$ is large and strong differential effects produce a fragmented spiral structure (e.g. NGC 7793, $L_c = 6.5$), whilst the larger spirals extend over a region where $d(V(R)/dR)dR$ has a lower variation and the spiral structure is best preserved (e.g., NGC 5457, $L_c = 1$).

In most SB galaxies the bar rotates as a solid body, the evidence coming from both emission – and absorption – line spectra. Streaming motions along the bar can be deduced from optical observations (Pismish, 1979).

The magellanic Irr I galaxies usually present a discrepancy in position between the center of rotation and the mass center (assumed to be in the bar-like feature characteristic of these systems). This effect was first noticed in the LMC, but now is known in several other late-type galaxies. Streaming motions away from the center of the bar have been also detected.

A discussion of the available information on rotation curves with regard to whether a global spiral pattern is observed in galaxies was made by Kormendy and Norman (1979). Their data show that global spiral patterns, if present, are associated either with very differentially rotating disks or the presence of a bar of a companion galaxy. A great many galaxies have no global spiral pattern and consequently the density-wave theory does not apply to them. This is particularly true for early-type spirals, which rotate very differentially e.g. NGC 2841 whose whole disk rotates differentially and does not show any coherent spiral pattern (Figure II.15). This is consistent with the spiral pattern predicted by stochastic star formation (Gerola and Seiden, 1978). Thus, filamentary arms would be explained, whilst the density-wave theory would account only for the very regular patterns.

Departures from Circular Motion

Pismish (1979) has called attention to the fact that rotation curves in many spirals exhibit a wavy structure which seems not to be an observational artifact. She inter-prets that phenomenon as a kinematic difference between the inter-arm and arm regions which respectively denote the slower rotation of the disk and the faster one of the arm.

Fig. II.15.   NGC 2841. A tightly wound spiral galaxy. (*Courtesy of Hale Observatories.*)

The outer parts of some galaxies have radial velocity residuals, obtained after discounting the smooth rotational field, which exhibits a double symmetry around the center of rotation. The residual distribution implies a systematic change in the line of nodes, as well as of the apparent major axis of the mass distribution. These warps in the outermost H I distribution question the identification of the observed velocities with the rotational ones. For many galaxies, however, the warps can be explained on basis of the entire information available in two dimensions whilst the flat rotation curve still persists (Faber and Gallagher, 1979).

Rotation in Early-type Galaxies

There are several cases of elliptical galaxies where rotation has been detected, e.g. NGC 4621 and NGC 4697 where inclined absorption lines indicating rotation were observed by King and Minkowski (1966). Bertola (1972) has extended these observations up to the turning point of the observed curve. Capaccioli (1979) and Illingworth and Field (1979) also contributed to this field. The difficulties of these observations have been discussed by Bertola (1972): because the measurements are made on absorption lines which are very weak outside the nuclear region and blended with the night sky lines, precision is far lower than in the case of late type galaxies. Since the rotational velocity $V(R)$ is the mean velocity of stars in an element of volume at distance R from the center, in the symmetry plane of the galaxy, the observed velocity V_R has to be deconvolved from the contributions by stars in the line of sight. This presupposes a previous knowledge of the local velocity dispersion in the system. The situation is complicated when two subsystems coexist as is the case of of the SO galaxies.

A first approach to the problem (Bertola and Cappaccioli, 1975) is to assume the weighting function to be the space density of light in the galaxy, as follows from the photometric profiles, although the difficulty still remains of extrapolating $V(R)$ beyond the last observed point.

In spite of all difficulties, there is an even increasing body of data about rotation curves in early type galaxies (see e.g., Capaccioli, 1979). Elliptical galaxies seem to rotate at slower rates than the standard isotropic oblate models predict. SO galaxies rotate faster than ellipticals, their rotation curves are more structured, steeper in the central regions, which suggests that more than one subsystem contributes to the observed rotation. There is also evidence of a different kinematical behavior of the gaseous component, if any, which rotates usually faster than the stellar one.

Velocity Dispersion in Early-type Galaxies

Presently the best procedure to obtain the integrated value of the velocity dispersion in early-type galaxies and nuclear bulges of early spirals in the radial direction σ_R is to evaluate it by comparing the spectrum of the galaxy with that of a giant K-type star, similar to those which dominate the light of early type galaxies (Section II.1.2). In this way the difficulties arising from blending of the absorption lines and instrumental resolution are avoided. The first attempts in this line were due to Minkowski (1961) through an analog device which widened the star spectra in order to produce artificial dispersions which, after comparison with the un-widened galaxy spectrum allowed interpolating the value of σ_R. A numerical procedure due to Richstone and Sargent (1972) leads to substantially improved results: Let f_λ be the observed star spectrum. Assume a Maxwellian velocity distribution for the stars contributing to the integrated light of the galaxy. A numerically broadened spectrum is possible to be computed through

$$F_\lambda(\sigma_c) = \frac{C}{\lambda(2\pi\sigma_R)^{1/2}} \int_{-\infty}^{+\infty} f_\lambda \exp\left[-\frac{1}{2}\left(\frac{C(\lambda'-\lambda)}{\sigma_R\lambda}\right)^2\right] d\lambda'$$

and to be compared with the observed spectrum of the galaxy. This procedure has been succesfully applied to M32.

Simkin (1974) and Illingworth *et al.* (1979) have used the Fourier transform of the former expression to determine an optimum value for σ_R through the convolution theorem.

A correlation between intrinsic luminosity L and the central velocity dispersion σ_R was found. This is an expected relationship, since masses, luminosities, and dispersions are proportional to each other.

II.2.2. MASSES OF GALAXIES

The basic knowledge underlying all methods to determine the mass of a galaxy is that of the gravitational force acting on a test particle. Since the force is impossible to obtain because we measure velocities only once, we are forced to assume a stationary state for the system. In this way the possibility of mass determination becomes restricted to a definite class of objects. Since symmetry and stationarity are conditions fairly satisfied by normal galaxies (Section I.1) we are able to know, at least in principle, the masses of normal systems. Within these limitations there are several ways to proceed with various degrees of certainty, namely: (i) from curves of rotation, (ii) from the measurement of the velocity dispersion of stars, (iii) through the statistics of binary galaxies, (iv) through the dynamics of systems of galaxies.

Masses from Curves of Rotation

When only gravitational forces are considered, the stationary circular motion of a test particle in the potential ϕ of the system is given by the equation

$$V^2(R) = R\frac{\partial\phi}{\partial R} \tag{9}$$

where R is the distance to the center in the equatorial plane and

$$\nabla^2\phi = 4\pi G\rho, \tag{II.10}$$

the local density being ρ. The first equation states the equilibrium between centrifugal and gravitational accelerations in the plane of rotation, but nothing is said regarding the conditions along the vertical coordinate besides the postulated symmetry with respect to the plane. We are then free to assume the way mass is distributed in that direction, and several methods have been proposed to give a reasonable vertical mass distribution whilst keeping the problem under tractable form (see e.g. Perek, 1962; Burbidge *et al.*, 1959; Schmidt, 1956; Brandt *et al.*, 1962; Toomre, 1963).

The general solution of Poisson's equation (II.10) for the potential in a point on the plane of the galaxy at a distance R form the center is

$$\phi(R) = -G\int\frac{dM(r)}{|R-r|},$$

where the integral extends to all mass elements $dM(r)$ of the distribution.

A point-like distribution has $dM(r) = M\delta(r)$, where $\delta(r)$ is Dirac's three-dimensional delta function. We obtain $\phi = -GM/R$ and $V^2(R)R = GM$. This is the so-called keplerian approximation, where the test particle is so far away as to 'feel' the galaxy as a mass-point. In practice, keplerian masses given only an order-of-magnitude lower bound for the mass of a glaxy.

Let us consider now homogeneous spheroidal distribution of mass of density ρ and equatorial radius R_0. The potential of such a distribution on a point of the equatorial plane at distance R from the center is given by $\phi(R) = -AGM(R)/R$, where $M(R)$ is the mass inside the homogeneous spheroid with equatorial radius R. The coefficient A is a function of the eccentricity which assumes values of the order of the unity. The circular velocity associated to this particular distribution is given by Equation (II.1), $V^2(R) = AGM(R)/R$ and has a different dependence on R according to whether the test particle is interior or exterior to the distribution of mass. Since $M(R) \propto \rho R^3$ we have $V(R) \propto \sqrt{\rho}\,R$ for $R \leqslant R_0$, whilst $V^2(R) \propto M(R_0)/R$ for $R > R_0$, The circular velocity results proportional to the distance inside the homogeneous distribution, whilst it has a keplerian fall outside it. These results give some insight about the meaning of the properties of the rotation curves. The central linear increase of the circular velocity is related to a homogeneous spheroidal solid rotation, whilst outside it the curve is expected to show a keplerian fall from the maximum at the edge of the distribution. However, this is not observed: the dependence of the mass with radius $M(R) \propto V^2(R).R$ varies as R for those cases and regions where the rotation curve has a constant circular velocity (Section II.2.1).

The evidence for flat rotation curves on larger extension than previously detected implies that galaxy masses are much larger than suspected, perhaps the new values double the previous optical estimates (Table II.4).

TABLE II.4

Average mass of Galaxies*

Type	M(in 10^{10} solar units)
gE	100
dE	0.3 to 5
$Sa^{(+)}$	100:
$Sb^{(+)}$	9
$Sc^{(+)}$	5.4

*   Adapted from Burbidge and Burbidge (1975).
$^{(+)}$ Out to the last measured point in the rotation curve.

Indicative Masses

The large body of H ɪ-observations with beamwidths larger than the apparent diameter of a galaxy have led to the introduction of a statistically defined 'indicative' mass (Heidmann, 1969) $\mathfrak{M}_i = KV^2(R_M)R_M/G$, where R_M is the distance at which the maximum rotational velocity is attained, and K a numerical coefficient. The H ɪ observations provide the global profile of the 21-cm line integrated over the whole of the galaxy, so that the line half-width (usually taken at 20% maximum detection

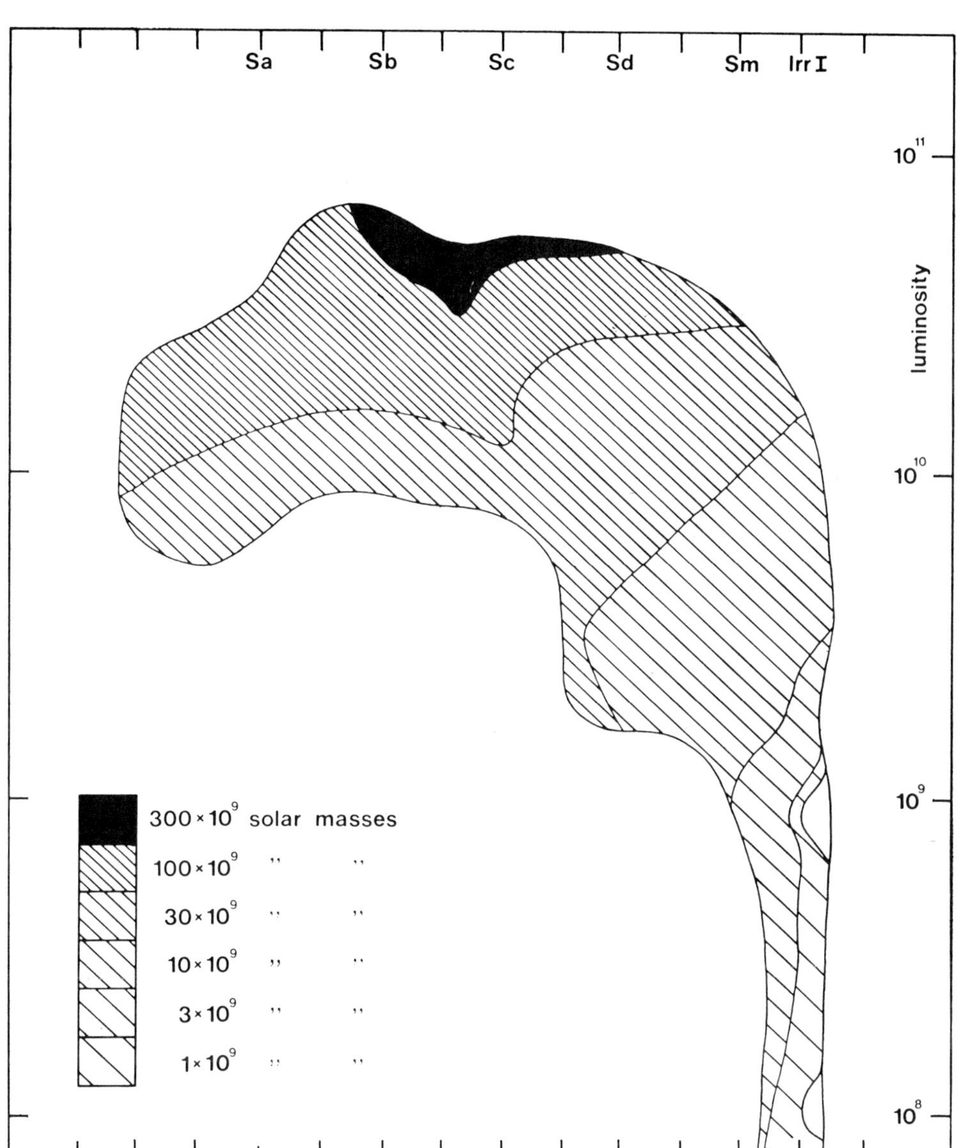

Fig. II.16.   Masses and luminosities of galaxies. (*Adapted from Heidmann* (1969).)

level) is proportional to $V(R_M) \sin i$, whilst R_M is assumed to be related to the Holmberg R_H isophotal radius. The coefficient K was then determined in order to get the best agreement with the masses obtained through rotation curves, resulting $\mathfrak{M}_i = 3 \times 10^4 (W/\sin i)^2 R_H$ when R_H is given in kpc and W in km s^{-1}. The indicative masses are particularly suited for statistical studies (Figure II.16).

Mass Distribution

In order to gather information on the mass distribution from the rotation curve, let us assume now an inhomogenous spheroidal distribution. The model was developed by Burbidge *et al.* (1959) and makes use of the property of the homeoids, thin ellipsoidal shells of constant density, of not producing attraction on their interior points.

Let $\phi\,(R, \lambda)$ be the potential of a homogeneous spheroid of unit density on a test particles at R, and R_0 be the equatorial radius of the spheroid. The potential on the same point of a distribution $\rho(\lambda)$ of similar shells now becomes

$$\phi(R) = \int^R \rho(\lambda)\, \frac{\mathrm{d}\phi(R, \lambda)}{\mathrm{d}\lambda}\, \mathrm{d}\lambda$$

the integration being extended up to $\lambda = R$ if $R < R_0$, because of the mentioned property of the homeoids, or up to R_0 if the test particle is exterior to the ellipsoidal distribution ($R > R_0$). Equation (II.9) can be written now

$$V^2(R) = \int^R \rho(\lambda)K(R, \lambda)\, \mathrm{d}\lambda \tag{II.11}$$

as in integral equation for $\rho(\lambda)$, with kernel

$$K(R, \lambda) = R\frac{\mathrm{d}^2\phi}{\mathrm{d}R\,\mathrm{d}\lambda} = 4\pi G(1 - e^2)^{1/2}\, \lambda^2(R^2 - e^2\lambda^2)^{-1/2}, \tag{II.12}$$

where $e = (1 - c^2/\lambda^2)^{1/2}$ is the eccentricity of the spheroid with major axis λ. Equation (II.12) can be solved by power series,

$$\rho(\lambda) = P_0 + P_1\lambda + P_2\lambda^2 + \cdots$$

and

$$K(R, \lambda) = (\lambda^2/R)(K_0 + K_2\lambda^2 + \cdots)$$

provided $V^2(R)$ be expressed in the same way. This can be made through a polynomial representation of the observed curve of rotation,

$$B(R) = R(A + BR + \cdots)$$

which carried to Equation (II.11) together with the series for $\rho(\lambda)$ and $K(R, \lambda)$ provide a set of equations for $P_0, P_1, \ldots$ etc. after equalization of the coefficients with the same power of R. From these equations we obtain

$$P_0 = 3A/K_0, \qquad P_1 = BAB/K_0, \quad \text{etc.}$$

which solves, in principle, the problem of finding the density distribution $\rho(R)$. The coefficients K_0, K_1, etc. contain the eccentricity e of the spheroidal distribution. There is no clear-cut procedure to define the value of e in a particular case from the observations. The best procedure is to use the eccentricity related to the intrinsic flattening associated to the morphological type of the galaxy (Section I.3).

The assumption of similar spheroids for the density distribution has no theoretical or physical ground besides being a convenient computational device and not contradicting seriously with what we would expect from the meagre available information.

Brandt and Belton (1962) have shown that Equation (II.11) reduces to Abel's equation when the spheroids degenerate into disks. The volume density ρ is replaced by the surface density $\sigma(R)$ and, because Abel's equation can be inverted, $\sigma(R)$ is expressed as an integral on $V^2(R)$.

Angular Momentum from Rotation Curves

To estimate angular moments of spiral galaxies, a model for the density distribution is needed. Takase and Kinoshita (1967) have proposed to do that through projecting the mass within cylindrical shells of radius R and width dR, on the symmetry plane. Each ring is assumed to rotate with circular velocity $V(R)$.

Marton instead (1978) has estimated the angular momenta under the assumption that each homeoid with semi-major axis R and width dR rotates with the same angular velocity as the matter in the symmetry plane at distance R from the center, which is equivalent to a distribution of rigidly rotating homeoids.

The moment of inertia dI of the homeoid of eccentricity, e, density $\rho(\lambda)$ and equatorial radius is

$$dI = \frac{8}{3}\pi(1 - e^2)^{1/2}\,\rho(\lambda)\lambda^4\,d\lambda$$

because the angular velocity is $w(\lambda) = V(R)/R$, we have

$$J = \int dJ = \int w(\lambda)\,dI(\lambda) = \frac{8}{3}\pi(1 - e^2)^{1/2}\int \rho(\lambda)v(\lambda)\lambda^3\,d\lambda$$

for the angular momentum. The substitution of the polynomial series found for the density distribution and the fit of the rotation curve $V(R)$ gives

$$J(R) = J_0 R^5 + J_1 R^6 + \cdots$$

the angular momentum as a function of the distance R to the center in the plane of rotation.

An order-or-magnitude relationship between J and the mass $\mathfrak{M}$ of the galaxy can be derived noticing that

$$\mathfrak{M} \propto V^2(R)R, \qquad J \propto \mathfrak{M}\,RV(R),$$

where R is a suitable characteristic length. A third relationship comes from the photometric properties of disk distribution (Section II.1.4). The approximate constancy of the peak surface brightness I_0 found by Freeman, required the total disk luminosity $L_D \propto I_0 R^2$, where R is the characteristic length of the brightness distribution. For disk galaxies $L_T \approx L_D$ and the mass-luminosity ratio can be assumed to be the same. So that we would expect

$$\mathfrak{M} \propto R^2.$$

Since I_0 is constant. From these three relations and the assumption R is the same characteristic length, we obtain

$$V^2 \propto R \propto \mathfrak{M}^{1/2} \quad \text{and} \quad J \propto \mathfrak{M}^{7/4}.$$

Takase and Kinoshita (1967) have carried out extensive calculations for several galaxies finding

$$\log J = 1.76 \log \mathfrak{M} - 3.92$$

(c.g.s. units) for the expected relationship.

Masses from the Velocity Dispersion of Stars

The virial theorem for a non-rotating stationary-state system is written

$$2\langle T \rangle + \langle W \rangle = 0,$$

where the brackets mean time-average values of the total kinetic energy T and the gravitational potential energy W, which is given by

$$W = - G \int \frac{\mathfrak{M}(s) \, d\mathfrak{M}(s)}{s},$$

where $d\mathfrak{M}(s) = 4\pi s^2 \rho(s) \, ds$ is the mass element in the shell between the radii s and $s + ds$. The customary procedure to compute W for elliptical galaxies is due to Poveda (1961), who assumed the projected radial distribution of mass to be the same as the projected brightness distribution, which is equivalent to postulate a constant mass-luminosity ratio through the galaxy.

The projected brightness distribution for spheroidal subsystems is given by de Vaucouleurs law (Equation (II.7), Section II.1.1). Poveda has shown that

$$W = -0.33GM^2/R_e$$

for such mass distribution. Here R_e is the effective radius associated with the brightness distribution. The kinetic energy T of the stellar motions is given by

$$T = 3 \, \mathfrak{M}\sigma_R^2/2,$$

where the dispersion velocity is derived from the Doppler width of absorption lines (Section II.2.1). Since elliptical galaxies are relaxed systems, T and W may be used in the virial theorem instead of its average values, so that the resulting mass is given by

$$\mathfrak{M} = 9(\sigma_R^2 R_e/G)$$

in terms of the two observable quantities σ_R and R_e.

Statistical Masses from Galaxy Pairs

There is a strong tendency between galaxies to form pairs (Section IV.1.1). Statistical analysis shows they are more frequent than chance associations. Holmberg (1937) published a catalogue of 827 pairs and multiple systems. From these almost 700 entries are simple pairs with separations S smaller than 10 arc min. The frequency function for S has a definite maximum for $S \to 0$. Holmberg has demonstrated that less of 12% of double galaxies are optical pairs.

Although the orbital motion of galaxies in pairs is not observable, the measurements of radial velocities give statistical information about their masses. Assume that two galaxies with a combined mass M move in circular orbits one around the other, with relative velocity V and separation R. If we consider them as point masses, then we have $V^2 = G\mathfrak{M}/R$. The relative velocity cannot be directly measured, only the difference of radial velocities ΔV_R. The same occurs with the distance R, for which we only observe the apparent separation r normal to the line of sight. Let be θ, ϕ the two angles which define the direction of the radius vector towards the observer, taking as origin one of the galaxies. We have

$$r = R \sin \theta$$

$$\Delta V_R = V \sin\theta \cos\phi$$

and thus

$$(\Delta V_R)^2\, r = G\mathfrak{M} \sin^3 \theta \cos^2 \phi.$$

ΔV_R is directly observed, whilst r follows from the observed angular separations of the galaxies and the distance D. Since we ignore the orientation of the orbital plane with respect to the line of sight, we do not know the two angles θ, ϕ. It is possible, however, to know the mean value of the projection factor $\sin^3 \theta \cos^2 \phi$ in the reasonable assumption that the orbits of binary galaxies are random oriented, in order to obtain masses of statistical significances.

The probability of these angles to be within a given cell $\Delta\theta\, \Delta\phi$ is proportional to the solid angle corresponding to that cell, so that the mean value results from integration over the sphere,

$$\langle \sin^3 \theta \cos^2 \phi \rangle = \int \sin^3 \theta \cos^2 \phi\, \frac{d\Omega}{4\pi}$$

$$= \frac{1}{4\pi} \int_0^{2\pi} \sin^2 \phi\, d\phi \int_0^{\pi} \sin^4 \theta\, d\theta = \frac{3\pi}{32} \approx 0.296$$

and consequently

$$\langle (\Delta V_R)^2\, r \rangle = 0.296 G\mathfrak{M}.$$

Holmberg (1954) and Page (1961, 1962) have followed in essence this procedure for a great number of pairs.

The later author has given the following values:

$$\mathfrak{M}_S = (2.1 \pm 1.5) \times 10^6 \text{ solar masses}$$

$$\mathfrak{M}_E = (59.3 \pm 15) \times 10^{10} \text{ solar masses}$$

for pairs formed by spirals and ellipticals, respectively.

Statistical Masses from Systems of Galaxies

The virial theorem

$$\langle\!\langle 2T \rangle\!\rangle + \langle\!\langle W \rangle\!\rangle = 0$$

relating the time averages of the kinetic and potential energies T and W of a dynamical stationary system, can be applied to groups and clusters of galaxies provided the ergodic hypothesis that the time averages can be replaced by space averages $\langle T \rangle$ and $\langle W \rangle$ taken over the volume of the system, is adopted.

The astronomical observations provide us with the magnitudes, relative positions on the sky, and radial velocities for each member of the system. We are not in a position to know the space distribution from either the galaxies, nor their velocity vectors, so we have to rely on the calculation of the statistical projection factors for rondom distributions of mutual distances and velocities (Limber *et al.*, 1961). Let θ be the angle between the velocity vector v and the line of sight of a given galaxy, we have

$$\sum_{i}^{N} m_i v_i^2 \cos^2 \theta_i = \sum_{i}^{N} m_i v_{iR}^2,$$

where m_i and v_{iR} are the mass and radial velocity associated to the ith galaxy, the summation being performed through the N objects of the system. After taking spatial averages the preceding expression becomes

$$\langle \sum_{i}^{N} m_i v_i^2 \rangle \langle \cos^2 \theta_i \rangle = \frac{1}{3} \langle \sum_{i}^{N} m_i v_i^2 \rangle = \langle \sum m_i v_{iR}^2 \rangle$$

because the average of $\cos^2 \theta$ over the sphere is $\frac{1}{3}$. In this way we obtain the statistical estimate of $\langle T \rangle$.

$$2 \langle T \rangle = 3 \langle \sum_{i}^{N} m_i v_{iR}^2 \rangle.$$

For the calculus of the potential energy estimate $\langle W \rangle$ we proceed in an analogous fashion, after introducing the angle η between the vector joining two galaxies and the line of sight. Then we have

$$G \sum{}' \frac{m_i m_k}{r_{ik} \cos \eta_{ik}} = G \sum{}' \frac{m_i m_k}{\delta_{ik}},$$

where $\sum'$ means $i = k$ is not considered in the summation, and $\delta_{ik} = r_{ik} \cos \eta_{ik}$ is the projected distance over the sky between the galaxies m_i, m_k. The space average of the foregoing expression is

$$\langle G \sum{}' \frac{m_i m_k}{r_{ik}} \rangle \langle \frac{1}{\cos \eta_{ik}} \rangle = G \sum{}' \frac{m_i m_k}{\delta_{ik}}$$

and $\langle 1/\cos \eta \rangle = \pi/2$, so that we obtain

$$\langle W \rangle = - \frac{\pi}{2} G \sum{}' \frac{m_i m_k}{\delta_{ik}}$$

the statistical estimate of the potential energy. Many approaches to the handling of the virial

$$\langle 2T \rangle + \langle W \rangle = 0$$

have been made in the literature (Page *et al.*, 1961). The following analysis due to Rood *et al.* (1970) is particularly suitable for further discussions of the subject (Section IV.2). These authors introduce the 'virial mass' $\mathfrak{M}_{VT}$ of the system through

$$\mathfrak{M}_{VT} = V^2 R/G,$$

where

$$V^2 = 3 \langle T \rangle / \mathfrak{M}_L \quad \text{and} \quad R = \mathfrak{M}_L^2 / \langle -W \rangle$$

$\mathfrak{M}_L$ is the 'luminosity mass' obtained from luminosities L_i and mass-luminosity ratios f_i for each member of the system. Notice $\mathfrak{M}_L$ is not necesarily known to compute V^2 and R, because they only depend on the relative weights $f_i L_i$, which can be calculated with the apparent luminosities instead of the intrinsic ones. Once $\mathfrak{M}_{VT}$ is known, the estimates for the individual masses follow from their relative weights.

The problems arising in applying this method to groups and clusters will be discussed in Section IV.2.

II.2.3. Mass-Luminosity Ratios

The data about masses can be combined with luminosities to obtain the mass-luminosity ratio $\mathfrak{M}/\mathfrak{L}$ a very useful parameter for our understanding of the evolution of the galaxy content.

Values of $\mathfrak{M}/\mathfrak{L}$ are dependent on D^{-1}, the inverse of the distance of the galaxy, because the mass is αD and the intrinsic luminosity is αD^2. In order to compare diverse $\mathfrak{M}/\mathfrak{L}$ ratios it is necessary to reduce all the data to a common scale of distances. Since $\mathfrak{L}$ can be obtained through different photometric windows, it is also convenient to define an intrinsic luminosity associated to a definite window, such as $\mathfrak{L}_B$, the intrinsic luminosity resulting from using corrected total B-magnitudes B_T^0 (Section II.1). It is customary to express $\mathfrak{M}$ and $\mathfrak{L}_B$ is solar units, which requires to define $\mathfrak{M}_\odot = 2 \times 10^{33}$ g and $\mathfrak{M}_{B\odot} = +5.48$ according to Allen (1973).

Integrated Values

The $\mathfrak{M}/\mathfrak{L}_B$ values can be given in an integrated way, valid for a galaxy as a whole, or in a local form, for different parts of the object. In the first case it is convenient to use a method for mass estimates which insures the whole mass of the system has been detected, otherwise a deficiency in $\mathfrak{M}$ will appear, leading to smaller values for $\mathfrak{M}/\mathfrak{L}_B$. This problem is critical in the case of masses derived from rotation curves: in these cases the values of $\mathfrak{M}$ and $\mathfrak{L}_B$ should be restricted to a definite volume of the galaxy, avoiding in this way unsafe extrapolations. This is usually done through the isophotal diameters given by Holmberg. Faber and Gallagher (1979) have given $\mathfrak{M}/\mathfrak{L}_B$ for galaxies with rotation curves extended at least to one-half of the Holmberg radius.

TABLE II.5

Mass-luminosity ratios for normal galaxies*

Type	$\mathfrak{M}/\mathfrak{L}_B^{(+)}$
E–SO	10–20
Sa/O	12
Sab–Sbc	13
Sbc–Sc	9
Scd–Sd	8
Sdm–Irr I	3

* After Faber and Gallagher (1979). $H = 100$ km s^{-1} Mpc^{-1}.
(+) Within a Holmberg radius.

They simply compute a sort of 'indicative mass' $\mathfrak{M}_{H0} = V^2(R_{H0})R_{H0}/G$ assuming a spherical homogeneous distribution, which could harmonize with the possible existence of extended halos around spirals (see below). A similar procedure was followed with indicative masses from H I observations. The resulting figures are summarized in Table II.5. A trend of the $\mathfrak{M}/\mathfrak{L}_B$ ratios along the sequence is then put in evidence. Since color indices correlates with the morhpological type (Section II.1) it is natural also to expect a correlation between $\mathfrak{M}/\mathfrak{L}_B$ and colors in the sense that redder galaxies have the larger $\mathfrak{M}/\mathfrak{L}_B$-values, as mentioned by de Vaucouleurs (1974).

Most of the present information regarding $\mathfrak{M}/\mathfrak{L}_B$ ratios in early-type galaxies is due to Faber *et al.* (1976) and Sargent *et al.* (1977), with masses derived from central velocity dispersions. The agreement between both groups is good and a value of $\mathfrak{M}/\mathfrak{L}_B \approx 10$–20 is found for E–SO galaxies. There are, however, suggestions of larger values in cases of giant cD systems.

From statistical masses in galaxy pairs Page (1961, 1962) has derived $(\mathfrak{M}/\mathfrak{L}_B) \approx 1.5$ for pairs of spirals and $(\mathfrak{M}/\mathfrak{L}_B) \approx 90$ for pairs containing early type (E, SO) galaxies.

Variation of the Local Value of the $\mathfrak{M}/\mathfrak{L}_B$ Ratio

Because of the uncertainties of surface photometry, but mostly because of the model-dependent nature of the density distribution results, the variation of the local value of $\mathfrak{M}/\mathfrak{L}_B$ with distance to the center in spiral galaxies is still subject to debate. Nordsiek (1973) has derived mass distributions for several nearby spiral galaxies and found that some large, well observed objects, like M31, M51, and M81 have surface-brightness and projected-density profiles which are fairly similar, suggesting a uniform $\mathfrak{M}/\mathfrak{L}$ ratio over the galaxy disks. Moreover, a correlation in the expected sense between the local values of $\mathfrak{M}/\mathfrak{L}_B$ and the color index was suggested in Nordsiek's data, with $\mathfrak{M}/\mathfrak{L}_B \approx 2$ for $B - V = 0.5$ and $\mathfrak{M}/\mathfrak{L}_B \approx 10$ for $B - V = 0.8$. An increase of $\mathfrak{M}/\mathfrak{L}_B$ towards the outer parts of the galaxy has been reported however by Roberts (1975) as well as by several other authors (Baldwin, 1975; Bosma and van der Kruit, 1979; Agüero, 1980).

Massive Halos

Ostriker *et al.* (1974) have suggested the spiral galaxies may be surrounded by very extended halos populated by low-luminosity stars, which makes the total mass an order of magnitude larger and increases the $\mathfrak{M}/\mathfrak{L}_B$ ratios over the accepted current values for the visible bodies of the galaxies. Evidence for the existence of these massive haloes is to be found in the flat rotation curves (Section II.2.1), in the low-luminosity haloes observed far away from the main bodies of elliptical (Section II.1.4) and apiral galaxies (Spinrad and Ostriker, 1974). If these halos were due principally to stars their red color would mean they must be dwarfs, so that $\mathfrak{M}/\mathfrak{L}_B$ should increase towards the outer parts of the system. Further evidence is to be found in the systematically larger mass and $\mathfrak{M}/\mathfrak{L}$ ratio estimates derived from larger systems (binaries, groups and clusters). Ostriker *et al.* (1974) have shown that the 'excess mass' over the rotation-curve masses linearly increases with the size of the system which suggests that the farther away the 'test particle' is placed, the larger is the detected mass.

Ostriker and Peebles (1973) gave theoretical reasons for the existence of massive halos: a rotating disk of stars cannot be stable and favors bar-like perturbations if the ratio of rotational kinetic energy T_R versus potential energy $|W|$ exceeds the value 0.14: for pure rotation this ratio is 0.5 from the virial theorem ($2T_R + W = 0$) and either the random motions in a disk are considerable (which is not supported by the observations), or the disk is immersed in a spherical mass distribution or corona which does not partake of rotation but contributes to increase $|W|$ and then to reduce the $T_R/|W|$ ratio. That massive subsystem has to be populated by stars: such a massive halo, if gaseous, would contribute a X-ray flux which is not observed.

The consequence of massive halos populated by low-luminosity stars for groups and clusters of galaxies will be seen in Section IV.1.2.

II.3. Models for Normal Galaxies.

The critical evaluation of the obtained data is a primary intellectual process done step by step. In the preceding sections the diverse observations were presented in an ordinate way, stressing regularities and abstracting common patterns. Now we have to go a further ahead by means of the modeling process.

We describe in this section two approaches to our understanding of normal galaxies. The first considers a galaxy as a N-body problem and tries to give an explanation of their forms and motions in terms of the Newtonian attraction as the sole interaction.

In the second approach the evolutive processes taking place between the different components of galaxy are considered, but no dynamical considerations are made.

Both approaches should converge to more unitary models, like those proposed by Larson (1974), but the difficulties are still large.

II.3.1. DYNAMICAL MODELS

The main purpose of stellar dynamics is to derive the relation between the velocity and density distributions of stars in a galaxy, and to follow the evolution of such relation in time. Because of their different behavior, the stellar and gaseous components have been usually treated separately, although we known that stars and gas interact strongly in a galaxy (Section II.3.2).

The construction of dynamical models of galaxies is based on the following assumptions:

(i) Conservation of the number of particles. Once the stars are formed, new ones can not be created, and none can be destroyed. This is equivalent to disregarding the dynamical consequences of stellar formation and evolution.

(ii) There are no stellar encounters. The time of dynamical relaxation, due to binary interactions between stars (T_R) is, for the system as a whole, greater than its age. This justifies neglecting the effect of encounters for time scales shorter than T_R.

(iii) The system is self-gravitating. Each star moves in the collective gravitational field of the others.

Although models based on (i) and (ii) have existed, considering (iii) makes the problem substantially more difficult to treat and advances in this direction have only been possible after the on coming of high velocity computers.

The statistical description of a stellar system is carried out by means of the distribution function $f(r, v, t)$ defined as the number of stars of mass m per unit volume in the (r, v) space at the instant t. The evolution of f in time is given by Liouville's equation

$$\frac{\partial f}{\partial t} + \sum_{1}^{3}\left(\nu_i\,\frac{\partial t}{\partial x_i} - \frac{\partial \phi}{\partial x_i}\,\frac{\partial t}{\partial v_i}\right) = 0 \tag{II.13}$$

which expresses the conservation of f in the co-moving element of volume. From the definition of f we get that the density of mass $\rho = m \int f\, d^3v$ and therefore Poisson's equation gives us

$$\nabla^2 \phi = 4\pi\, Gm \int f(r, v)\, d^3v. \tag{II.14}$$

Any function $f \geqslant 0$ satisfying the integro-differential system (II.13)–(II.14) represents a possible stellar system. Poincaré's theorem, which in stellar dynamics is known as Jeans', states that the general solution of Equation (II.13) can be written

$$f = f(I_1, I_2, I_3, I_4, I_5, I_6), \tag{II.15}$$

where the set of functions $I_i = cte$ are the six integrals of the equations of motion for a single star $(dx_i/dt = v_i; dv_i/dt = - \partial\phi/x_i)$.

Thus we have two depending variables, ϕ and f, and the independent variables r, v. A particular solution of the system (II.13)–(II.14) is said to be a self-consistent model. The boundary conditions for a physically realistic solution of the problem are: $f \geqslant 0$; $\phi(r) \to r^{-1}$ with $r \to \infty$ and $\phi(r = 0) = $ minimum. The first excludes solutions with negative densities and the two others are the usual conditions for the Poisson equation. The limits of integration in the space of velocities also must be established. Although generally it is acceptable to integrate for $|v| \leqslant \infty$, this leads to difficulties of interpretation. A more realistic approach for finite systems requires us to write $E = u^2/2 = \phi(r) \leqslant E_0 < 0$, which implies a cutoff slightly below the escape velocity in the distribution. This defines a limit for the system and, if no escape of stars exists, fits the assumption (i).

Elliptic Galaxies

Due to their smooth appearance and evident structural simplicity, these galaxies seem to be the natural candidates for the application of the methods of stellar dynamics. At the same time this very structural simplicity puts a limit to the kind of observations one can carry out. Precision photometry is necessary to differenciate between the various theoretical models. The central regions are of high surface luminosity and therefore the most observed ones, but the outer envelopes can have a surface brightness so low that it is hard to detect it against the sky background.

Since the theory of formation and later violent relaxation (Chapter VII) of the elliptic galaxies is not yet well developed, we must limit ourselves to looking for the form of the distribution function f and the collective gravitational potential in the hyphothesis of a stationary regime. The application of the Poincaré–Jeans theorem requires thus the distribution function J to depend on the energy E and the component of the angular moment corresponding to a stellar orbit, parallel to the axis of symmetry, i.e. $f = f(E.\ J)$. Prendergast and Tomer (1970) have pioneered in the field of E-modeling. They have found a sequence of models which depend on two parameters: the value $\phi(0)$ of the potential in the center and the angular velocity. The deeper the potential well the larger the central condensation will be and when the parameter of rotation increases from zero, the models become progressively flatter. The isophotes computed on the basis of these models are, in general lines, very similar to those observed. It is characteristic for them to be come more spherical the farther out the isophotes are. The models do not reproduce systems with axial ratio a/c, larger than 3.

Wilson (1975) has computed models of elliptic galaxies with algorithms similar to those of Prendergast and Tomer but with a different form for the distribution function f. The models were projected on the plane of the sky and compared with observations of NGC 3379 (Section II.1). One was found agreeing reasonably well with the observations, including even the velocity dispersions. Wilson's models represent quite well the luminosity profiles but the large variety of curves that connect the flattening of the isophotes with their equivalent radii cannot be reproduced. These models show nearly circular isophotes around the center and the flattening increases to a maximum, to become circular again in the outer regions.

Flattening of Elliptical Galaxies

We have seen (Section I.3) that the flattest elliptical galaxies are of type E7, which corresponds to an axial ratio of 3.33:1, whereas SO and S galaxies have axial ratios of the order of 20:1. Thuan and Gott (1975) have given an interpretation of this property. These authors postulate that a protogalaxy can be described as a homogeneous sphere of mass M and initial radius R_i rotating uniformly as a whole with angular velocity w_i but also with local random velocities. Let t be the ratio of the kinetic energy of rotation R_T of the sphere at its potential energy W, that is $t = T_R/W_i = (w^2R^3/3GM)$ where the index i means that the initial values have to be taken. The binding energy is then $E = E_i(1 - t) = 3GM^2(1 - t)/5R$. If the formation of stars has been completed for the time in which the collapse begins, this will be dissipationless, the system relaxes violently and acquires an equilibrium configuration ofter a few damped oscillations. The violent relaxation thermalizes the distribution of velocities. The system can then be represented by a McLaurin ellipsoid in virial equilibrium, with isotropic pressure and uniform rotation. The energy of this spheroid is written now

$$E_f = \frac{1}{2}W_f = -\frac{3}{5}\frac{GM^2}{R_f}F(e),$$

where the index f denotes final values and $F(e)$ is a function of the eccentricity e of the ellipsoid. As there is no dissipation in the collapes, $E_f = E_i$ and we have

$$(1 - t_i) = \frac{R_i}{R_f}F(e). \tag{II.16}$$

The conservation of the angular moment requires, on the other hand, that

$$(T_R)_f = \frac{R_i}{R_f}(T_R)_i = \frac{3}{5}\frac{GM^2}{R_i}t\frac{R_i}{R_f}.$$

Again, the theory of the equilibrium figures in rotation gives for the kinetic energy of a McLaurin ellipsoid

$$(T_R)_f = \frac{3GM^2}{5R_f}G(a)$$

which finally gives, eliminating R_i/R_f in (II.16),

$$t(1 - t) = F(e) \, (G(e).$$

The function $F(e)G(e)$ increases with e, but the first member is symmetric in t and reaches its maximum at $t = \frac{1}{2}$, a value to which corresponds the maximum ellipticity of 0.709 25, which leads to a maximum flattening of $c/a = 0.705$. For $t < 0.5$, as it probably is in most cases, the galaxy flattens when its angular moment increases, but beyond $t = 0.5$ the increase of the angular moment makes it more circular so that no elliptic galaxy can be flatter than E7 independently of how fast it initially rotates. The reason for this interesting result is due, according to Thuan and Gott, to the fact that the McLaurian spheroids go to infinite for $t \to 1$, although the increase in angular momentum has been finite. With $t = 1$ the galaxy has only $\sqrt{2}$ times the angular moment of one with $t = 0.5$ but since the kinetic energy is equal in absolute value to the potential ($t = 1$), the binding energy is zero and R_f becomes infinite. The numerical experiences of Bouvier and Janin (1970) have demonstrated that the violent relaxation is minimum in the case of $t = 0.5$ (E7 galaxies) and increases symmetrically towards $t = 0$ and $t = 1$.

Triaxial Structures in Elliptical Galaxies

According to Miller (1978a, b), dynamical arguments exist suggesting that E4 and flatter galaxies are prolated objects rotating end-over-end in space. This leads to an interpretation divergent from the traditional standpoint which, as we have seen, interprets elliptic galaxies as oblate objects of axial symmetry with different degrees of intrinsic flattening q_0, due to rotation.

The dynamical arguments in favor of prolate or triaxial structures are the result of numerical experiments due mainly to Miller and Smith (1979) who have found that the bars are the natural form towards which stellar systems in rapid rotation evolve. Even systems collapsing from the most diverse initial configurations all end up in bars, which, once formed, are very stable.

On the other hand, the observational results of Bertola and Capaccioli (1975), Illingworth (1977), Peterson (1978) and others do not prove the fast rotation which should exist if elliptic galaxies really were objects with axial symmetry, flattened by rotation. Binney (1976) suggested this to imply that the structure of elliptic galaxies was due to an anisotropic velocity field.

Illingworth describes the rotations observed in terms of a diagram V/σ vs ε, where V is the maximum rotational velocity found in a galaxy, σ the velocity dispersion, and ε the ellipticity. Most models with axial symmetry and rotation fall around the line $V/\sigma \simeq 2\varepsilon$ in this diagram. The observations, on the other hand, show most elliptical galaxies well below this line, around $V/\sigma \simeq 2\varepsilon/3$. Depending on the direction from which they are viewed, prolated models may fall anywhere within the region between $V/\sigma \simeq 2e$ and $V/\sigma = 0$.

According to Miller and Smith, the prolate ellipticals, if viewed along the axis of rotation, would have small values of V/σ whereas if the line of sight goes through the equatorial plane, V/σ approaches values corresponding to models of axial symmetry in rotation.

In the last instance, the question of tridimensional figures of elliptic galaxies should be resolved by observations. The dependency of the surface brightness on the tridimensional figure has served Marchant and Olsen (1979) and independently Richstone (1979), as a statistical test for the tridimensional models.

Let us assume that an E galaxy of oblated figure is seen from two points of view: from the pole it is observed more circular and with lower mean surface luminosity than seen edge-on. The contrary occurs with a prolate galaxy. From the statistical analysis of the distribution of real objects on the plane surface brightness vs q, Richstone arrives at the conclusion that the sample used (35 elliptical galaxies brighter than $M_B = -21.5$) favors oblate galaxies.

II.3.2. EVOLUTION OF GALAXY CONTENT

Most of the theories about galaxy formation assume an initial gaseous state (protogalaxy) which collapses and fragments into stars (Chapter VII). The evolution of the content of such a system and its integrated properties can be followed in time through analytical or numerical models with independence, in principle, of the dynamical behavior. The backbone of these models is the knowledge of stellar evolution, which can be used as input on a set of differential equations supplemented by simple assumptions regarding the initial mass function (IMF) for stars and the rate of stellar formation. In this way we are able to compute the fraction of the total mass in the form of gas, the abundance of heavy elements, the total luminosity of the system, its colors, etc. as a function of time.

The concept of IMF was introduced first by Salpeter (1955). van den Bergh (1957) and Schmidt (1959) derived the basic equations for a time-dependent rate of stellar formation. A simple, analytical way to describe the general properties of the galaxy content at any time was given independently by Pagel (1973) and Searle (1973). Further refinements by several authors are described by Trimble (1975).

Basic Equations

Let $\xi(m, t)$ the mass of stars being created in a definit volume of space, per unit of time in the range of stellar masses $m, m + dm$ at time t. The function $\xi(m, t)$ is known as Salpeter's birth-rate function.

The assumption of separate variables, namely

$$\xi(m, t) = \xi(m) f(t)$$

is usually made. Here $f(t)$ is the overall birth-rate function, and $\xi(m)$ was found to be a power-law of stellar mass $\xi(m) \propto m^{-s}$ in the interval $m'' > m > m'$. For the solar neighborhood Salpeter found $s = 1.35$.

The IMF $\psi(m)$ is related to $\xi(m)$ through

$$\psi(m) = \xi(m)/m \propto m^{-(1+s)} \tag{17}$$

in the same interval m'', m'. Schmidt (1963) has given proof that $\psi(m)$ is a weakly time-dependent function, and the foregoing separation is only an analytical approximation which however leads to reasonable predictions regarding the chemical composition of the interstellar medium (Talbot and Arnett, 1973).

According to a simplified view of stellar evolution, stars spend a time $\tau(m)$ in the main sequence (MS) consuming up to 10% of their H content, to move then fast out of the MS towards the red-giant region of the H-R diagram. Their subsequent evolution is even faster and, after losing most of their mass, they end up as dark stellar remnants, mostly white dwarfs. Assuming the red-giant phase can be disregarded as short-lived, star evolution reduces to a long stay in the MS to end up as dark remnants of average mass w, after losing their excess over w solar masses.

The observed MS mass function $\varphi(m)$ is then given by

$$\varphi(m) = \phi(m) \int_{t_0-\tau(m)}^{t_0} f^*(t)\, dt \quad \text{if} \quad \tau(m) < t_0,$$
$$\varphi(m) = \phi(m) \qquad\qquad\qquad \text{if} \quad \tau(m) \geqslant t_0, \tag{II.18}$$

where $f(t)$ has been redefined to satisfy the normalization condition

$$\int_0^{t_0} f^*(t)\, dt = 1.$$

The factor of $\phi(m)$ in the definition (II.18) of $\varphi(m)$ is the fraction of stars still at the MS in time t_0. All stars formed with mass $m \leqslant m_0$, where $\tau(m_0) = t_0$, are still in the MS. In case of a uniform rate, $f^* = t_0^{-1}$, and we have $\varphi(m) = \phi(m)(\tau/t_0)$ for $m > m_0$.

Under homology conditions the luminosity of a MS star with mass m is $l(m) = m^\alpha$ and $\tau(m) = m^{-\alpha/\gamma}$ (Tinsley, 1973). The units of l and m are solar units, whilst that of τ is 10^{10} yr, the Sun's time-life in the MS. Typical values for the exponents are $\alpha = 4$, $\gamma = 1.3$, and $s = 1.35$ (cf. Chapter VII). The upper m'' and lower m' mass limits are usually $m'' = 50$ and $m' \approx 0.2$.

The total mass used to form stars up to time t is

$$M_{TS}(t) = \int_{m'}^{m''} \phi(m)m\, dm \int_0^t f^*(t)\, dt$$

and the mass ejected by stars in their evolution up to the same epoch is (Schmidt, 1959)

$$M_E(t) = \int_{m'}^{m''} \phi(m)(m-w) \int_0^{t-\tau(m)} f(t)\, dt\, dm.$$

Assuming now the system is closed, that is, matter neither enters nor leaves it, all the mass was in form of gas at $t = 0$ and $M_G(0) = M_T$ is the total mass. At time t we have

$$M_T = M_G(t) + M_{TS}(t) - M_E(t)$$

and

$$M_S(t) = M_T - M_G(t) = M_{TS}(t) - M_E(t). \tag{II.19}$$

The first relation expresses the conservation of the mass and the second defines the mass in the form of stars (luminous or dark) at time t (van den Bergh, 1957).

Chemical Evolution of the Interstellar Gas

Under a further assumption that the interstellar gas is well mixed, we may introduce the function $z(t)$ such that zM_G gives the mass of heavy elements in the interstellar gas. The rate at which this is lost to form stars, dM_G/dt, can be separated into a

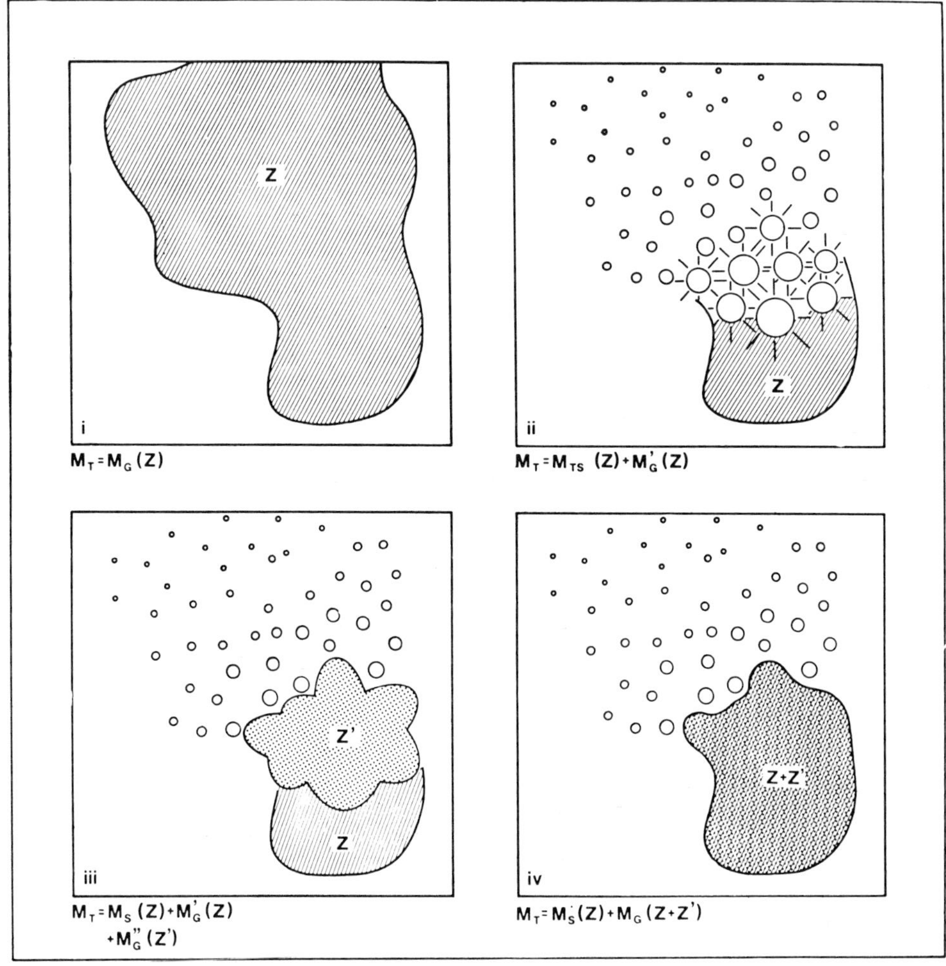

Fig. II.7. Evolution of the chemical content of the interstellar medium: (i) A gas cloud with a fraction z of heavy elements collapses. (ii) Star formation proceeds. (ii) Stars synthesize more heavy elements (z') and the massive ones explode as supernovas and turn back their content (iii) to the surrounding medium which results enriched ($z + z'$) after mixing (iv).

contribution $z(\mathrm{d}M_G/\mathrm{d}t)$ corresponding to the heavy elements and another $(1 - z)$ $(\mathrm{d}M_G/\mathrm{d}t)$ which will be processed inside the stars to produce further heavy elements. A balance equation can be written now for zM_G,

$$\frac{\mathrm{d}}{\mathrm{d}t}(zM_G) = z\,\frac{\mathrm{d}M_G}{\mathrm{d}t} - y(1 - z)\,\frac{\mathrm{d}M_G}{\mathrm{d}t},$$

where the second term at the right side describes the enrichment in heavy elements resulting from an efficiency or 'yield' (Sargent, 1977b) of massive stars in turning back heavy elements to the interstellar medium (Figure II.17). The preceding equation reduces to

$$\frac{\mathrm{d}z}{\mathrm{d}t} = - \frac{y}{M_G} \frac{\mathrm{d}M_G}{\mathrm{d}t_G} \quad \text{(for } y_z \ll 1\text{)} \tag{II.20}$$

and z can be computed for any t if the 'yield' y and $\mathrm{d}M_G/\mathrm{d}t$ were known. Talbot and Arnett (1973) has found $y \approx 0.01$ for a IMF with $s = 1.45$. With a constant y, Equation (II.19) admits the solution

$$z(t) = y \ln\left[\frac{M_T}{M_G(t)} \right] = y \ln\left[1 + \frac{M_S(t)}{M_G(t)} \right]$$

which is independent of the explicit form of $\mathrm{d}M_G/\mathrm{d}t$.

An inmediate interpretation of the metal abundances along the sequence of galaxies follows from Equation (II.20): the smaller the fractional content of gas is, the higher is the metal content in the interstellar medium of galaxies (Section II.1.3). On the other hand, the logarithm dependence on M_G/M_T accounts for the rather low variation of z between late-type galaxies. We had seen (Section II.1.3) that M_G/M_T lies between 0.05 and 0.25 for late spirals and irregulars, which leads to a change in z of only about a factor 2 for these objects.

The Rate of Stellar Formation

In order to have a solution of the basic equations we need a form for the rate-function $f(t)$. The usual assumption (Schmidt, 1959) is that in a closed volume

$$\frac{\mathrm{d}M_{TS}}{\mathrm{d}t} = KM_G^n \tag{II.21}$$

the rate of mass conversion into stars is proportional to a certain positive power of the gas density (or mass in the given volume). The gas mass M_G will be dependent both on star formation and the gas turned back by the evolving stars. Equation (II.19) gives

$$- \frac{\mathrm{d}M_G}{\mathrm{d}t} = \frac{\mathrm{d}M_{TS}}{\mathrm{d}t} \left(1 - \frac{\mathrm{d}M_E}{\mathrm{d}M_{TS}} \right) = (1 - \beta)KM_G^n$$

β is the fraction of matter which going through stars, comes back to the interstellar medium. Its value depends on the adopted form of the IMF, ranging between 0.12 to 0.25. In consequence we may write

$$\frac{\mathrm{d}M_G}{\mathrm{d}t} = - \frac{1}{\tau_*} M_G^n, \tag{II.22}$$

where $\tau_*^{-1} = K(1 - \beta)$ is the characteristic time-scale of the process. Equation (II.22) has different analytical solutions depending on the value of the exponent n.

Since the number of young massive ($m > m_*$) stars $N(m_*)$ correlates with the present rate-function $f(t_0)$,

$$N(m_*) = \int_{m_*} \varphi(m)\, \mathrm{d}m = f(t_0) \int_{m_*} \psi(m)\tau(m)\, \mathrm{d}m \propto f(t_0)$$

because $\tau(m) \ll t_0$ if $m_* > m_0$, we should expect from Equation (II.18) also a correlation between the number of bright stars $N(m_*)$ and $M_G^n(t_0)$. Several authors

(Sérsic, 1964; also Table in Madore, 1977) have estimated n on basis of star counts and H I mass distributions in nearby galaxies.

The H II regions result from the ionization of the interstellar hydrogen by the ultraviolet ($\lambda < 912$ Å) radiation emitted by young massive stars. This provides us with a natural indicator for the places where star formation is proceeding. Sérsic (1964) has made use of this technique to determine n.

The volume of ionized gas is $V_i \propto I/\rho_G^2$, where I is the ionizing radiation and ρ_G the gas density.

Assuming the UV luminosity of exciting stars is proportional to its bolometric luminosity $l(m)$, the ionizing radiation produced in a volume V by the exciting stars ($m > m_*$) is

$$I \propto V \int_{m_*} \varphi(m) l(m) \, \mathrm{d}m = V f(t_0) \int_{m_*} \psi(m) m \, \mathrm{d}m. \qquad \text{(II.23)}$$

Since $\tau(m) = m/l(m) \ll t_0$. Introducing now the ionized fraction of space $\xi = V_i/V$, Equation (II.23) and V_i together give

$$\xi = \frac{V_i}{V} \propto \frac{f(t_0)}{\rho_G^2} \int_{m^*} \psi(m) m \, \mathrm{d}m$$

or

$$\xi \propto \rho_G^{n-2} \int_{m''} \psi(m) m \, \mathrm{d}m \propto \rho_G^{n-2}$$

at least for the largest H II complexes ($m \gg m^*$).

The foregoing expression implies: (i) if $n < 2$, an anticorrelation between the space distribution of the larger H I regions and the gas density will be found; (ii) if $n = 2$ there will be no correlation; and (iii) if $n > 2$ the larger H II regions will correlate with the gas.

Radiofrequency observations, particularly in the case of our Galaxy and the nearby ones (Section II.1.3), suggest that $n > 2$ is required. This result applies to massive stars and agrees with Schmidt's (1963) value $n \approx 3$ for them, which on average gives $n \approx 2$.

Madore (1977) has called attention to the fact that Equation (II.21) is a causal one, relating the gas density at the start of the star formation process, to the number of stars which will be formed. By means of a simple numerical model he discusses the correlation between N and M_G in a given enclosure and found that $n \approx 2$ is a result compatible with a natural time scale $\rho^{-1/2}$ provided the efficiency of the process is low enough.

Integrated Colors as age Indicators

It is possible to compute the integrated color of a closed system in which stars are formed and evolve. The color index increases when the high luminosity stars in the upper MS are decimated by stellar evolution, so that the global properties of the system became determined mainly by the great number of red stars of low intrinsic brightness.

Searle *et al.* (1973) have computed the U, B, V colors of model galaxies characterized by three parameters: the exponent s of the power-law expression on for the IMF, τ_* the time scale for an exponential ($n = 1$) stellar formation rate, and time t.

When we make $\tau_* \to \infty$ (uniform rate) the model satifactorily represents the integrated colors of open clusters as a function of time, while the colors of the Sc and Irr I galaxies are obtained provided (i) all galaxies were about 10^{10} yr old; (ii) the IMF is a Salpeter-type power-law such that $s(\text{Irr I}) < s(\text{Sc})$, which means that massive stars are more abundent in Irr I than Sc galaxies; (iii) the mean rate of stellar formation, averaged on large volumes and long times, behaves as a decreasing function of time; (iv) galaxies of the same morphological type may be obtained from different time-scales τ_*, but many galaxies have gone instead through a uniform rate in the last 10^{10} yr. The same authors have suggested that very blue galaxies can be interpreted as systems in which sudden and intense star formation outbursts are produced in a random fashion (see below).

To achieve the preceding results, Searle *et al.* have resorted to the existing information on evolutionary tracks for stars with different masses ($0.3_\odot < M < 30_\odot$) taken from Stothers (1963). They computed for each time the luminosities L_B, L_V contributed by the stars on the whole H-R diagram, after assuming they were formed at $t = 0$ with a Salpeter-type IMF. Thus, in a semi-empirical way, they obtained the values of the second column in Table II.6 which also can be represented with

$$\langle B - V \rangle_{\text{SSB}} = - 0.97 + 0.12 \log t, \quad \log t < 9$$

and

$$\langle B - V \rangle_{\text{SSB}} = - 2.64 + 0.35 \log t, \quad \log t > 9.$$

The different time dependence is related to the CN-and PP-cycles which provide the energy source in stars.

In order to stress the essence of the procedure, we will develop next a much simplified model to show the time-dependence of the integrated color of an evolving population.

Let us assume the IMF given by Equation (II.17) and a uniform rate for stellar formation, so that for any time $t > 0$ we have from Equation (II.18)

$$\varphi(m) = \tau(m)\psi(m) \quad \text{if} \quad m > m_0,$$
$$\varphi(m) = t\psi(m) \qquad \text{if} \quad m \leqslant m_0,$$

where m_0 is such that $\tau(m_0) = t$. The zero-age main sequence (ZAMS) is defined in the luminosity-color plane through B–$V = -2.5 \log a + bM_V$, which is equivalent

TABLE II.6

Integrated colors and M/L_V ratios as a function of time*

Age (10^7 yr)	$(B - V)_i$	$(U - B)_i$	$M/L_V{}^{(+)}$
0.1	−0.22	−1.10	0.07
1	−0.18	−0.78	0.05
10	+0.19	−0.23	0.30
100	+0.46	+0.16	2.3
1000	+0.90	+0.36	10

* Adapted from Searle *et al.* (1973) for $\psi(m) \propto m^{-2\,45}$.
$^{(+)}$ Normalized to $M/L_V = 10$ for $t = 10^{10}$ yr.

to $l_B = al_V^{1+B}$, a relation connecting the B- and V-luminosities of a star of mass m in the ZAMs. The integrated luminosities L_V, L_B of the system at time t are now given by the expressions

$$L_V = \tau(m_0) \int^{m_0} \phi l_V \, dm + \int_{m_0} \phi l_V \, \tau \, dm,$$

$$L_B = a \, \tau(m_0) \int^{m_0} \phi l_V^{1+b} \, dm + a \int_{m_0} \phi l_V^{1+b} \, \tau \, dm,$$

which, because the mass-luminosity law $l_V = m^\alpha$, and $\tau(m) = m^{-(\alpha/\gamma)}$, jointly with a Salpeter power-law for the IMF $\phi(m) \propto m^{-(s+1)}$, reduce to

$$L_V(t) = L_V^0(\alpha, \gamma, s) m_0^{\alpha - [s + (\alpha/\gamma)]}$$

and

$$L_B(t) = aL_B^0(\alpha, \gamma, s) \, m_0^{\alpha(1+b) - [s + \alpha/\gamma]}$$

after integrating between a lower limit $m' \ll m_0$ and an upper limit $m'' \gg m_0$. The integrated color of the system is now given by

$$(B - V)_i = C + 2.5b \log t$$

after noticing that $t = m_0^{-(\alpha/\gamma)}$. The time-dependence of integrated color is now apparent: it arises from the decreasing mass m_0 at the upper end of the MS. The constant C is the zero age integrated color and should coincide with that of young clusters and associations.

The mass-luminosity ratio M/L_V can also be computed for these models. In our crude example we obtain

$$\log\left(\frac{M}{L_V}\right) = \log\left(\frac{M}{L_V^0}\right) + \frac{\alpha(\alpha - s) + \alpha}{\alpha} \log t$$

an increasing function of t. Through elimination of time between the last expression and that for the integrated color,

$$\log\left(\frac{M}{L_V}\right) = \text{const} + \frac{\gamma(\alpha - s) + \alpha}{2.5\alpha\gamma b} (B - V)_i$$

a relationship between both parameters is predicted. Nordsiek (1973) has shown that this is the case for spiral galaxies. Since the M/L ratio is very sensitive to the presence of faint red dwarfs, Sargent and Tinsley (1974) have used this technique to test whether or not the galaxies observed by Nordsiek have the same IMF in the whole range of masses. They obtained suggestions of a possible enhancement in the lower IMF in earlier Sc's.

The Nature of Dwarf Galaxies

When the morphological classification of galaxies was discussed in Section I.1, it was noticed that the surface brightness fluctuations conceals the underlying symmetries of normal Irr I galaxies. According to Searle *et al.* (1973) a galaxy is composed of a certain number ν of elementary cells with a characteristic size, where star formation proceeds at random bursts simultaneously in about $\sqrt{\nu}$ cells. For large galaxies ν

is also large and the effect of the bursts on global properties, such as luminosity or color, is small. The case of Irr I systems puts in evidence the cell structure, although the fluctuations are still restrained. The dwarf systems are thought to be comparable to an elementary cell and then fluctuations become wholly random: during a period which is comparable with the time-life of a burst, we observe a blue dwarf system (Section III.2).

During long times between active phases the galaxy is hardly visible or, if any, appears as a red dwarf surrounded by a H I envelope (Section II.1).

Though the application of the stochastic self-propagating model for star formation (Section II.2.1) Gerola *et al.* (1980) have suceeded to show that small systems can easily experience a fluctuation of the star formation, which naturally leads to a bursting mode. The transition size from continuous to bursting modes is 1 to 2 kpc.

Because all the mass of the dwarf system is available for star formation between active phases, the rate of formation in the bursts is larger than the one prevailing in the continuous mode.

Through the model developed by Gerola *et al.* (1980), these authors have been able to incorporate dwarf galaxies within an unifying concept and to show that the percolating nature of the self-propagating stellar formation model is the reason for the change in modes at a critical size. The energy liberated by the just formed massive stars, responsible for the propagation of the process, sweeps out the remaining gas and consequently prevents further local star formation. In a large system star formation can percolate into regions which have already recovered from earlier bursts. In small systems, however, this is not possible and some recovering time has to elapse till the next spontaneous burst.

In this way the dual behavior of dwarf galaxies, ranging from low surface brightness reddish objects to high surface brightness blue compact galaxies, can be understood.

References

Agüero, E.L.: 1980, *Astrophys. Space Sci.* **73**, 193.

Allen, R.J.: 1973, *Astrophysical Quantities*, London.

Allen, R.J.: 1975, in L. Weliachew (ed), 'La Dynamique des Galaxies Spirales', *Colloque CNRS* **241**, 157.

Aller, L. and Faulkner, D.J.: 1962, *Publ. Astron. Soc. Pacific* **74**, 219.

Arp, H. and Bertola F.: 1969, *Astrophys. Letters* **4**, 23.

Baade, W.: 1944, *Astrophys. J.* **100**, 137.

Baldwin, J.E.: 1975, in A. Hayli (ed.), 'Dynamics of Stellar Systems', *IAU Symp.* **69**, 341.

Baum, W.: 1959, *Publ. Astron. Soc. Pacific.* **71**, 106,

Bertola, F.: 1972, *Soc. Astron. It. Gionnate di Studio* **XV**, 199.

Bertola, F. and Capaccioli, M.: 1975, *Astrophys. J.* **200**, 439.

Bosma, A. and van der Kruit, P.C.: 1979, *Astron. Astrophys.* **79**, 281.

Bouvier, P. and Janin, G.: 1970, *Astron. Astrophys.* **5**, 127.

Brandt, J.C. and Belton, M.J.: 1962, *Astrophys. J.* **136**, 352.

Burbidge, E.M. and Burbidge, G.R.: 1962, *Astrophys. J.* **135**, 694.

Burbidge, E.M. and Burbidge, G.R.: 1965, *Astrophys. J.* **142**, 634.

Burbidge, E.M. and Burbidge, G.R.: 1975, in A.R. Sandage, M. Sandage, and J. Kristian (eds.), *Galaxies and the Universe*, Univ. of Chicago Press, p. 81.

Burbidge, E.M., Burbidge, G.R., and Prendergast, K.H.: 1959, *Astrophys. J.* **130**, 739.

Bulesteix, J., Courtés, G., Laval, A., Monnet, G., and Petit, H.: 1974, *Astron. Astrophys.* **37**, 33.

Burstein, D.: 1979, *Astrophys. J. Suppl. Sér.* **41**, 435.

Burstein, D.: 1979, *Astrophys. J.* **234**, 435.

Burstein, D.: 1979, *Astrophys. J.* **234**, 829.

Capaccioli, M.: 1979, in D.S. Evans (ed.), *Photometry, Kinematics and Dynamics of Galaxies*, Univ. of Texas Press, p. 165.

Coleman, G.D. and Worden, S.P.: 1977, *Astrophys. J.* **218**, 792.

Davies, R.D.: 1974, in J.R. Shakeshaft (ed.), 'The Formation and Dynamics of Galaxies', *IAU Symp.* **58**, 119.

Deutsch, A.J.: 1964, *Astrophys J.* **139**, 532.

de Vaucouleurs, G.: 1948, *Compt. Rend. Acad. Sci. Paris* **227**, 548.

de Vaucouleurs, G.: 1958, *Astrophys. J.* **128** 465.

de Vaucouleurs, G.: 1959, *Handbuch der Physik* **LIII**, 275.

de Vaucouleurs, G.: 1969, *Astrophys. Let.* **4**. 17.

de Vaucouleurs, G.: 1974. in J.R. Shakeshaft (ed.), 'The Formation and Dynamics of Galaxies', *IAU Symp.* **58**, 1.

de Vaucouleurs, G.: 1977, in B.M. Tinsley and R.B. Larson (eds.), *The Evolution of Galaxies and Stellar Populations*, Yale Univ. Press, p. 43.

de Vaucouleurs, G. and Capaccioli, M.: 1979, *Astrophys. J. Suppl. sér.* **40**, 699.

de Vaucouleurs, G.: 1979, in D.S. Evans (ed.), *Photometry, Kinematics and Dynamics of Galaxies*, Vol. 1, Univ. of Taxas Press.

de Vaucouleurs, G. and de Vaucouleurs, A.: 1964, *Reference Catalogue of Bright Galaxies*, Univ. of Texas Press.

de Vaucouleurs, G., de Vaucouleurs, A., and Corwin, H.G.: 1976, *Second Reference Catalogue of Bright Galaxies*, Texas Univ. Press.

Emerson, D.T.: 1974, *Monthly Notices Roy. Astron. Soc.* **169**, 607.

Evans, D.S. (ed.): 1979, *Photometry, Kinematics and Dynamics of Galaxies*, Univ. of Texas Press.

Faber, S.M.: 1973, *Astrophys. J.* **179**, 731.

Faber, S.M. and Gallagher, J.S.: 1970, *Astrophys. J.* **204**, 365.

Faber, S.M. and Gallagher, J.S.: 1979, *Ann. Rev. Astron. Astrophys.* **17**, 135.

Faber, S.M. and Jackson, R.E.: 1976, *Astrophys J.* **204**, 668.

Faulkner, D.J. and Aller, L.: 1965, *Monthly Notices Roy. Astron. Soc.* **130**, 393.

Freeman, K.C.: 1970, *Astrophys. J.* **160**, 811.

Gerola, H. and Seiden, P.E.: 1978, *Astrophys. J.* **223**, 129.

Gerola, H., Seiden, P.E., and Schulman, L.S.: 1980, *Astrophys. J.* **242**, 517.

Hanes, D.A.: 1977, *Monthly Notices Roy. Astron. Soc.* **179**, 331.

Hanes, D.A.: 1977, *Monthly Notices Roy. Astron. Soc.* **180**, 309.

Hanes, D.A.: 1977, *Mem. Roy. Astron. Soc.* **84**, 45.

Harris, W.E. and Racine, R.: 1979, *Ann. Rev. Astron. Astrophys.* **17**, 241.

Heidmann, J.: 1969, *Astrophys. Letters* **3**, 19.

Heidmann, J., Heidmann, N., and de Vaucouleurs, G.: 1971, *Mem. Roy. Astron. Soc.* **75**, 85.

Hodge, P.W.: 1961, *Astron. J.* **66**, 249, 384.

Hodge, P.W.: 1974, *Publ. Astron. Soc. Pacific* **66**, 845.

Holmberg, E.: 1937, *Lund. An.* **6**.

Holmberg, E.: 1954, *Medd. Lund. Ser. I* **86**, 1.

Holmberg, E.: 1958, *Lund. Obs. Medd. Ser. II*, **136**.

Illingworth, G.: 1977, *Astrophys. J.* **218**, L43.

Illingworth, G. and Field, J.: 1979, in D.S. Evans (ed.), *Photometry, Kinematics and Dynamics of Galaxies*, Univ. of Texas Press, p. 177.

Israël, F. P. and van der Kruit, K.C.: 1974, *Astrophys J.* **187**, L83.

Jaschek, C.: 1957, *Z. Astrophys.* **44**, 23.

Johnson, H.M.: 1959, *Publ. Astron. Soc. Pacific* **71**, 245.

King, I.R.: 1962, *Astron. J.* **67**, 471.

King, I.R. and Minkowski, R.: 1966, *Ann. Astrophys.* **23**, 385.

Kormedy, J.: 1977, *Astrophys. J.* **214**, 359.

Kormendy, J.: 1977, *Astrophys. J.* **217**, 406.

Kormendy, J.: 1977, *Astrophys J.* **218**, 333.

Kormendy, J.: 1977, *Astrophys. J.* **227**, 714.

Kormendy, J. and Bahcall, J.: 1974, *Astron. J.* **79**, 671.

Kormendy, J. and Norman, C.A.: 1979, *Astrophys J.* **233**, 539.

Larson, R.B.: 1974, *Monthly Notices Roy. Astron. Soc.* **166**, 585.

Limber, N.: 1961, in T.L. Page, J. Neyman, and E. Scott, (eds.), *Santa Barbara Conference on Galaxy Instability.*

Madore, B.: 1977. *Monthly Notices Roy. Astron. Soc.* **178**, 1.

Marton, S.: 1978, Seminar paper, Observatorio Astronómico, Univ. Nacional, Córdoba.

Mayall, N.U.: 1958. in N.G. Roman (ed.), 'Comparison of the Large-Scale Structure of the Galactic Systems with that of other Stellar Systems', *IAU Symp.* **5**, 23.

Mayall, N.U.: 1960, *Ann. Astrophys.* **23**, 344.

Marchand, B. and Olsen, D.W.: 1979, *Astrophys. J.* **230**, L157.

Mathis, J.S.: 1965, *Publ. Astron. Soc. Pacific* **77**, 90.

Mathews, W.G. and Baker, J.C.: 1971, *Astrophys. J.* **170**, 244.

McClure, R. and van den Bergh, S.: 1968, *Astron. J.* **73**, 313.

Metzger, P.G.: 1975, in G. Setti (ed.), *Structure and Evolution of Galaxies,* D. Reidel Publ. Co., Dordrecht, Holland, p. 143.

Miller, R.H.: 1978, *Astrophys. J.* **223**, 192.

Miller, R.H.: 1978, *Astrophys. J.* **224**, 32.

Miller, R.H. and Smith, B.F.: 1979, *Astrophys. J.* **227**,785.

Minkowski, R.: 1961, in McVittie (ed.), 'Problems of Extragalactic Research', *IAU Symp.* **15**, 112.

Minkowski, R. and Osterbrock, D.: 1959, *Astrophys J.* **129**, 583.

Monnet, G.: 1971, *Astron. Astrophys.* **12**, 379.

Nordsiek, K.H.: 1973, *Astrophys. J.* **184**, 719.

Osterbrock, D.: 1960, *Astrophys. J.* **132**, 325.

Ostriker, J.P. and Peebles, P.J.E.: 1973, *Astrophys. J.* **186**, 467.

Ostriker, J.P., Richstone, D.O. and Thuan, T.X.: 1974, *Astrophys J.* **188**, L87.

Page, T.L.: 1948, *Astrophys J.* **108**, 157.

Page, T.L.: 1952, *Astrophys. J.* **116**, 63.

Page, T.L.: 1961, *Proc. Verk. Symp. Math. Stat. Prob.* **IV**, 111.

Page, T.L.: 1962, *Astrophys. J.* **136**, 685.

Pagel, B.: 1973, *ICTP Trieste It. Rept. IC* **173**, 169.

Peimbert, M. and Spinrad, H.: 1970. *Astron. Astrophys.* **7**, 311.

Perek, L.: 1962, *Adv. Astron. Astrophys.* **1**, 165.

Peterson, C.J.: 1978, *Astrophys. J.* **207**, 382.

Pismish, P.: 1979, in D.S. Evans (ed.), *Photometry, Kinematics and Dynamics of Galaxies*, Univ. of Texas Press, p. 235.

Poveda, A.: 1961, *Astrophys. J.* **134**, 910.

Prendergast, K. and Tomer, E.G.: 1970, *Astron. J.* **75**, 674.

Reynolds, J.H.: 1913, *Monthly Notices Roy. Astron. Soc.* **74**, 132.

Roberts, M.S.: 1963, *Ann. Rev. Astron. Astrophys.* **1**, 149.

Roberts, M.S.: 1969, *Astrophys. J.* **158**, 123.

Roberts, M.S.: 1972, in D.S. Evans (ed.;, 'External Galaxies and QSOs', *IAU Symp.* **44**, 12.

Roberts, M.S.: 1975a, in A. Hayli (ed.), 'Dynamics of Stellar Systems', *IAU Symp.* **69**, 331.

Roberts, M.S.: 1975b, in A.R. Sandage, M. Sandage, and J. Kristian (eds.), *Galaxies in the Universe*, Chapter IX, Vol. IX, Chicago Univ. Press.

Roberts, M.S. and Rots, A.H.: 1973, *Astron. Astrophys.* **26**, 483.

Rood, H.J., Rothman, V.C., and Turnrose, B.E.: 1970, *Astrophys. J.* **162**, 411.

Richstone, D.O.: 1979, *Astrophys. J.* **234**, 825.

Richstone, D.O. and Sargent, W.L.W.: 1972, *Astrophys. J.* **176**, 91.

Rubin, V.C., Ford, W.K., and Thonnard, N.: 1979, Ann. Rept. DTM 1978/9, Carnegie Inst., Washington.

Rubin, V.C., Ford, W.K., and Thonnard, N.: 1980, *Astrophys. J.* **238**, 471.

Sandage, A.R.: 1958, *Astrophys. J.* **127**, 513.

Sandage, A.R.: 1961, *The Hubble Atlas of Galaxies*, Carnegie Inst., Washington, D.C.

Sandage, A.R.: 1971, in D.O' Conell (ed.), *Nuclei of Galaxies*, North-Holland, Publ. Co., Amsterdam, p. 271.

Sargent, W.L.W.: 1977, in G. Setti (ed.), *Structure and Evolution of Galaxies*, D. Reidel Publ. Co., Dordrecht. Holland, p. 265.

Sargent, W.L.W. and Tinsley, B.: 1974, *Monthly Notices Roy. Astron. Soc.* **168**, 19p.

Sargent, W.L.W., Schechter, P.L., Boksenberg, A., and Shortridge, K.: 1977, *Astrophys. J.* **212**, 326.

Salpeter, E.E.: 1955, *Astrophys J.* **121**, 161.

Schmidt, M.: 1956, *Bull. Astron. Inst. Neth.* **13**, 15.

Schmidt, M.: 1959, *Astrophys. J.* **129**, 243.

Schmidt, M.: 1962, J. Sahade (ed.), *Symp. Stellar. Evolution*, La Plata Observatory, p. 61.

Schmidt, M.: 1963, *Astrophys. J.* **137**, 758.

Schweitzer, F.: 1976, *Astrophys. J. Suppl. Sér.* **31**, 313.

Searle, L.: 1973, *Stellar Ages LII*, Obs. Paris-Meudon, p. 1.

Searle, L. and Sargent, W.: 1972, *Astrophys. J.* **173**. 25.

Searle, L., Sargent, W.L.W., and Bagnuolo: 1973, *Astrophys. J.* **179**, 427.

Sérsic, J.L.: 1959, *Observatory* **79**, 54.

Sérsic, J.L.: 1960, *Z. Astrophys.* **50**, 168.

Sérsic, J.L.: 1964, *Z. Astrophys.* **58**, 259.

Sérsic, J.L.: 1968, *Atlas de Galaxias Australes*, Univ. Nacional Córdoba.

Sérsic, J.L. and Calderón, J.H.: 1979, *Astrophys Space Sci.* **62**, 211.

Shane, W.W.: 1975, in L. Weliachew (ed.), 'La Dynamique des Galaxies Spirales', *Colloque CNRS* **241**, 217.

Shu. F.H.: 1974, in K. Pinkau, (ed.), *Interstellar Medium*, D. Reidel, Publ. Co., Dordrecht, Holland, p. 219.

Simkin, S.M.: 1974, *Astron. Astrophys.* **31**, 129.

Spinrad, H.: 1961, *Publ. Astron. Soc. Pacific* **73**, 336.

Spinrad, H.: 1962, *Astrophys. J.* **135**, 715.

Spinrad, H. and Ostriker, J.P.: 1974, *Bull. Am. Astron. Soc.* **6**, 332.

Stothers, R.L.: 1963, *Astrophys. J.* **138**, 1074.

Takase, B. and Kinoshita, H.: 1967, *Publ. Astron. Soc. Japan* **19**, 409.

Talbot, R.J. and Arnett, W.: 1973, *Astrophys. J.* **186**, 51, 69.

Thuan, T.X. and Gott, J.R., III; 1975, *Nature* **257**, 774.

Toomre, A.: 1963, *Astrophys. J.* **138**, 385.

Trimble, V.: 1975, *Rev. Mod. Phys.* **47**, 877.

Turnrose, B.E.: 1976, *Astrophys. J.* **210**, 33.

van den Bergh, S.: 1957, *Z. Astrophys.* **43**, 236.

van den Bergh, S.: 1965, *Astron. J.* **70**, 124.

van den Bergh, S.: 1974, in J.R. Shakeshaft, (ed.), 'The Formation and Dynamics of Galaxies', *IAU Symp.* **58**, 15.

van den Bergh, S.: 1975a, *Ann. Rev. Astron. Astrophys.* **13**, 217.

van den Bergh, S.: 1975b, *J. Roy. Astron. Soc. Can.* **69**, 57.

van den Bergh, S.: 1977, in B.M. Tinsley and R.B. Larson, (eds.), *The Evolution of Galaxies and Stellar Populations*, p. 32.

van der Kruit, P.C. and Allen, R.J.: 1976, *Ann. Rev. Astron. Astrophys.* **14**, 417.

Visvanathan, N. and Sandage, A.R.: 1977, *Astrophys. J.* **216**, 214.

Wakamatsu, K.I.: 1977, *Publ. Astron. Soc. Japan* **89**, 504.

Young, P.J.: 1976, *Astron. J.* **81**, 807.

Yohizawa, M. and Wakamatsu, K.I.: 1975, *Astron. Astrophys.* **44**, 363.

Warner, P.J., Wright, M.C.H., and Baldwin, J.E.: 1973, *Monthly Notices Roy. Astron. Soc.* **163**, 163.

Wilson, C.P.: 1975, in A. Hayli, (ed.), 'Dynamics of Stellar Systems', *IAU Symp.* **69**, 209.

CHAPTER III

ACTIVE GALAXIES

From the knowledge of the spectral distribution of energy radiated by extragalactic objects we can infer the emission mechanisms which generate the spectrum, although not necessarily the ultimate nature of the energy source. If by any reason this were known, it would be necessary to establish the chain of processes which transform the energy released in it until it is radiated. Three problems have to be distinguished: the origin of the energy, the mechanism of transformation, and that of the emission.

In order to get an idea of the nature of the emission mechanism it is necessary to compute the total energy spent by the system (E). We already know the radiated power (L) and if we could estimate the mean lifetime of the process (t) we would be able to calculate roughly the total energy $E = L \times t$.

The major difficulty lies in the estimation of the lifetime of the process so that the results obtained up to now should be accepted only as orders of magnitude. The interpretation of the observations shows that most of the different categories of extragalactic objects consume energy of the order of 10^{62} erg during intervals that go from 10^{10} yr for normal galaxies down to 10^6 yr for QSO's and intense radio galaxies.

Cosmically speaking, an energy release of the order of from 10^{60} to 10^{62} erg in a time scale of the order of 10^6 to 10^7 yr can be considered as a genuine explosion.

III.1. Classification

Ozernoy (1970) has proposed an ordering of the galaxies according to their degree of activity or excitation. This author takes as a starting point the observation that the difference existing between objects such as the nucleus of our galaxy, a Seyfert nucleus or a QSS, lies only in the degree of intensity of the process of liberation of internal energy.

The duration of the active phase in these galaxies may be estimated from its frequency (2%), provided we assume 'activity' occurs in normal galaxies ramdomly. The harmonic mean of the activity phases is then of the order of 10^8 yr. This allows us to conjecture about the direction of galactic evolution towards denser states (Section III.3) and it induced Ozernoy (1970) to introduce an enlarged Hubble sequence with two additional hypothetical classes C_1 and C_2 which assimilate Zwicky's more massive and compact galaxies (Figure III.1). At the other end of minimum compactness, we shall add the so-called intergalactic H II regions observed by Sargent and Searle (1970) (IG H II).

The Hubble sequence thus generalized (IG H II, Irr I, S, SO, E, C_1 and C_2) constitutes what Ozernoy (1970) calls the 'ground state' of galaxies. From the energetic point of view, the ground level is characterized because its members emit very little or no non-thermal energy relative to their luminosity of thermal origin, that is to say $L_{NT} \ll L_T$.

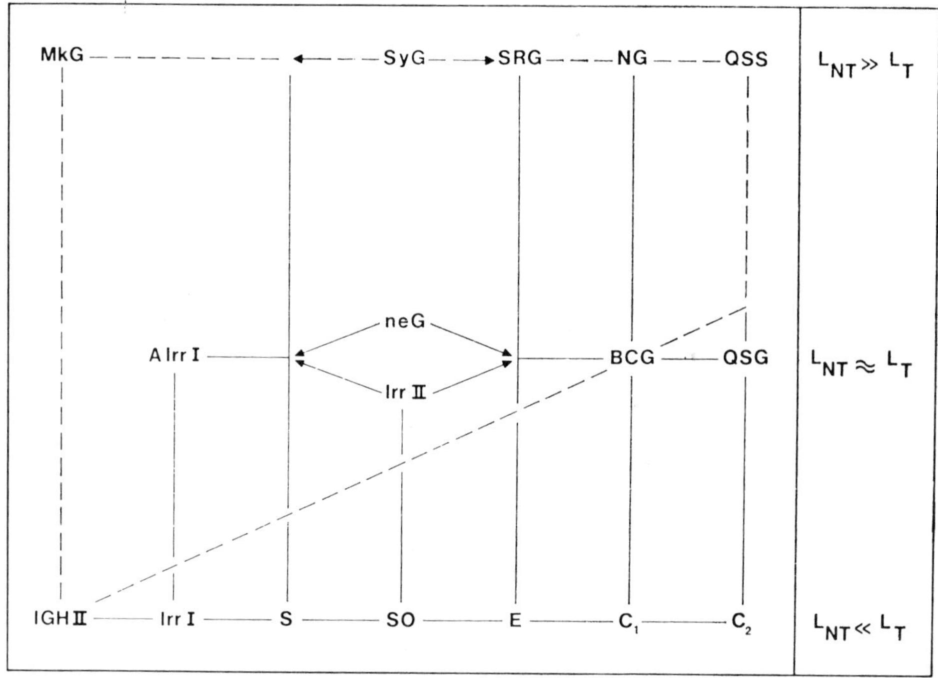

Fig. III.1. Classification of active galaxies. (*Adapted from Ozernoy*, 1970.)

On the other hand, Seyfert galaxies (SyG), strong radio galaxies (SRG), N galaxies (NG), and quasars (QSO) are considered the 'ceiling' of excitation activity wherein the luminosity of non-thermal origin (L_{NT}) exceeds that of thermal origin (L_T). This scheme is completed by introducing a sequence of moderate excitation $L_T \leqslant L_{NT}$ formed by the quasi-stellar galaxies (QSG), blue compacts (BCG), galaxies with nuclear emission (neG), with nuclear radio emission (NRG), and finally by Irr I objects of moderate activity (AIrr I) at the end of least compactness.

The galaxies in the fundamental state ($L_{NT} \ll L_T$) are not connected genetically among themselves and differ only by their initial conditions, mainly their mass and angular momentum (cf. Section II).

The vertical lines in the Ozernoy diagram suggest probable directions for the excitation of galaxy nuclei. On the contrary, this author holds that a transition in the horizontal sense (from QSS to SRG or viceversa, e.g.) is improbable. The ground states of the QSO's and NG's are, according to Ozernoy (1970), galaxies of extreme compactness whose spatial density should be very small compared to that of the BCG's and QSG's if their mean lives are of the order of 10^8 yr.

In Figure III.1 we have summarized Ozernoy's diagram which, in the last instance, is equivalent to establishing the natural groups in the plane defined by their 'activity' L_{NT}/L_T and central concentration (or stage of the extended Hubble sequence).

Markarian UV galaxies (MkG): Since 1967 B. Yu. Markarian has published lists of galaxies with an unusually intense UV continuum spectra. These objects were found in survey plates obtained with a Schmidt camera and an objective prism which produce a dispersion of 1800 Å mm^{-1} in the Hα region.

On the average, 0.1 MkG have been found per square degree. Detailed studies of spectra and photographs with larger dispersions and scales made by Weedman, Sargent, Kachikian and others, have shown the great variety of morphologies which fit the Mk type.

Markarian has also furnished an estimate of the degree of concentration of the ultraviolet continuum for the objects in his lists. Objects classified as '*s*' have the sharpest and best defined images whereas those of class '*d*' are diffuse and ill-defined. Intermediate classes have also been used. According to Markarian the most active galaxies with nuclear emission would be the '*s*' type while young objects and galaxies with disks formed by hot stars and gas would be classified as '*d*'.

A statistical investigation carried out by Sulentic (1976) shows that a third of the Mk galaxies are spirals, a fifth irregulars and another third is composed indistinctly by elliptics (*E*), N galaxies and other compact objects. The absolute magnitude range of the *S* and *I* Markarian objects goes from $M_B = -23$ to -14. Between Mk galaxies exists an apparent excess (30%) of pairs relative to normal galaxies, 9% are radio-sources with a mean spectral index $\alpha = 0.75$ and continuum class '*s*'. Most of them are Seyfert galaxies.

From the spectroscopic point of view we distinguish two groups, namely:

– Diffuse galaxies (*d*) with intense UV continuum emission and lines extending all over the object. The spectrum shows no difference from that of the H II regions, Within this group the members with smaller intrinsic luminosity ($M < -15$) form a well characterized class identified by Sargent and Searle (1970) as genuine intergalactic H II regions (IG H II).

The objects of higher luminosity present a larger diversity of forms. Some are irregular galaxies of high surface brightness, possibly erupting stellar formation (AIrr I) (Figure III.2). Other objects are spirals with emission in the nucleus (neS) and among the nearest we find galaxies with peculiar nuclei characterized by Sérsic and Pastoriza (1967). The Haro galaxies (found by G. Haro in Mexico with a technique similar to that of Markarian but a decade before him) also belong to this group. Their spectra are of moderate excitation, typical for giant H II regions (in some cases) superposed to the stellar background of a SO galaxy (Irr II).

– Galaxies with strong nuclear emission (S): the energy source is of small dimensions, variable intensity and non-thermal origin. Ten percent of the Mk galaxies are Seyfert galaxies (SyG). We also find here most of the N galaxies with associated radio emission, the BL Lac objects and the QSS.

To sum up, the Mk objects constitute a heterogeneous group which comprises a wide range of activity and concentration such as shown in Figure III.1.

– Zwicky Compact Galaxies (ZCG): Zwicky's survey was based on the morphological appearance on plates of the Palomar Sky Survey and revealed an aboundant population of compact objects in space.

Zwicky's lists, as well as Markarian's contain a great variety of objects many of them in common but, as the basic criterion was compacticity, they are partially complementary. Among those objects not yet described as Mks, we find objects which have spectra with only absorption lines, typically Ca II, the G-band and the Mgb

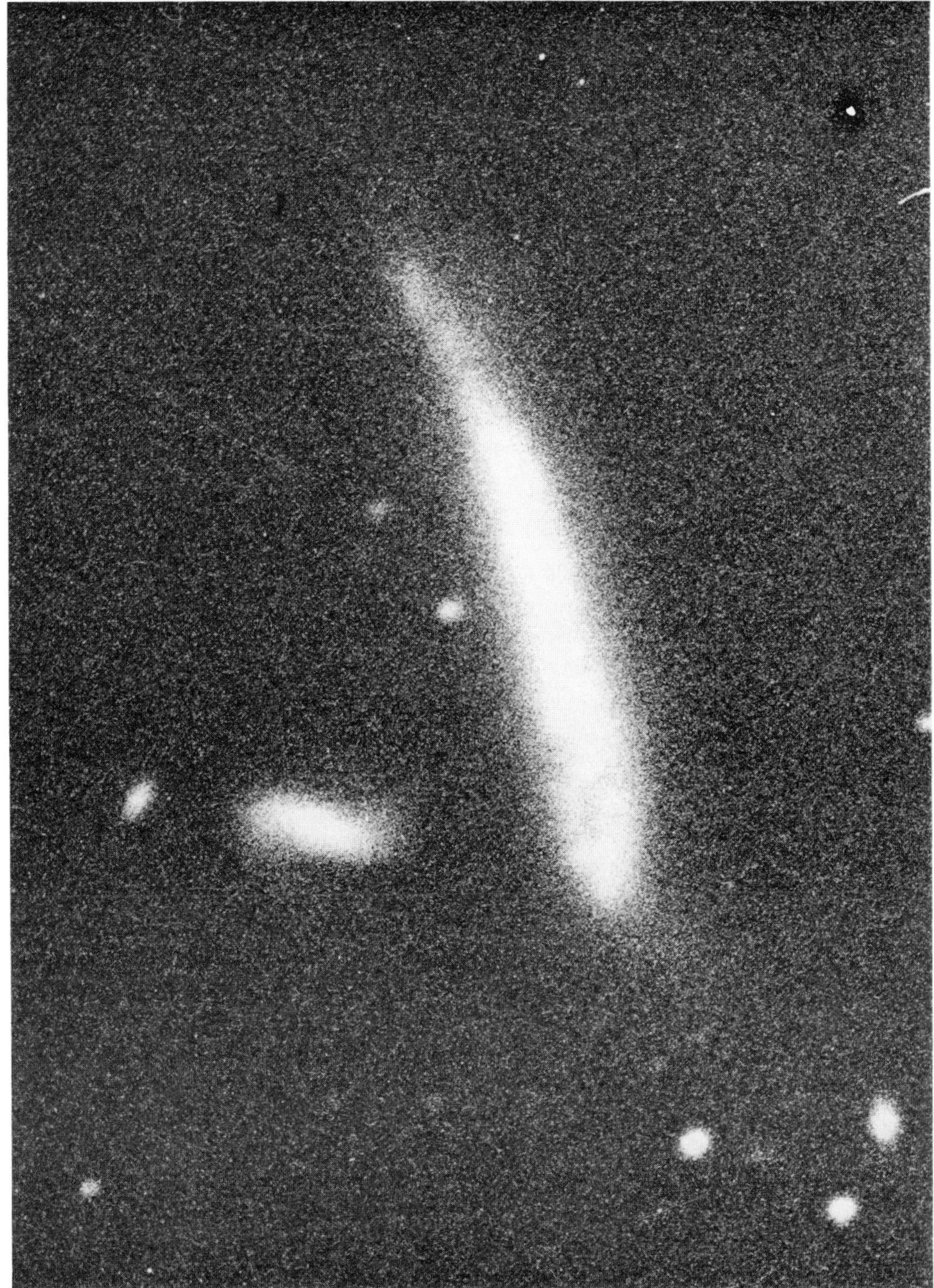

Fig. III.2.   The anonimous Object in 0^h05^m–41° is an example of AIrr I galaxies.
(Sérsic *et al.*, 1968.)

feature. Very few show Balmer and Ca II emission lines. Their colors are similar to those of the elliptic galaxies so that they can be interpreted as dense groups of old stars. This raises the problem of why they have developed a more compact structure than the normal elliptics.

A typical example for this class of galaxies is NGC 4486-B. From its velocity dispersion and brightness distribution a mass of the order of 5×10^{10} suns is inferred and a M/L ratio of the order of 80. Although Rood (1965) suggests that its dimensions might be the result of a tidal limitation caused by its giant neighbor NGC 4486, Zwicky (1965, 1966) has found several other objects of this class that could not be explained by tidal limitation.

III.2. Intergalactic H II Regions (IG H II)

These objects are located at the end of the least degree of concentration in the sequence of galaxies and their emission is only thermal ($L_{NT} \ll L_T$). Their extragalactic character is clearly indicated by their redshifts. Investigations by Sargent and Searle (1970) (Table III.1) show their characteristics to be comparable to the large emission complexes in spiral galaxies (Section II.1).

TABLE III.1

Properties of the IG H II

Dimensions	0.2
Luminosity (visual suns)	10^7
Colors $(B - V)_0 = 0$, $(U - B)_0 = -0.6$	
Total mass (suns)	10^8
H I mass (suns)	10^8
H II mass (suns)	10^6
Examples	IZw18, IIZw40.

III.3. Galaxies with Nuclear Emission (neG)

Various surveys have led to the discovery of galaxies with nuclei conspicuous in various aspects. When Morgan introduced the Yerkes classification (Section I.1) he called attention to some galaxies whose nuclei are formed by a certain number of bright H II regions ('hot spots'). Sérsic and Pastoriza (1967) have published lists of galaxies with peculiar nuclei in the sense that "there is a sharp change in the luminosity gradient in the central regions, and some structure due to the existence of high excitation clouds around the nucleus itself". These lists show some overlap with Morgan's work. Vorontsov-Velyaminov, Zaitseva and Lyutyi also have done similar surveys in larger samples of galaxies.

Among 174 galaxies selected as brighter than magnitude 11.0 in the Shapley–Ames catalogue and inclinations large enough as to allow the unobstructed inspection of their nuclei, Sérsic and Pastoriza (1967) have found 20 with peculiar nuclei. All these objects belong to de Vaucouleurs SAB and SB families (Figure III.3) which means that this kind of nuclei seems to be associated with the bar phenomenon. Although this result has been confirmed by Velyaminov and his group, Heckman (1978) has

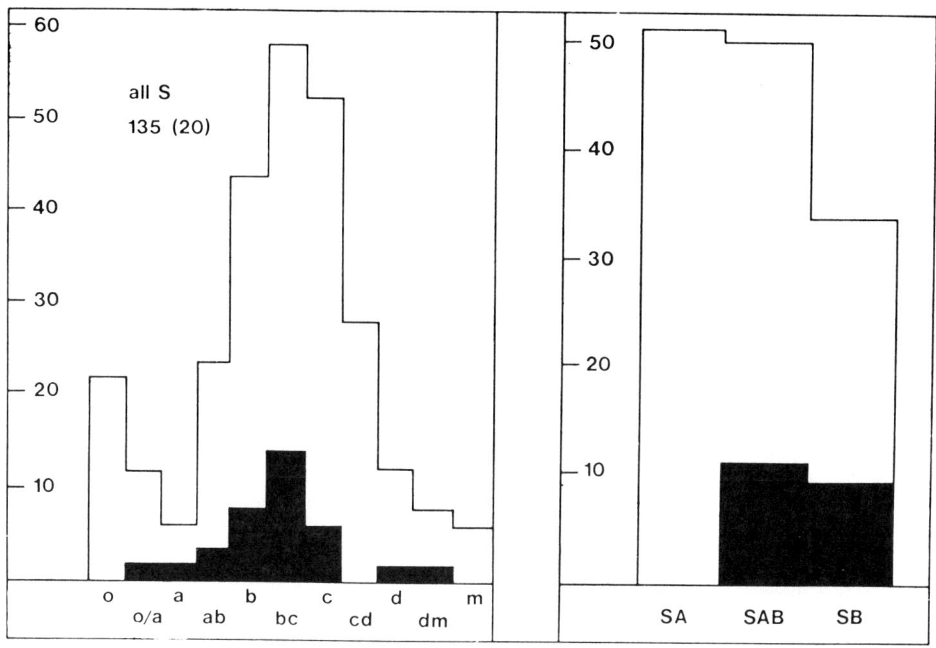

Fig. III.3. Frequency of galaxies with peculiar nuclei. The histogram on the left shows the correlation with the presence of the bar phenomenon (black). On the right, the histogram does not evidence any correlation with the degree of concentration (place along the sequence) of spirals. Below: structure of the peculiar nucleus of NGC 1672. (*After Calderón et al.*, 1980.)

cast doubts on it. His definition of a peculiar nucleus is more general because he also considers absorption features which naturally leads to different results.

Table III.2 lists the 20 bright galaxies with peculiar nuclei.

The histogram of Figure III.3 shows that there is no correlation between the presence of these nuclei and the stage along the Hubble sequence. The numerical density of these objects has been estimated by Sérsic (1968) at $10^{-3.5}$ Mpc^{-3}, a number of the same order as that for the Seyfert galaxies.

TABLE III.2

Galaxies with peculiar nuclei brighter than $m_T = 11.0$*

NGC	Type	L_C
613	SB(rs)bc	2
925	SAB(s)d	4
1097	SB(s)b	2
1365	SB(s)b	2
1433	SB(r)a	—
1672	SB(s)b	2
1808	(R)SAB(s)o/a	—
2903	SAB(rs)bc	2
2997	SAB(rs)c	1
3310	SAB(r)bcp	3
3351	SB(r)b	3
3359	SB(rs)c	3
4051	SAB(rs)bc	3
4151	SAB(rs)ab	3
4303	SAB(rs)bc	1
4321	SAB(s)bc	1
5236	SAB(s)c	2
5248	SAB(rs)bc	1
7424	SAB(rs)bc	2
7552	(R')SB(s)ab	—

* Source RCGII

According to Prabhu (1979), the 'peculiar nuclei' are redder than the main body of the galaxy. The difference between the luminosities in the infrared ($\lambda_e = 7600$ Å) and the blue ($\lambda_e = 4200$ Å) goes from $0\overset{m}{.}4$ to $0\overset{m}{.}8$ relative to the adjacent galaxy background. Holmberg (1960) finds a mean nuclear color excess of the order of $-0\overset{m}{.}06$ in these galaxies, which he attributes to the emission of $\lambda 3727$ (O II). Tifft (1961) and de Vaucouleurs (1961) also have found blue and UV excess in galaxies of this type. The mean absolute magnitude of *ne* galaxies is $M_B = -19.5$ equivalent to a mean luminosity class $L_c = 2.2$ whilst the corresponding value for the nuclei found by Pastoriza (1967) is $M_B = -18$, brighter in the blue than the Seyfert nuclei (Section III.5). The same author has found a mean nuclear diameter $d_N \approx 0.8$ kpc. There are fewer than ten hot spots in a nucleus with a mean diameter of from 0.1 to 0.2 kpc which makes them comparable to the giant H II regions found in the spiral arms of Sc galaxies (Section II.). The hot spots have non-rotational motions (streaming and or z-motions) of the order of a few hundreds of km s^{-1} as follows from spectroscopic observations by Burbidge *et al.* (1962) Pastoriza (1967), Arp and Bertola (1970), Simkin (1971), Oke *et al.* (1974), etc. Burbidge and Burbidge (1960) have estimated the masses of the nuclei of NGC 1097 and 1365 to be of the order of 10^{10} solar masses.

Although Lequeux seems to have found weak radio emission in 75% of these objects, Crane does not confirm these results. Thus, radio evidence does not support the idea of strong activity in galaxies with nuclear emission.

Spectroscopic observations by various groups of observers show that the degree of excitation in hot spots is similar to that of H II regions with $T \approx 8000\ K$ and N_e 10^3 cm^{-3}. The stellar contribution to the continuum comes from F stars, although Pastoriza (1967) finds an appreciable contribution of K and M stars in NGC 1097 and NGC 1808.

The absolute fluxes emitted by hot spots in $H\alpha$ are explained by the presence of numerous OB stars, whilst the intensity ratio $H\alpha/H\beta$ requires that there should be reddening due to absorbing matter (Osmer *et al.*, 1974). All this leads to an interpretation of peculiar nuclei as examples of clouds excited by clusters of young giant stars. According to Prabhu (1979) these clusters originated from sporadic outburst of a star formation, similar to those proposed by Sargent and Searle (1970) in the case of IG H II and AIrr I galaxies.

III.4. Irr II Galaxies

Introduced by Holmberg (Section I.2), they have the spectroscopic and morphological characteristics of late objects – at least in blue plates – but their color index corresponds to early galaxies.

Krienke and Hodge (1974) have classified these objects in three categories:

– Irr I galaxies resulting from tidal interaction (such as NGC 5195, the companion of NGC 5194 = M51) (See Section IV.1).

– Explosive galaxies.

– Post eruptive galaxies.

Explosive Galaxies

Within this group we must distinguish those with normal morphological type E, SO, and those of normal S type. Among the first we find NGC 3077, NGC 2915 and NGC 5253. They have a spectrum of sharp emission lines and moderate excitation, typical of giant H II regions. The nuclear region is dominated by a complex of clouds, filaments, and jets which extend to well outside the main body (Figure III.4) (NGC 5253, NGC 2915).

Moderate nuclear radio emission has been detected and it is usual to find an excess of neutral hydrogen (H I) content relative to their basic morphologic types. Some of these objects have also been detected in the far infrared. The usual interpretation of these galaxies is that they are going through a violent outburst of star formation originated in a sudden contribution of gas which falls into the deep potential well of the galaxy, fragments, condenses into stars, which in their turn feed-back the dynamics of the remaining gas, generating in this way the large nuclear complexes of H II regions and the short-lived unstable structures. According to van den Bergh (1971), NGC 5253 might have accreted its gas passing close to NGC 5236 = M83, but the situation of NGC 2915 is not quite clear because no plausible partner can be found. The existence of H I intergalactic clouds (Section IV.3) could explain this case as a capture of a H I cloud by NGC 2915. The explosive galaxies of type S have as a classical example

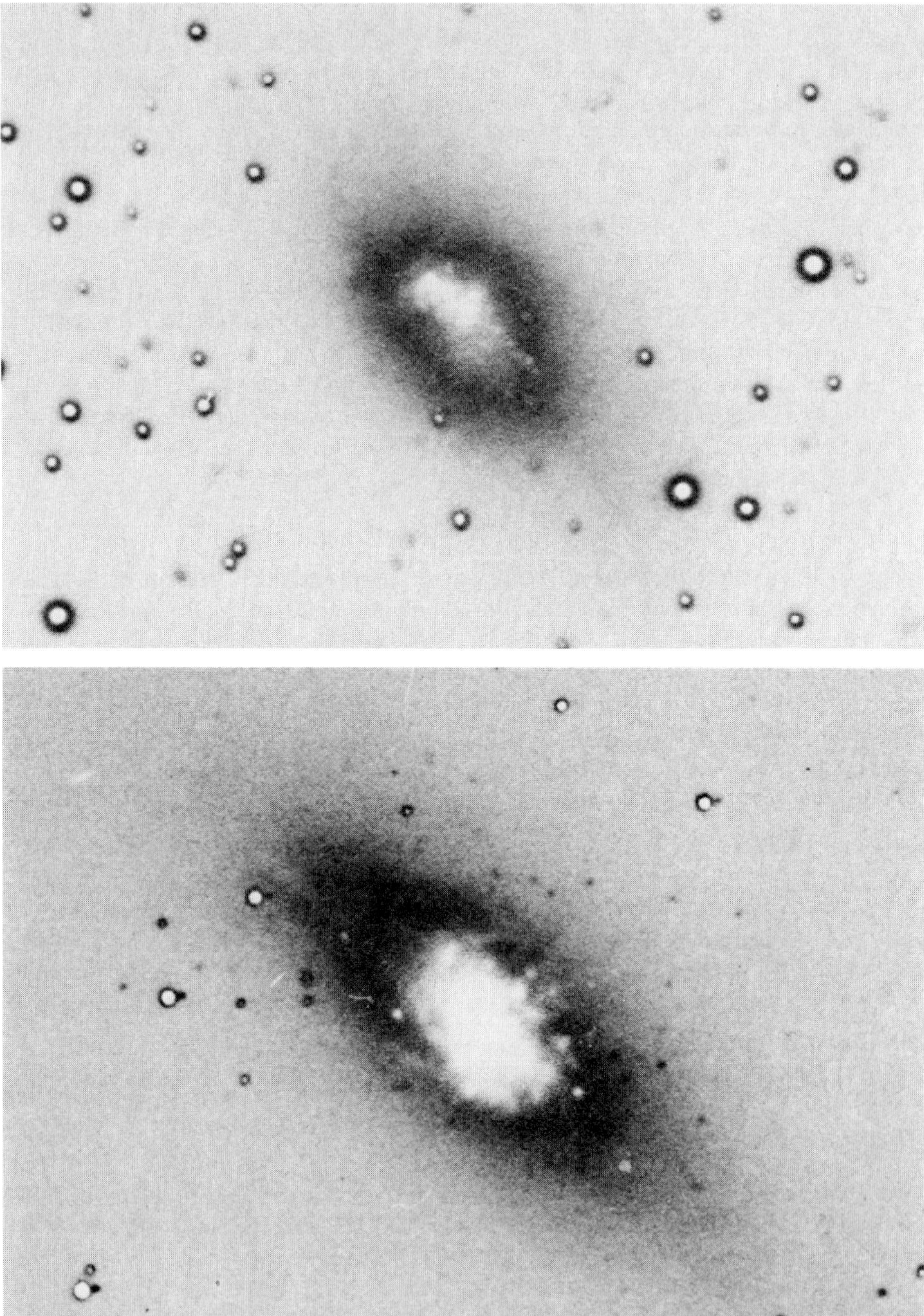

Fig. III.4.   Composite prints of NGC 2915 (above) and NGC 5253 (below). These Irr II southern galaxies display gigantic nuclear complexes from which jets and plumes are issued. Hα emission is shown here in white, whilst the yellow continuum coming from the underlying early-type galaxies is seen in gray. The original plates were taken with the 4 m CTIO telescope (see Sérsic *et al.*, 1972, 1977).

the case of NGC 3034 = M82. This galaxy has an early AF spectrum with strong emission lines in Hα arising from a complex system of Hα filaments. The radio emission is quite strong and its spectrum is flat (with slope $\alpha = 0.2$). The rotation curve of the underlying galaxy has been detected spectroscopically and it was found to correspond to that of a normal spiral galaxy.

Two different interpretations of this object have been advanced: A violent enough explosion in the nucleus of the galaxy can expel hot gases. If such an explosion takes place in the center of an S galaxy, the ejection would be along two oposite cones with their common axis coincidental with that of the rotation of the galaxy. That was the model proposed by Sandage and Lynds (1963) for M82 in order to give an interpretation of the filamentary Hα structure observed above and below the plane of that galaxy. The existence of a velocity gradient would suggest, in the frame of this model, that the matter had been expelled with velocities of the order of 1000 km s^{-1} from the nucleus. Later observations, however, showed that the Hα line as well as the continuum were strongly polarized, which would suggest that the radiation coming from them was not intrinsic but originated in the nucleus and scattered by the dust present on both sides of the plane of the galaxy. If the dust were in expansion from the nucleus, as it is assumed in the model by Sandage and Lynds, the scattered light would be redshifted on both sides of the plane, which is in contradiction with the observations, since it is redshifted on the north side of M82, whereas on the south it is blue shifted, These observations led to an alternative hypothesis: M82 would be moving within a gas cloud and capturing part of it. It is well known that H I studies in the region show the M81 = NGC 3031 group to which M82 belongs, to be surrounded by a hydrogen cloud. From the asymmetry of the gas distribution around M81 it is concluded that M82 quite probably had a close passage with capture of material 10^8 yr ago.

As a consequence of those difficulties in the interpretation of M82 it has been suggested that the peculiar structure of M82 was caused by an event (or events) of rapid and intense stellar formation in its central region (as in the cases of NGC 5253 and NGC 2915). van den Bergh (1971) found in plates taken in the near infrared a dozen of bright condensations dominating the central region of the object. Spectroscopic observations suggest them to be super-associations similar to those studied by Sabhazian but located in the central region of the object. van den Bergh (1971) suggests these objects were formed as a consequence of the events already described.

Recent optical observations of the splitting of the Hα lines in emission from the filaments on the south side of M82 hardly harmonize, however, with the idea of scattering by dust. So it seems after all that an explosion also took place in M82 since the filaments would be expanding at the surface of two hollow cones with common vertex at the nucleus and whose axis coincides with the axis of the galaxy. The ejection velocity implies this explosive event took place 10^7 yr ago, long after M82 began collecting gas from the envelope or M81. (Figure III.5).

Post Eruptive Galaxies

These are objects typically classified as Irr II. According to Krienke and Hodge (1974), because of the presence of a dust envelope these galaxies are much reddened in contrast to their early type spectrum.

Fig. III.5.   The M81 group. The brightest galaxy is M81, a Sb object whose distance is 2.7 Mpc (see Chapter V). The spindle-shaped object above it is the Irr II galaxy M82.
(*Courtesy of Hale Observatories.*)

Examples of this group of galaxies: NGC 972 (spectrum type B), NGC 3067 (F2), NGC 3955, etc. Nearly all these galaxies show normal H II regions and regular rotation curves which indicate that dust is the factor altering their morphology. There are also early galaxies with excess of dust, such as NGC 4753 and 5363.

According to Krienke and Hodge (1974), the post eruptive objects would be the result of ancient explosive events, that is to say, later stages of explosive galaxies.

III.5. Seyfert Galaxies (SyG)

A Seyfert galaxy is a spiral object with a compact nucleus of stellar appearance, much brighter than the nuclei of normal galaxies. The spectra of Sy galaxies may contain features which usually are not observed in other galaxies, especially a degree of excitation higher than the one required to produce [O II] and [Ne III]. The emission lines, at least those of hydrogen, have large widths, equivalent to Doppler velocities going from 500 to 10^4 km s^{-1}.

Table III.3 summarizes some characteristics of the seven SyG with $z < 0.009$.

It is hard to distinguish and classify the structure of the disk underlying the bright nucleus in the more distant Seyfert galaxies ($z > 0.1$) so that many objects classified as N according to their appearance (Section III.7) can be Sy because of their spectrum.

An early discussion by A. and G. de Vaucouleurs about the abundance of bright Sy galaxies in relation to those catalogues by Shapley–Ames showed that 1 % of spirals are SyG, which is equivalent to a spatial density of the order of 5×10^{-4} Mpc^{-3}. These authors also found that, compared with galaxies of the same morphological type, Sy galaxies are 1.5 to 1 magnitude more luminous. These statistical results are confirmed by a larger sample discussed some years later by Huchra and Sargent (1973): 5 % of the galaxies are Mk objects of which 10 % are Seyfert galaxies.

The scarcity of the E type among the Sy is a noticeable fact, shared with the galaxies of moderate nuclear emission (Section III.4). Most of Sy galaxies are of a variety of the Sc, Sb types, but only 5 % of them can be considered elliptic.

TABLE III.3

Characteristics of Seyfert Galaxies*

Object	Type	Class	m_T	cz	$U - B$	$B - V$	M_B	M_N	d_N
NGC 1068	(R)SA(rs)b	2	10.5	1134	1.0	0.0	−20	−18	0.5
1566	SABbc	1	10.3	1178	0.8	0.0	−20	−16	0.3
3227	SAB(s)a:	2	11.7	1050	0.8	−0.1	−19	–	0.3
3516	(R)SBO	1	12.5	2710	0.8	−0.2	−20	–	–
4051	SAB(rs)bc	1	11.0	726	0.7	−0.4	−19	−15	0.3
4151	SAB(rs)ab pec	1	11.1	1002	0.5	−0.7	−19	−18 −16	0.5
6814	SAB(rs)bc	1	12.0	1578	1.1	+0.4	−20	–	–

* Data from the Reference Catalogue of Galaxies II, Weedman (1976) and de Vaucouleurs (1968).

III.5.1. SPECTRUM OF EMISSION

The relative width of the emission lines in the spectra of Sy nuclei leads Weedman to distinguish two subtypes:

- Sy1: with broad hydrogen lines and narrower forbidden lines.
  Example: NGC 4151.
- Sy2: both the forbidden and the hydrogen lines are of equal width which varies between 500 and 1000 km s^{-1}. Their profiles are generally asymmetrical.

The most direct approach to estimate the Sy sub-type consists in comparing the widths of Hβ and [O III] λ 5007, adjoining in the spectrum. At low dispersions it can be difficult to distinguish some Sy2 from non-Sy galaxies with nuclear emision, as the difference of line widths can not be distinguished (500 km s^{-1} in Sy2 vs 200 km s^{-1} in a galaxy with nuclear emission). Another criterion of comparison is to use the intensities of these same lines as they are of the same order for the Sy1, whereas Hβ is one order of magnitude more intense than [O III] λ5007 in the Sy2. Finally, a two color $U - B$, $B - V$ diagram permits separation of both types of galaxies from each other and all Sy's from the neG and the objects N and QSS (Figure III.12).

TABLE III.4

Emission lines in IC 4329*

Ion	λ	I/Hβ	10^{-3} FWHM (km s^{-1})	10^{-3} FWOI (km s^{-1})
[Ar III]	7135	0.42	–	–
He I	7091	0.42	–	–
[S II]	6030–17	0.88	–	–
[N II]	6583	2.38	1.3	3.5
Hα	6563	3.53	1.6	3.6
[N II]	6548	1.59	1.3	4.1
O I	6300	0.67	1.6	3.7
[O III]	5007	5.11	1.3	4.1
[O III]	4958	1.99	1.5	3.1
Hβ	4861	1.00	1.2	3.7
[O III]	4363	0.55	1.1	2.4
Hγ	4340	0.60	2.0	3.8

* Data by M. Pastoriza (1979).

A characteristic phenomenon of the Seyfert galaxies is that the Balmer decrement is larger than predicted by the theory of recombination. The easiest explanation for this is the reddening.

Table III.4 lists identified lines in IC 4329 Å, their relative intensities and widths in km s^{-1}.

Direct information on the temperature T and the electronic density n_e can be deduced from the relative intensities of the forbidden lines. The intensity ratio of the components of the doublet [S II] λ 6717/6730 reflects conditions in the regions of low ionization. The mechanism responsible for the widening of the lines seems to be the Doppler effect caused by the motion of discrete clouds moving in a very rarified medium where forbidden lines are produced. The cloud motions are actually observed in the fine structure of the spectra. In the case of IC 4329 Å, Pastoriza (1980) finds that the Balmer lines in these spectra arise from three regions: the nucleus itself, where also the forbidden lines are originated, and two hydrogen clouds, one with a large dispersion of velocities (5000 km s^{-1}) coming towards the observer and another receding from the nucleus.

III.5.2. CONTINUUM SPECTRUM

The continuum spectrum also differs significantly from that of normal galaxies as it is much flatter and resembles the composite stellar spectra prevailing in the nuclei of normal galaxies (Section II.1.2). To give an interpretation of them we have to resort to three mechanisms, each one being responsible for the spectrum in different frequency ranges. The nature of the radio waves is non-thermal synchrotron radiation, which also may be responsible for the infrared excesses, although most of them can be explained as a result of re-emission by heating of dust caused by the optical and ultraviolet radiation coming from the nucleus. In the optical range the continuum results from the superposition of thermal emission of stars and gas, plus non-thermal emission. The linear polarization (of up to 5% in NGC 1068) observed both in IR and in the optical range seems to justify the hypothesis of non-thermal radiation. Some bands and absorption lines of stellar origin have been identified.

From them attempts have been made to infer the underlying stellar population. However no unequivocal solution could be obtained on the basis of the equivalent width of so few lines.

Two dozen Sy galaxies have been detected with varying degrees of confidence as X-ray emitters. Type 1 Sy's are intrinsically more powerful X-ray sources than type 2 Sy's. According to Wilson (1979), the X-ray power appears to correlate with the optical and IR continuum-luminosities and also possibly with the Balmer line width.

The optically brightest are NGC 4151, 3227, 5548, 6814 and 3C120. Their luminosities are of the order of $L_x \approx 10^{43}-10^{44}$ erg s^{-1} in the 2 to 10 eV energy range, which makes these objects about two orders of magnitude more intense in X-radiation than in the optical range.

The infrared emission is a universal characteristic in Seyfert galaxies. Their flux is responsible for practically all the luminosity of the Sy2 and for a significant part in the Sy1. The IR luminosity of NGC 1068 is two orders of magnitude higher than the optical one and half of it arises from a compact region with a diameter of about 90 pc.

III.5.3. VARIABILITY

Variations of the optical and IR flux have been detected, confirmed in the cases of NGC 1566, 3C120, and NGC 4151. In NGC 1566 variations in the spectrum have also been found (Figure III.6). The time scale varies between one month and one year and there is evidence favoring the fact that variations in the infrared follow with a certain delay those produced in the optical and ultraviolet range. This is consistent with the idea of re-emission of radiation by dust, because then the delay would reflect the dimensions of the cloud complex of surrounding the nucleus.

III.5.4. RADIOEMISSION

Sy galaxies have nuclear radio emission which on the average is more intense than can be detected in normal spirals. The Sy2 on the average, have stronger emission than the Sy1. The dimensions of the regions responsible for radio emission vary considerably from one object to the other, but in general it can be said that they go from 10 pc up to various kpc. No bilobulate structures are observed as in the case of intense radiosources (Section III.6).

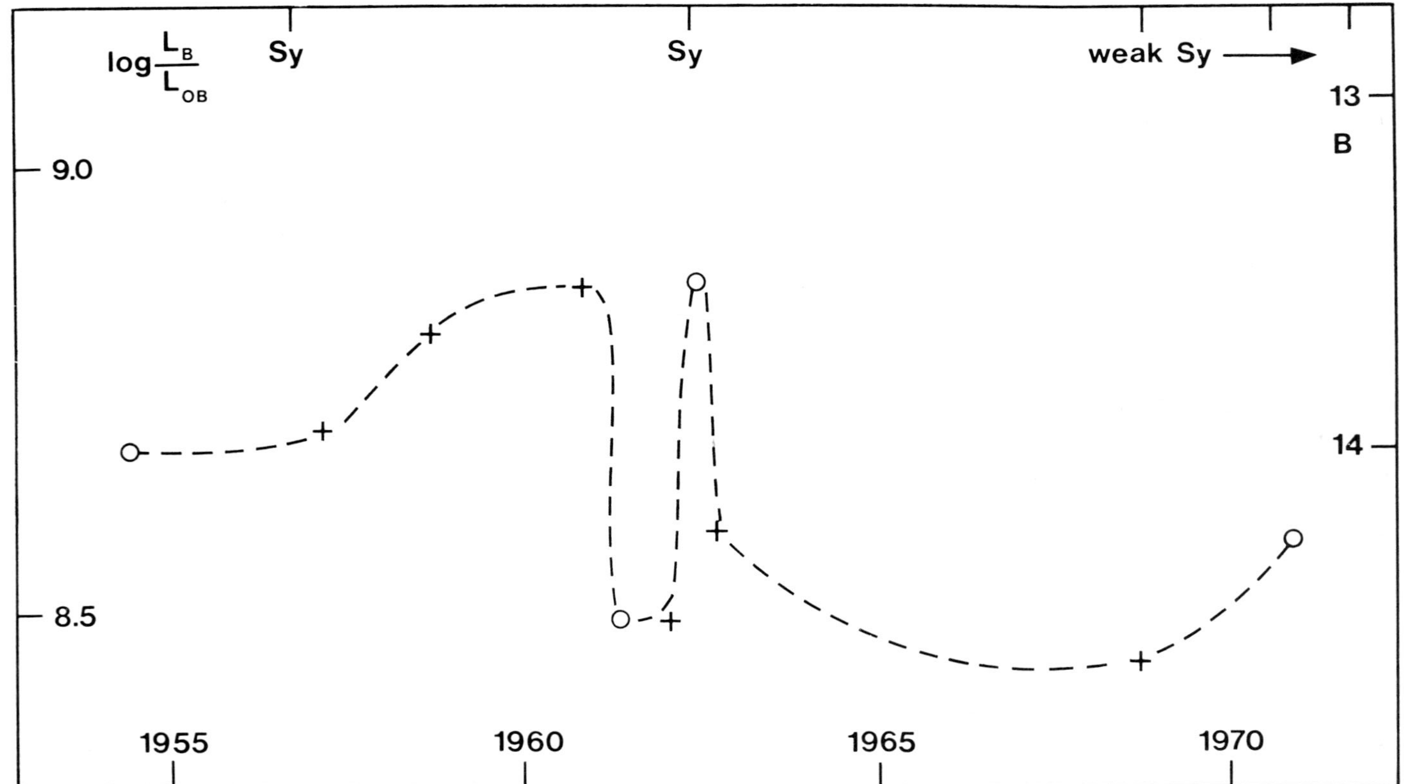

Fig. III.6.   NGC 1566 is a southern Sy1 galaxy which has a spectacular double arm system. The nucleus has suffered variations both in spectrum (Pastoriza and Gerola, 1970) and brightness (de Vaucouleurs, 1973; Quintana *et al.*, 1975). From the last paper we reproduce here the brightness variation of the nucleus in the last years.

Recent observations in 21 cm of the galaxy Mk 238, which is a Sy2, carried out at Arecibo by Morris and Wannier (1980), have revealed an extensive H I cloud whose dimensions surpass by one order of magnitude the optical dimensions of the object. Its discoverers maintain that it is composed of primordial material.

III.5.5. CORRELATIONS

There exist various possible correlations between the nuclei and the global properties of Sy galaxies. The dimensions of Sy2 nuclei are somewhat larger than those of Sy1 and optically weaker in relation to the luminosity of the galaxy as a whole. According to Kachikian and Weedman (1974) the Sy1 galaxies seem to have better developed spiral structures, something not to be found among the Sy2, which are somewhat peculiar in their morphology. This would suggest a connection between the activity of the nucleus and the dynamics of the disk.

The spectra of the regions immediately outside the nuclei show only the typical emission lines of normal galaxies, such as $H\alpha$, [N II] and [O II]. They are narrow and tilted. Some show deviations from rotational symmetry towards the inner part of the galaxy due to non-circular motions around the nuclei but the inclination is due basically to rotation and leads to estimates of mass comparable to those of normal galaxies of the same morphological type.

The disorderly motions around the nucleus amply exceed the local escape velocity. In this sense the presence of faint and extensive annular structures around NGC 1068 and 7469 should not be forgotten.

III.6. Strong Radio Galaxies (SRG)

Many discrete sources of radio emission are identified or associated either with peculiar type galaxies or they are of rather normal appearance although the features of their optical spectra show a certain degree of 'activity', usually both these circumstances appear together. An extragalactic object of these charactcristics is called 'radio galaxy'. Radio emission can come from a small centrally located source or from diffuse regions which extend much beyond the visible limits of the optical objects. The rate of energy emitted in radio frequencies is equal or larger than the optical emission. In this latter case we have a strong radio galaxy (SRG).

The difference between normal radio galaxies and SRG's arises naturally from a clearcut separation of the radio sources into two groups. That of low radio luminosity ($L_R < 10^{41}$ erg s^{-1}) is populated by normal spirals, whereas that of high luminosity ($L_R > 10^{41}$ erg s^{-1}) corresponds to the SRG's and is populated by a variety of optical objects which, on general lines, are associated with the E type galaxies. Detailed optical studies of radio sources identified as SRG can be made only for the nearest ones. In many cases these objects already had been noticed in the optical range because of their peculiar appearance. In this group we can name as examples NGC 1275 = Per A; NGC 1316 = For A; NGC 4486 = M87 = Vir A; NGC 5128 = Cen A.

Radio and optical studies of the structure and the spectra of these galaxies have shown evidence of explosive events having taken place in their nuclei, involving

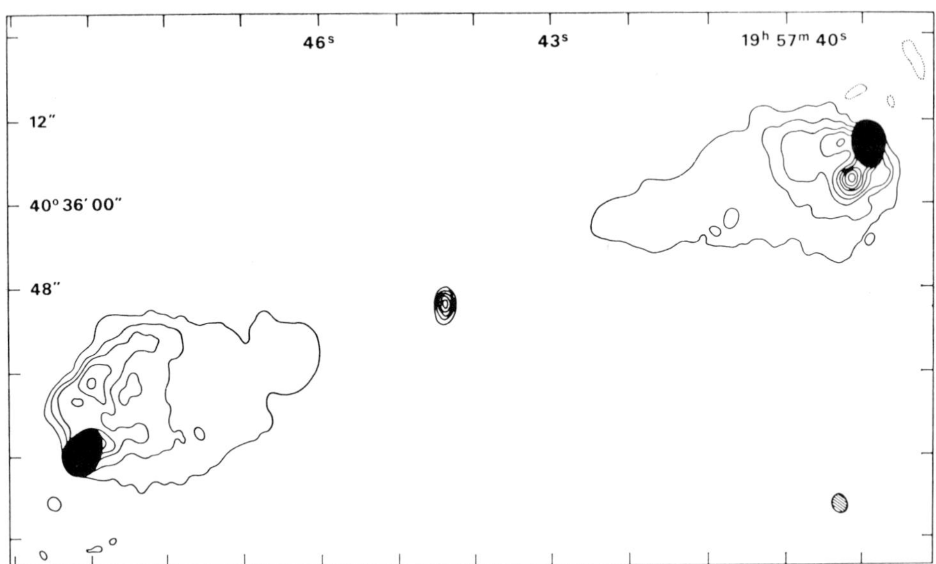

Fig. III.7.   Radio structure of Cyg A. Radio map in the 6 cm wavelength made with the Cambridge
5 km interferometer. (*Adapted from van der Laan*, 1977.)

energies from 10^{55} to 10^{61} erg in intervals of only 10^6 yr. In some of these objects
more than one explosive event has taken place, such as in NGC 5128.

III.6.1. RADIO STRUCTURES

In marked contrast to 'normal' radio galaxies, SRG's have double structures in
radio. Cyg A is a typical example: both components of its radio source (Figure III.7)
are separated from each other by about 100 seconds of arc (0.08 Mpc at a distance
of 170 Mpc). They are of the order of 60 kpc of diameter. The compact central source
is weaker and does not exceed 40 kpc, which is less than the dimensions of the optical
object (48 × 32 kpc).

Most of the energy emitted in radio frequencies comes from the symmetrical
components. The brightness temperature of the SRG's goes from 10 K in the extended components of Cen A, up to 7×10^6 K in the cases of Cyg A and 3C295,
although according to Moffet a typical value is 10^4 K. The fine structure of intense
radio sources may have hot-spots with temperatures up to 10^6 K. Some cases of
distant SRG's have been found recently which posess highly collimated radio jets
connecting the central compact radio source with the distant outer lobes. In some
cases only one jet exists.

Observation of linear polarization as high as a 30 % implies the existence of regular
and unidirectional magnetic fields in extensions of the order of tens to hundreds of
kiloparsec.

The details of the distribution of the magnetic field inferred from the polarization
measurements in more than two frequencies show it to be tangential in spheroidal
radio sources but along the main direction in elongated structures.

III.6.2. Spectrum of Radio Frequencies

Most radio frequencies have spectra which obey the power law

$$S \sim \nu^{-\alpha} \tag{III.1}$$

which connects the flux density S with the frequency ν. The exponent α is called 'spectral index' and is positive for non-thermal spectra.

The spectrum is interpreted as the result of the synchrotron radiation emitted by relativistic electrons in the magnetic field associated with a plasma. The index α increases with frequency in the spectra of many radio sources. This is due to energy loss on the part of the electrons. The synchrotron radiation theory shows that the spectrum radiated by a single electron of energy E covers a wide range of frequencies, with a maximum in $E_c = 0.3\ \nu_c$ where the critical frequency is

$$\nu_c \sim H_\perp E^2$$

and $H_\perp$ is the component of the magnetic field normal to the line of sight. The shape of the spectrum depends only on the ratio ν/ν_c through the function $P(\nu/\nu_c)$. Assuming now that the electron population $N(E)$ follows the power law

$$N(E)\,\mathrm{d}E = KE^{-\gamma}\,\mathrm{d}E \tag{III.2}$$

the spectrum $J(\nu)$ resulting from adding the individual contributions $P(\nu/\nu_c)$ can be written now

$$J(\nu) = \nu^{1/2(1-\delta)} \int_0^\infty P(x)\,\frac{\mathrm{d}x}{x^2}$$

after having introduced the variable $x = \nu_c/E^2$ and used relation (III.1).

Since the observed flux $S(\nu)$ is related to $J(\nu)$ through a geometrical factor of dilution, we finally get

$$S(\nu) \infty\ J(\nu) \propto \nu^{-1/2(\gamma-1)}$$

and the spectral index α is related to γ by means of $\gamma = 1 + 2\alpha$.

The mean spectral index for radio galaxies (in the interval from 10 to 1400 MHz) is $\alpha = 0.79$ with a dispersion $\sigma = 0.15$ due mainly to observational errors. This constancy of α suggests the universality of the mechanism responsible for the spectrum.

We have already seen that radio emission of these objects is strongly polarized and this is precisely what must be expected from the synchrotron mechanism. With $\alpha = 0.8$ the value of $\gamma = 2.6$ is very close to that observed for the energy spectrum of cosmic radiation.

III.6.3. Time Scale

The mean life of the active phase of a SRG can be estimated in three ways:

(1) Computing the mean life of the radiation associated with the electrons emitted in a given frequency interval.

$$\frac{\mathrm{d}E}{\mathrm{d}t} = -\,\beta H^2 E^2 \tag{III.3}$$

so that between the critical frequency ν_c, the magnetic field H, and the age t of the process, the following relation exists

$$\nu_c = \frac{10^9 \text{ Hz}}{H_\perp^3 t^2}.$$ (III.4)

When t increases, ν_c shifts towards lower frequencies and the whole spectrum shifts without changing its shape. In every particular case the critical frequency can be found by making a theoretical curve coincide with the observed spectrum. In such a way we can establish t, if H were known by some other means.

(2) An inferior limit for t can be estimated in the case of a symmetrical source, recurring to the separation of the outer lobes. If their velocity of ejection does not surpass $0.1c$, no relativistic effects modifying the symmetry can be expected. The age thus obtained is $\geqslant 10^6$ yr.

(3) Since all galaxies identified with intense radio sources are intrinsically bright elliptic galaxies, a relation will exist between their spatial densities and their ages. Schmidt (1966) has estimated that 5% of the bright ellipticts are SRG's, if their age coincides with that of the universe ($\approx 10^{10}$ yr), the mean harmonic life for SRG's will be of the order of 5×10^8 yr.

From (1) we can deduce time scales so short as 1 yr for very compact radio sources, up to 10^8 yr for extensive radio sources, whilst (2) and (3) give a range of from 10^6 to 10^9 yr.

III.6.4. COMPACT SOURCES

The spectra in radio frequency of some SRG's and QSO's show a curvature such that $\alpha < 0$. This always implies the presence of complex structures in the radio source which cause superposed spectra of varied characteristics.

The synchrotron radiation of compact sources produces self-absorption in the spectrum. This self-absorption produces a maximum in the frequency spectrum, below which the flux varies as $S \sim \nu^{2.5}$. This is the kind of maximum usually observed in the centimeter wave region of radio sources. Figure III.8 shows the composite spectrum of the SRG Per A = 3C84 = NGC 1275. Three components can be distinguished in it. In those sources with a well defined self-absorption frequency the flux density at maximum S_p, and the angular dimension θ, the intensity of the magnetic field $H_\perp$ can be determined with the expression

$$H_\perp \approx 2.4 \times 10^{-5} S_p^{-2} \theta^4 \nu_c^5 (1 + z)^{-1} G$$

predicted by the synchrotron radiation theory.

The analysis of the spectra allows the identification of their components. For those which are variable its time scale enables us to estimate their linear dimensions, and from their distance the angular separation θ from which the magnetic field H can be inferred.

The theory of the spectrum of a cloud of relativistic electrons and its magnetic field in adiabatic expansion was successfully developed by van der Laan (1977) (Figure III.9). According to this model the compact sources should have such an intense radiation that a great part of the energy of the electrons would be dispersed by the Inverse Compton Effect in much shorter time scales than their observed mean

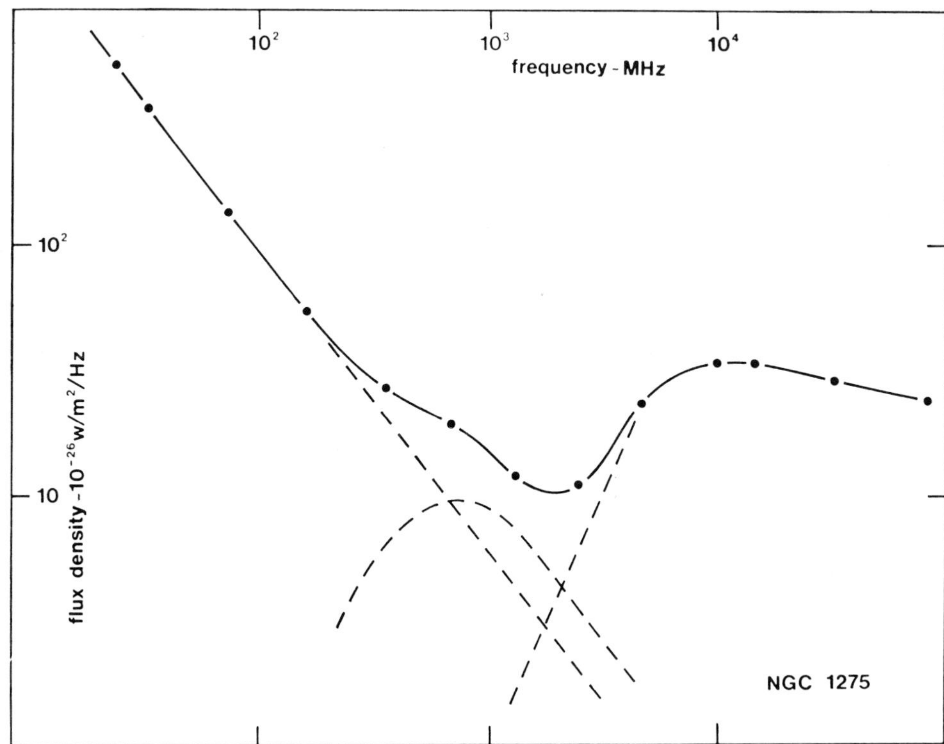

Fig. III.8.   Composite radio spectrum of NGC 1275.

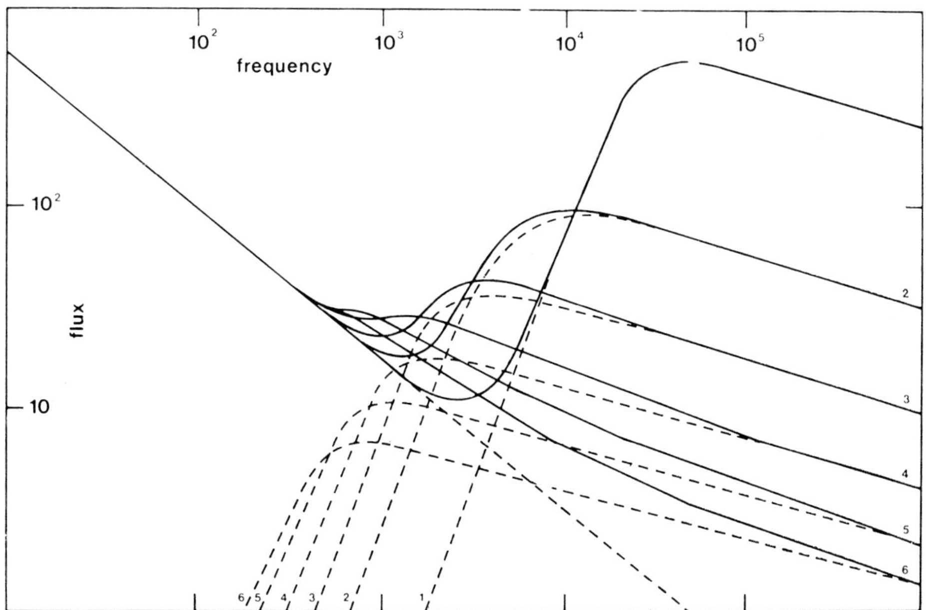

Fig. III.9.   Evolution of the radio spectrum of an expanding cloud of relativistic electrons, according to van der Laan (1977).

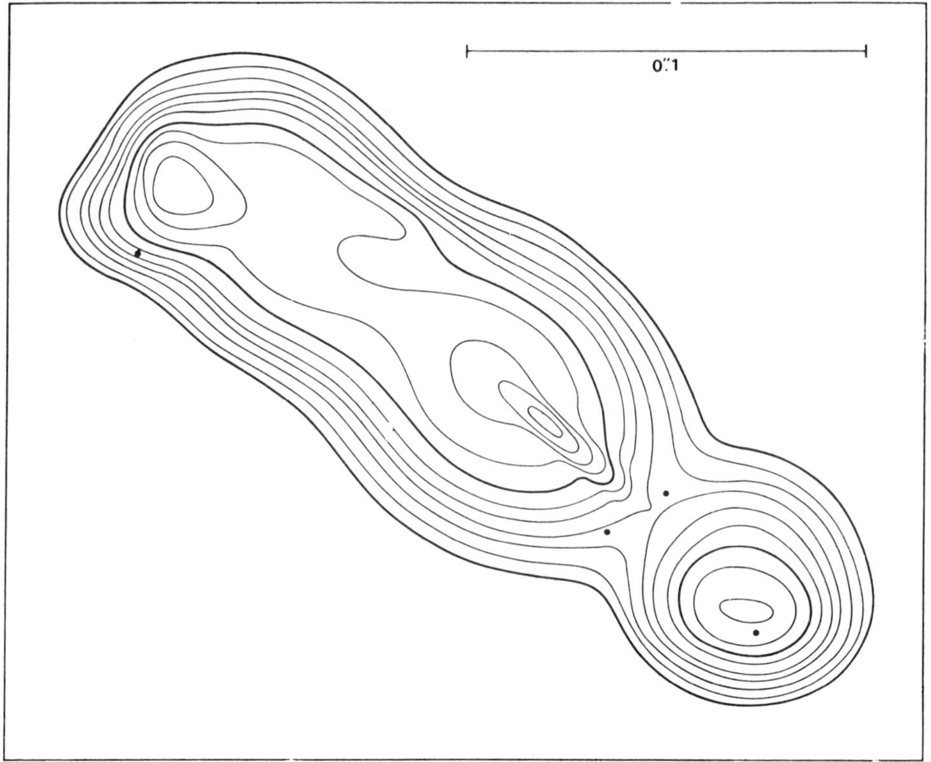

Fig. III.10.   Structure of 3C273 in radio emission deduced from VLBI.

lives. However, Rees and Simon (1968) have pointed out that if the expansion velocities are relativistic, the delay with which we receive the contributions to the flux from different parts of the source will produce a dilatation in the evolution of the spectrum in agreement with what has been observed.

The VLBI techniques have confirmed the existence of compact sources by direct measurements. In the case of NGC 1275, interferometry reveals the existence of the three components responsible for the spectrum: a halo of 5′ a component of 0″02 and a third more compact one of only 0″0025 in 1969. At the distance of NGC 1275 this is equivalent to a diameter of 1 light year, but only a year before, in 1968, the diameter of this source was 0″0017, which leads to a velocity of expansion of 0.35c.

In the case of QSO 3C273 three components are found: a jet, a halo of 0″022, a compact component of 0″0020 and another at the limit of resolution with 0″0004 which corresponds to only three light years at the distance of the QSO (Figure III.10).

III.6.5. Optical Forms Associated with SRG's

The forms of the optical objects associated with the SRG's have been classified by Morgan in the Yerkes system (Section I.2). In a now classical work, this author

Fig. III.11.   NGC 5128 = Cen A, a SRG with a complex optical structure which suggests recurrent activity has taken place in it. Color 4 m – CTIO photograph Copyright by AURA Inc. (cf. Figures IV.5 and IV.6)

together with Matthews *et al.* (1964) have shown that the SRG's appear in the following family forms:

Ep – Elliptical peculiars. They have absorption features which give them a peculiar character (e.g. NGC 5128 = Cen A) (Figure III.11).

D – Systems with rotational symmetry lacking a spiral or elliptic structure. Some cases can be considered SO Hubble galaxies. The D objects differ from elliptical giants (gE) in having very extended halos (e.g. 3C449).

db – Double D type galaxies. Both nuclei are comparable in dimensions and luminosity but they are surrounded by a common extensive envelope (e.g. IC 2082).

The objects belonging to these families are of high luminosity. According to Sandage (1968) the mean absolute magnitude $M = -22.0$ and color $B - V = 0.9$ make the SRG indistinguishable from the central objects in clusters of galaxies (Section IV.1).

III.6.6. OPTICAL SPECTRA

Schmidt (1969) did a detailed study of the SRG spectra based on 31 objects with redshift $z \leqslant 0.46$ and apparent magnitudes between 14th and 20th. Their spectra were taken in the UV-blue range with a dispersion of 400 Å mm^{-1}.

The galaxies observed show a variety of spectra, with and without emission lines, with $\lambda 3727$ as their only feature. The most intense identified lines are [O III] $\lambda 5007$, Hβ, and [O II] $\lambda 3727$. On a second plane of importance appear [O III] $\lambda 4959$, [O III] $\lambda 4363$, Hγ, [Ne III] $\lambda 3869$, [Ne V] $\lambda 3426$, and [Ne II] $\lambda 4686$.

The absorption lines H and K of Ca II are observed when the spectra are sufficiently exposed. In some cases the G band, the sodium lines, and Fe were found in the ultraviolet. Schmidt (1969) has discussed several correlations between the intensity of the emission spectrum and parameters such as the radio flux, the features of radio emission distribution (dimensions, duplicity) with negative results. Nor did he find any relation with the spectral index or the presence of absorption features.

III.6.7. ENERGETICS OF SRG'S

The total energy of the electrons responsible for syncrotron radiation

$$E_e = \int_{E_1}^{E_2} EN(N)\,\mathrm{d}E \tag{III.5}$$

is determined by observing that the radiated power

$$L_{NT} = \int_{\nu_1}^{\nu_2} J(\nu)\,\mathrm{d}\nu$$

by the radio source in a given frequency interval must be equal to the radiative loss in the corresponding energy range,

$$L_{NT} \propto H_\perp^2 \int_{E_1}^{E_2} EN(E)\,\mathrm{d}E. \tag{III.6}$$

By using Equation (III.2) and eliminating K between Equations (III.5) and (III.6) we obtain

$$L_{NT} \propto H_\perp^2 E_e \frac{\int_{E_1}^{E_2} E^{2-\gamma}\,\mathrm{d}E}{\int_{E_1}^{E} E^{1-\gamma}\,\mathrm{d}E}$$

an expression in which we can replace E by the critical frequency by means of Equation (III.1). This way we get

$$E_e = 4 \times 10^7 \, H^{-3/2} L_{NT},$$ (III.7)

where we have put $\gamma = 2.6$ ($\alpha = 0.8$), selected $\nu_1 = 10$ MHz and $\nu_2 = 10^4$ MHz as frequency limits, and introduced the numerical constants.

The energy of the magnetic field H in the volume V in its turn comes from the well known expression

$$E_H = \frac{H^2}{8\pi} \, V$$

and the total energy of the radio source is obtained then adding the energy E_p of the protons and heavier nuclei. The value of E_p is unknown but it is assumed to be proportional to E_e, that is $E_p = \lambda E_e$, so that the total energy can be written

$$E = 4 \times 10^7 (1 + \lambda) H^{-3/2} L_{NT} + \tfrac{1}{6} H^2 R^3$$

after remembering that $V = 4\pi R^3 / 3$.

The minimum E^* of the total energy can be estimated with respect to the magnetic field H, which gives

$$E^* = \tfrac{7}{3} E_H^* = \tfrac{7}{4}(1 + \lambda) E_e^*$$

and leads to the following equipartition values

$$E^* = 2 \times 10^4 (1 + \lambda)^{4/7} \, L_{NT}^{4/7} R^{9/7}$$

$$H^* = 230(1 + \lambda)^{2/7} \, L_{NT}^{2/7} R^{-6/7}.$$

The values of E^* vary between 10^{56} and 10^{62} erg with $\lambda = 0$, while H^* is generally found between 10^{-3} and 10^{-6} G.

There are no compelling reasons for accepting the validity of the conditions of equipartition, all the more so taking into account the non-equilibrium character of the phenomenon and its proven recurrency.

Although both E^* and H^* strongly depend on the value assumed for λ, the main cause for uncertainty lies in the luminosity L_{NT}, so that errors of one order of magnitude should not be surprising.

TABLE III.5

Minimum energies and magnetic fields for SRG ($\lambda = 0$)

	Cyg A	Cen A*	Cen A**
Distance (Mpc)	170	5	5
L_{NT}(erg s^{-1})	10^{45}	2×10^{41}	2×10^{41}
R (Mpc)	0.08	0.004	0.10
E^* (erg)	3×10^{58}	3×10^{56}	3×10^{58}
H^* (Gauss)	4×10^{-5}	2×10^{-5}	10^{-6}
T^* (yr)	3×10^5	5×10	8×10^9

  * Core.

  ** Halo.

The time scale of the process can be obtained by means of $t^* = E^*/L_{NT}$ and is written as

$$t^* = 6 \times 10^{-4} (1 + \lambda)^{4/7} L_{NT}^{-3/7} R^{9/7} \text{ yr}.$$

This leads to values of t^* in the order of 5×10^7 yr with $\lambda = 0$. If we remember that the time scales estimated on basis of the dimensions of the radio sources are one order of magnitude larger, it means that either the conditions of equipartition are not suitable or that $\lambda \approx 100$.

One of the biggest problems in the energetics of SRG's is the estimation of $\lambda = E_p/E_e$. Observations give at most E_e and there are no valid arguments to adopt $\lambda = 100$ as has been done in previous computations, except that $\lambda \approx 100$ is the value found for primary cosmic radiation in the Galaxy, at least at present.

Table III.5 sums up the typical information for some SRG's assuming $\lambda = 0$.

III.7. N Galaxies

This type of galaxy was introduced originally by Morgan as a 'morphological family' in the Yerkes classification (Section II.2). The objects of this class resemble a small bright nucleus superposed on a very faint background.

According to Sandage (1968), N galaxies are somewhat similar to the QSS, inasmuch as most of their luminosity comes from a point source. Nevertheless he points out that all N galaxies show a nebulous envelope, something not observed in most QSS, which conserve their stellar appearance. The two-color diagram $U - B, B - V$ clearly separates both classes of objects and Sandage has shown that the colors of an N galaxy result of the superposition of a giant elliptic galaxy (gE) with a blue central nucleus or source similar to a QSS.

In other cases, possibly the nearest ones, the intense nuclear region is seen surrounded by neat spiral arms. Many N galaxies have been identified as radio sources and their spectra are similar to Sy nuclei, although with somewhat narrower emission lines. There are exceptions (e.g. 3C177) which have a normal excitation spectrum.

The similarity with the Sy galaxies exists also in the variability in the optical range,

TABLE III.6

Properties of N Galaxies.

Object	z	M_B^*	Spectral lines**
3C390.3	0.057	-20.8	[Ne v] λ 3426, [O ii] λ 3727, [Ne iii] λ 3869, λ 3968, Hδ Hγ, Hβ, [O iii] $\lambda\lambda$4363 4959, 5007, Hα, No abs. Balmer lines very broad, two maxim.
3C371	0.051	-21.1	[O ii] λ3727, [O iii] $\lambda\lambda$4959 5007, Ca ii, H and K abs.
3C459	0.221	-21.5	[Ne v] λ3426, [O ii] λ3727, [Ne iii] λ3869, [O iii] $\lambda\lambda$4959, 5007 weak em. abs.

   * From apparent magnitudes given by Arp (1974).
** From Burbidge (1970).

caused by fluctuations of the non-thermal continuum. This is evidence of the smallness of the nuclear region.

The redshifts and luminosities of N galaxies place them in an intermediate position between the Sy's and the QSS's (Table III.6).

Both the Sy and N galaxies have similar features and their differentiation depends somewhat on the criteria used in their classification. Two classical Sy galaxies, NGC 4152 are considered N galaxies by Morgan (1958).

III.8. Quasi Stellar Objects (QSO)

Even with the largest telescopes, a QSO produces a stellar image on the photographic plate. Only in very few cases have there been found traces of nebulosity or other features. Spectroscopic research after the discovery of QSO's have shown that (1) the most typical feature of its emission line spectrum is its large redshift and (2) that less than 10 % of these objects are detectable as radio sources.

Due to the selection effect caused by their identification via radio sources, most of the QSO's known at present are radio sources but it is well known now that two intrinsically similar QSO's nevertheless, can differ greatly in their power emitted in radio frequencies.

The most outstanding feature of the radio emitting QSO's is the extreme compactness of the source, to such an extent that an important fraction of the flux received comes from an area in the sky whose diameter is less then one second of arc. Nevertheless, some QSO's also show a structure in radio emission similar to that of the SRG's.

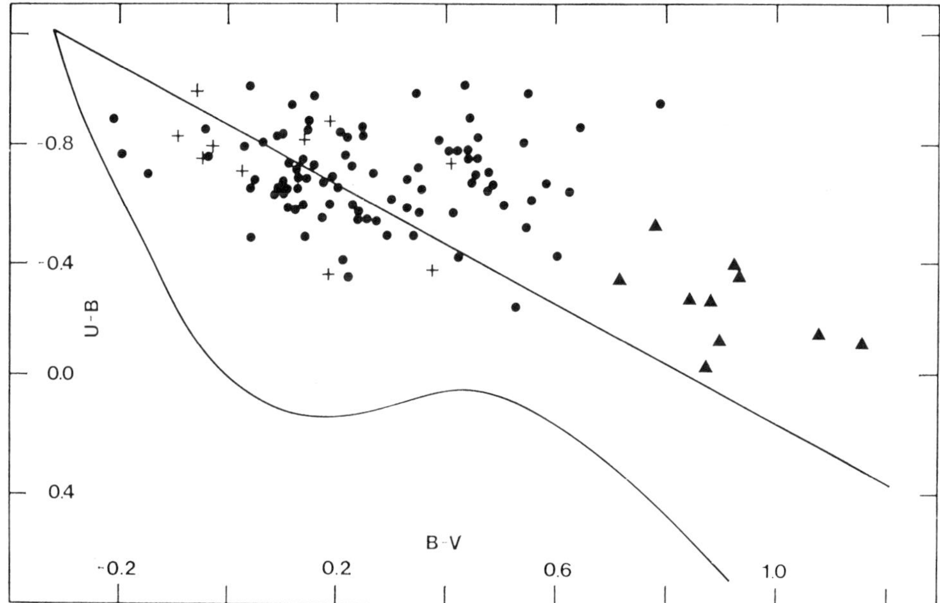

Fig. III.12.   On a two-color diagram it is possible to discriminate between QSR's (•), QSO's(+) and NG's(▲). The normal stellar relation and the black body line are also shown as a comparison. (*Adapted from Sandage*, 1968.)

UBV photometry shows strong ultraviolet excess in the QSO's (Figure III.12). The spectral emission lines are responsible for the correlation observed between $U - B$, $B - V$ colors and the redshift z. Sandage (1968) has constructed a composite $J(\nu)$ curve for a 'mean' QSO based on multicolor photometry at different redshifts.

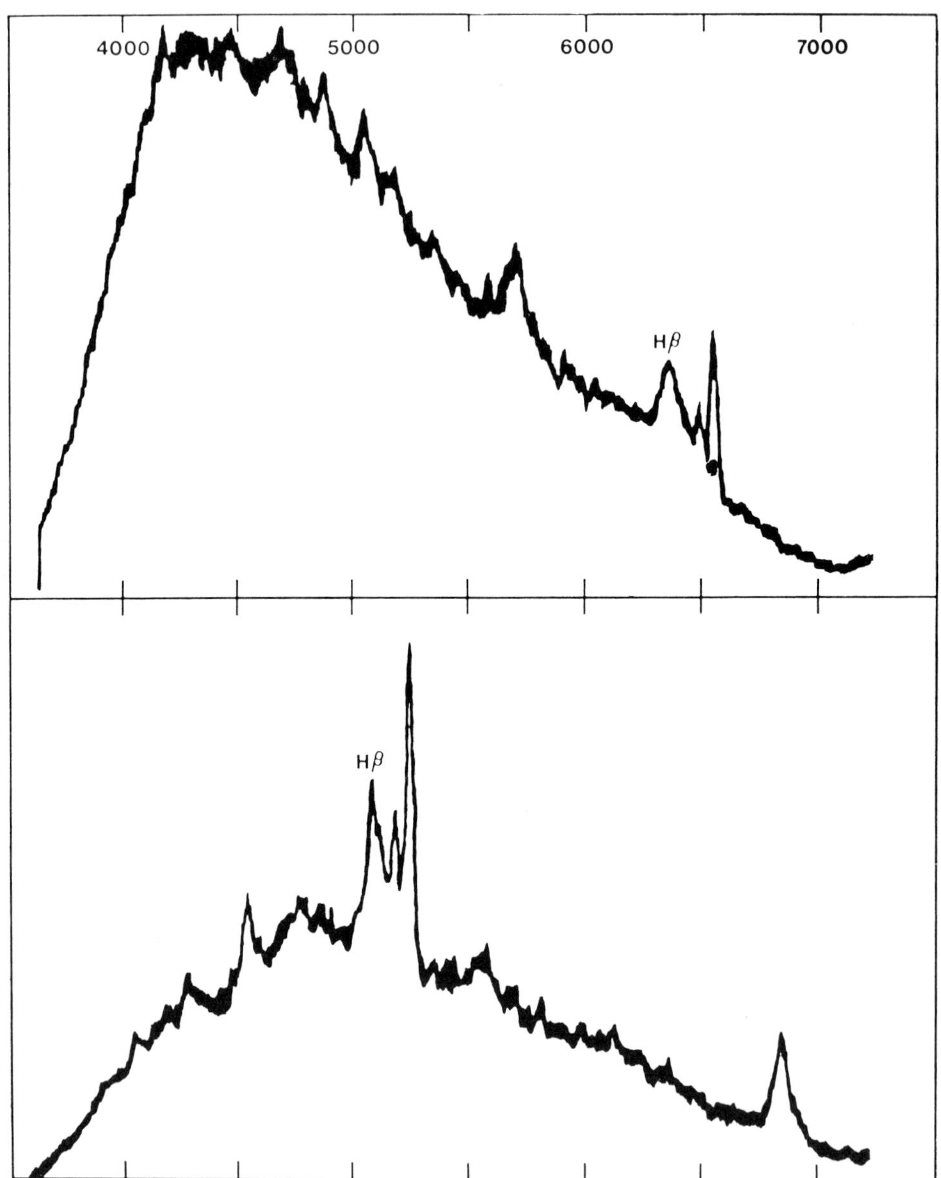

Fig. III.13.   The similitude between the spectra of QSO's and Sy galaxies is shown here in the case of the quasar 3C249.1 ($z = 0.31$) at the top of the figure and the Sy galaxy Mk 374 ($z = 0.04$). (*Adapted from Weedman*, 1976).

The QSO's show complex phenomena of variability in the optical range such as brightness fluctuations in time scales going from several years to a few days and amplitudes from a few tenths to more than one magnitude. This violent optical variability is generally accompanied by a flat continuum and the presence of structures such as jets (e.g. 3C273).

Most of the QSO's have a small intrinsic polarization of up to 10% in the visible, although much higher values have been detected in optically variable objects.

III.8.1. CONTINUUM SPECTRUM

This is similar to that of the SRG and Sy1 galaxies, (Figure III.13) following a law of the type $S \propto \nu^{-\alpha}$ with a mean value $\alpha = 0.81$ (in the range of 178 to 1400 MHz) but with larger dispersion ($\sigma = 0.25$). In the centimetric region the spectrum shows some complexities: the maxima seem to be due to a contribution from self-absorbing compact sources of synchrotron radiation. Time-variations and the very-long base interferometry (VLBI) confirm the existence of components with dimensions smaller than a light year.

Some QSO's show non-thermal continua which extend from the visible to the IR. The observations made by Oke and Shields (1976) or Oke et al (1974) between 0.3 μm and 2.2 μm of various QSO's show the optical continuum to be very flat ($\alpha = 0.2$) although curving towards the IR up to values of α of the order of 1.6. There is also variability in this part of the spectrum in time-scales ranging from years to days.

One of the nearest and intrinsically brightest QSO (3C273) was detected as an X-ray emitter in the energy range from 1 to 10 keV. There is no evidence of variations in the flux. 3C273 being up to now the only QSO with reliable associated X-ray emission, it is not yet possible to advance general statements for QSO properties in this spectral domain.

III.8.2. LINE SPECTRUM

The spectrum of emission lines is characterized by the high level of excitation of the atoms which are ionized up to five times. The various redshifts displace different parts of it towards the visible range so that the spectrum of these objects is known in a larger frequency range than that of any other class of celestial objects. In spite of the differences between the QSO's, it has been possible to compose a standard spectrum extending from the classical lines as far as $L\beta$. Table III.7 lists the most prominent lines together with an estimate of their intensities.

The characteristics of the emission lines in QSO's of low redshift ($z < 0.3$) have been studied in detail by Baldwin, who finds them very similar to those of the Sy1 and N objects. The QSO's spectra, on the contrary, differ from those of the Sy2 galaxies in the fact that their emission lines are weaker with respect to the continuum than in Sy2's. This difference also exists between the spectra of Sy1 and Sy2, which, according to Weedman, emphasizes the uniqueness of the QSO phenomenon.

Although the continuum of many QSO is variable, there is no evidence yet about changes in the emission line spectrum.

The presence of one or various absorption lines systems with different z_{abs}, usually smaller than z_{em}, has been found in a great number of QSO's. Most of these systems

TABLE III.7

Standard QSO emission spectrum*

Wavelength(Å)	Line	Strength
1026	Lβ	Absent Medium
1033	O VI	–
1216	L	Strong
1240	N V	Absent to Medium
1397	Si IV	Absent to weak
1406	[O IV]	–
1500	C IV	Strong
1640	He II	Absent to Medium
1909	[C III]	Medium to Strong
2798	Mg II	Strong
3426	[NeV]	Absent to weak
3727	[O III]	Absent to weak
3869	[Ne III]	Absent to Medium
4686	He II	Absent to Medium
4861	H	Medium
5007	[O III]	Absent to Strong
5876	He I	Absent to Medium
6562	H	Strong

* Adapted from Stritmatter and Williams (1976).

of lines when well established, are close to the emission redshift with $(z_{em} - z_{abs})/$ $(1 + z_{em}) < 0.03$. There are however, noteworthy exceptions which require relative velocities as high as $0.3c$.

III.8.3. BL LACERTAE OBJECTS

These are objects of stellar appearance which have all the characteristics of the QSO's excepting that they do not show spectral lines and therefore their redshifts are unknown.

Their prototype is the object BL Lac with apparent magnitude $V = 14.5$, which was taken for a variable star during many years.

The evidence of polarization, variability, and distribution of spectral index suggests they are a kind of QSO which do not show emission lines. However, there exist some differences in details: the non-thermal optical continuum is steeper ($\alpha \approx 2$), although many variable QSO's and N galaxies also share this feature. The brightest and most variable BL Lac objects are not associated with radio sources, a property also shared by the QSO's and N galaxies. Moreover, the time scale for the variability tends to be systematically shorter (of the order of 10 days).

There is a strong evidence that these objects are found in the nuclei of galaxies or near them (e.g. OXO29 near II Zw 129). There is only indirect evidence of large redshifts, thanks to the observation of intergalactic absorption lines. For BL Lac itself, Oke and Gunn claim $z = 0.07$ but this would require relativistic expansion velocities in the source.

Sulentic (1974) has suggested that the condensations emitted by SRG's like M87 or NGC 5128 could be BL Lac objects, in which case they would be very massive, up to 10^7 solar masses.

III.8.4. Redshifts and Nature of the QSO's

The QSO's show redshifts going from $z = 0.1$ up to $z \approx 4$, whereas the SRG's have only been observed up to $z = 0.8$. The giant elliptics (gE) and the SRG's fit nicely the Hubble law because of the uniformity of their intrinsic luminosities, as Sandage has shown (Chapter V). For the QSO's, however, the Hubble diagram is one of scatter. This has given rise to the thought that the QSO's could be objects of local origin, moderately distant and having a considerable intrinsic redshift z_i which, when combined with the cosmological redehift z_c typical for these distances, will produce the large values observed. If $z = 0.8$ as for a very distant E galaxy, $z_i = 1.5$ would be enough to produce an observed redshift as large as $z = 3.5$.

The absence of a consistent interpretation of the origin and nature of the intrinsic redshift within the frame of conventional physics, has become a serious difficulty for the non-cosmological hypothesis of the QSO's to such a degree that at present most astrophysicists are inclined to consider, at least as a working hypothesis, the cosmological nature of the QSO's. This rather conservative point of view is founded on the following arguments:

(1) Evidence that QSO's occur in the nuclei of galaxies. Due to the fact that identification of the QSO's always has been made on prints of the Palomar Sky Survey, the limitations of scale and seeing do not allow differentating the image of a very far galaxy from a stellar image. This gives rise to a definite frontier at a certain z in the Hubble diagram: for larger z's only QSO's are found and for smaller z's only galaxies. Kristian (1973) discussed a sample of 26 objects which at one time or another had been considered genuine QSO's. Four of them are located in the region dominated by galaxies and they all show an underlying galaxy on 5 m Palomar telescope plates. In the intermediate or transition region, five out of eight have underlying galaxies, whereas the 14 remaining ones in the high redshift area do not show diffuse images.

In the cases of association with galaxies, the QSO's satisfy the criteria of being considered N type galaxies, but for historical reasons they were called something else. Their apparent diameters are those to be expected from N galaxies and bright galaxies in clusters (cf. Section IV) at those distances. The QSO's appear in the nuclei and the galaxies as extended images with elliptic outlines.

(2) The spectra of the QSO's show these objects to contain heavy elements such as C, N, O, Ne, Mg, Si, and Ar. These elements are synthesized in the interior of the stars. Therefore the QSO's are formed, or take place in stellar environments, i.e., in galaxies.

(3) The spectroscopic evidence that QSO's, N and Sy1 galaxies are only manifestations of different degrees of the same phenomenon (cf. Section III.9) (Figure III.13).

(4) The existence of a redshift-magnitude correlation for the brightest QSO's. Bahcall and Hills (1973) have suggested considering the brightest QSO in a given interval of z in order to plot the Hubble relation, assuming that the optical luminosity function is always the same at different redshift and that its upper end is thus better defined. This leads to a high luminosity envelope parallel to the Hubble relation for galaxies.

(5) The double and symmetric radio sources associated with the QSO's show angular separations (θ) whose upper envelope in the diagram $\log \theta$ vs $\log z$ is compatible with the one to be expected if z is proportional to the distance.

III.9. Activity in Compact Objects

The active extragalactic objects described before (Sections III.5 to III.8) have various characteristics in common:

(1) Evidence of non-circular internal motions, generally of expansion, which can amount to several thousand km s^{-1}.

(2) Intrinsic luminosities going from 10^{38} to 10^{46} erg s^{-1} mostly contributed by non-thermal mechanism.

(3) Variability which implies sudden oubursts of non-thermal radiation and linear scales of the responsible objects which can be as small as 0.1 light years.

(4) An energy reserve which can be as large as 10^{62} erg.

(5) The motions, the absorption lines in the spectra, the observed structures (jets, plumes) suggest the ejection of mass, either in a continuous way or in the form of coherent masses.

(6) The time scale of these processes is as short as 10^5 to 10^8 yr, as the case may be.

(7) Recurrence.

It is convenient to take into account that, excepting scale factors, this phenomenology is not exclusive of extragalactic objects: Similar properties are also found in galactic objects such as η Carinae, the Crab Nebula or 30 Doradus in LMC. The essential property common to these phenomena is the compactness of the object or objects where they take place.

Various models have been proposed to explain the phenomenology of the activity of the compact objects but an explanation is still lacking about the ultimate energy source which feeds the whole process of activity (Figure III.14).

III.9.1. Excitation Mechanism

Continuous emission would have its origin in one or several compact sources with dimensions in the order of from 0.1 to 0.01 light years. The ultraviolet radiation of these non-thermal sources could be responsible for the ionization of a net of filaments and the jets surrounding the central source (or sources) occupying a much larger volume. In this way the line spectrum would be originated, which would reflect then the motion of expansion of the matter in the broadening of the lines. The uniformity of the hydrogen equivalent widths in all the compact objects (between 20 and 100 Å, according to Sargent and Searle (1970)) suggests a 'standard' mechanism of excitation such as non-thermal radiation. Other models, such as that proposed by Osterbrook, put the origin of excitation energy in collisions between clouds at high velocities, which leads to temperatures of the order of 10^6 K. This might explain the presence of high excitation coronal lines in some particular cases although the universality of this mechanism is not clear.

In the case of an explosive event, a large part of the energy would be dissipated throughout the surrounding gas in the form of a diluted and very hot medium, besides the non-thermal radiation originated in the accelerated particles. The temperature could reach 10^6 to 10^7 K and the supersonic expansion of this 'bubble' would generate, according to Wolfe (1974), a hydrodynamic flux similar to the solar wind, until it would reach a pressure equilibrium with the surrounding gas.

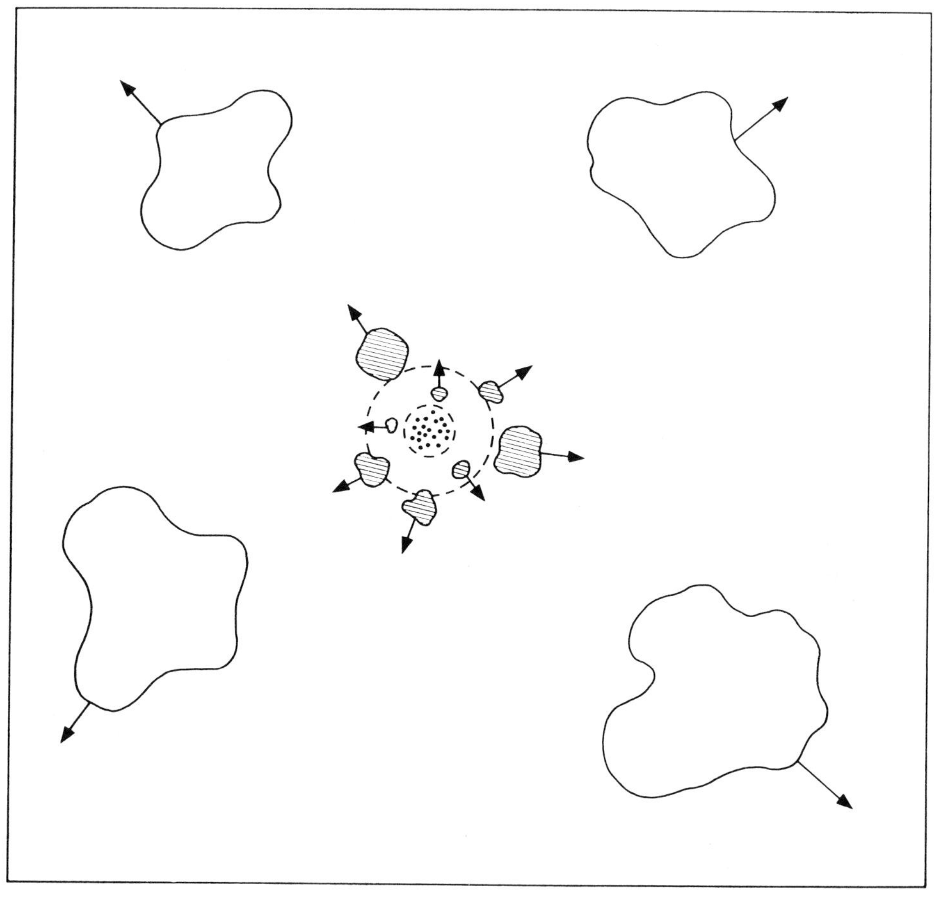

Fig. III.14.   Model for a QSO or a Seyfert nucleus.

The absorption spectrum has its origin in material belonging to the compact object, which is found in the form of clouds, shells, or jets moving relative to the central source. In certain cases, such as QSO's, the possibility cannot be ruled out that some absorption lines originate in galactic or intergalactic material with which radiation interacts on its long way to the observer.

On the other hand, the emission of energy in radio frequencies is produced in an extensive plasma which may have hot spots and which is associated with the magnetic field where relativistic electrons injected by the central source lose their energy, either in the form of low frequency synchrotron radiation, or by inverse Compton effect interacting with the primary radiation. The scale of this process seems to be related to the morphology of the region responsible for the radio emission. If the power is not large enough to produce an SRG or a QSO, as in an Sy galaxy, the structure of the radio source is simply one of core-halo. If, on the other hand, the power in question surpasses 10^{42} erg s^{-1}, bimodal structures will prevail.

Hot-spots are not only found associated with the denser parts of the compact objects but also at large distances in the radio-lobes.

The IR emission can be produced in a similar way, which means that it would be of non-thermal origin, but other factors such as the presence of dust can also be responsible for it. Similarly, an extension of the non-thermal spectrum towards very high frequencies could be the cause of X-emission, although thermal bremsstrahlung is also mentioned as a source. It is not clear where energetic enough electrons should come from.

III.9.2. THE SOURCE OF ENERGY

The compactness of the active regions and the persistence of their characteristics in a vast range of systems are unavoidable data for the construction of models to interpret the liberation of massive quantities of energy. In a system of mass M and radius R, the available nuclear energy is $E_N \propto M$ while the reservoir of gravitational energy is $E_G \propto M^2/R$, so that $E_G/E_M = M/R$ indicates that in compact objects (large M and small R) gravity is the dominant energy source. This reasoning, originally due to Fowler, underlies all proposed models. According to Rees (1977) the models can be grouped in three categories: (1) Dense stellar clusters; (2) Massive stars, massive rotators, magnetoids; and (3) Accretion in black holes.

(1) Dense stellar clusters. According to Spitzer, the ejection of mass in the course of evolution of the 10^9 old stars that populate the nucleus of a galaxy forms a dense cloud in the center which becomes unstable, fragments, and originates a cluster of new stars. The dynamical evolution of this cluster through inelastic collisions could then lead to phase (2).

First Ulam and then Spitzer and Saslaw have developed the concept of coalescence, a mechanism by which two stars when colliding finally form a new object with larger mass and cross section. Gold and Axford have developed the same subject but in the wider context of the irreversible contraction of large stellar systems (e.g. elliptical galaxies) as a consequence of the escape of stars.

(2) In one way or another material should collect in the central potential well of the galaxy. If the gas falls with sufficient velocity it creates turbulency, the dynamic dissipation heats the gas and its entropy increases. The opacity, assumed to be high, inhibits the escape of radiation, excepting from the surface layers so that the pressure increases enough as to support a spherical structure of up to 10^8 solar masses, as Fowler and Hoyle proposed. The energy losses through radiation induce this massive star to contract till its nuclear reactions stabilize it. Apparently a massive star would be built in this way, yielding part of its energy during the process of contraction and sustaining itself with marginal stability through the nuclear reactions.

If the fall of the gas were slow, there would be no possibility to generate a high entropy and the structure would not be spherical, but a rotating disk.

The coupling energy of a thin disk in uniform rotation increases monotonously with its degree of compactness up to a maximum of 0.3 Mc^2 if the gravitational redshift in the center is infinite. When a disk of these characteristics contracts, it creates a region from which energy in the form of radiation (ergosphere) can be extracted. Bardeen has pointed out that although the configuration of the rotating disk remains stable with respect to the collapse, it is not so with respect to fragmentation. As the disk looses energy, it becomes thinner. When it reaches a critical thickness, instabilities arise and it fragments into a few pieces. These presumably reproduce this process again and again, radiating, becoming flatter and again being fragmented. Although the whole of the fragments fill approximately the original volume, they are individually denser which indicates that they have lost energy. Since each generation of fragments looses the same fraction of total energy, the efficiency of this process is very great.

The rotation of massive objects with magnetic fields has been proposed mainly by Woltjer. This model which may provide a very efficient means of energy conversion, is inspired in the electrodynamics of the pulsars. A massive object is assumed in rotational equilibrium, which in its motion drags along a magnetic field. The electromagnetic stresses produce the loss of rotational momentum at the expense of which energy is radiated. The basic mechanism is the coupling between radiated energy and rotation via the magnetic field and, consequently, it can be applied to stars and the aforementioned massive rotators as well. In the latter case the fragments would behave like pulsars and the compact central region would be a cluster of them, as Rees has proposed.

The irreversible fate of all these configurations is to arrive finally at a situation of gravitational collapse.

(3) Rees has proposed as 'primum mobile' a black hole of $M \approx 10^7$ solar masses, fed by accretion of the surrounding gas (as previously suggested by Lynden–Bell), or even by stars. If the particles surrounding a black hole approached the smallest minimum allowed orbit, through a process of mutual collisions their capture would liberate 6% of the rest-mass as radiation. Lynden–Bell showed that if the black hole were rotating with the maximum angular momentum allowed by GR, an even greater efficiency could be achieved.

To attain an energy production along 10^7 yr with a 10% efficiency, the black hole would have to capture some tens of solar masses a year. This mass can be provided by the material ejected by the stars in the course of their evolution, particularly in early stages of the galactic evolution and through tidal disruption or collisions in star encounters. For the accretion to be effective, it has to take place near the minimum possible orbit, i.e. at a few Schwarzschild radii from the black hole.

Within the context of this model and according to statistical evidence that SRG's and QSO's seem to have been more abundant in the past ($z \approx 2$), it is admissible to think that the decay of their activity is due to the disappearance of more propitious conditions prevailing in galaxies of more recent formation and that the black holes can exist contemporaneously as passive entities which only occasionally revive their activities as a consequence of a random accretion process.

References

Arp, A.: 1974, K. Brecher and G. Setti (eds.), *High Energy Astrophysics*, MIT Press, Table VI.6.
Arp, A. and Bertola, F.: 1970, *Astrophys. Letters* **6**, 65.
Bachall, J.N. and Hills, J.G.: 1973, *Astrophys. J.* **179**, 699.
Burbidge, G.R.: 1970, *Ann. Rev. Astron. Astrophys.* **8**, 369.
Burbidge, E.M. and Burbidge, G.R.: 1960, *Astrophys. J.* **132**, 30.
Burbidge, E.M., Burbidge, G.E., and Prendergast, K.: 1962, *Astrophys J.* **136**, 119.

Calderón, J.H. and Sérsic, J.L.: 1980, *Observatory* **99**, 215.
de Vaucouleurs, G.: 1961, *Astrophys. J. Suppl.* **5**, 233.
de Vaucouleurs, G.: 1973, *Astrophys. J.* **181**, 31.
Heckman, T.: 1978, *Publ. Astron. Soc. Pacific* **90**, 241.
Holmberg, E.: 1960, *Ark. Astron.* **5**, 305.
Huchra, J. and Sargent, W.L.: 1973, *Astrophys. J.* **186**, 433.
Khachikian, E. and Weedman, D.W.: 1974, *Astrophys. J.* **192**, 581.
Krienke, O.K. and Hodge, P.W.: 1974, *Astron. J.* **79**, 1242.
Kristian, J.: 1973, *Astrophys. J. Letters* **179**, L95.
Matthews, T., Morgan, W.W., and Schmidt, M.: 1964, *Astrophys. J.* **140**, 35.
Morgan, W.W.: 1958, *Publ. Astron. Soc. Pacific* **70**, 364.
Morris, M. and Wannier, P.G.: 1980, *Astrophys. J.* **238**, L7.
Oka, S., Wakamatsu, K., Sakka, K., Nishida, M., and Jugaku, J.: 1974, *Publ. Astron. Soc. Japan* **26**, 280.
Oke, J.B. and Gunn, J.E.: 1974, *Astrophys. J.* **189**, L5.
Oke, J.B. and Shields, G.A.: 1976, *Astrophys. J.* **207**, 713.
Osmer, P., Smith, M.G., and Weedman, D.: 1974, *Astrophys. J.* **189**, 187.
Ozernoy, L.M.: 1970, *Circular Astr. Acad. de Ciencias, Moscú* **58**, No. 581.
Pastoriza, M.G.: 1967, *Observatory* **87**, 225.
Pastoriza, M.G.: 1979, *Astrophys. J.* **234**, 837.
Pastoriza, M.G. and Gerola, H.: 1970, *Astrophys. Letters* **6**, 155.
Prabhu, T.P.: 1979, Thesis.
Quintana, H., Kaufmann, P., and Sérsic, J.L.: 1975, *Monthly Notices Roy. Astron. Soc.* **173**, 57p.
Rees, M.J.: 1977, *Quarts. J. Roy. Astron. Soc.* **18**, 429.
Rees, M.J. and Simon, M.: 1968, *Astrophys. J.* **152**, L145.
Rood, H.J.: 1965, *Am. Astron. J.* **70**, No. 9.
Sandage, A.R.: 1968, in L. Perek (ed.), *Highlights of Astronomy*, 45.
Sandage, A.R. and Lynds, C.R.: 1963, *Astrophys. J.* **137**, 1005.
Sargent, W.L. and Searle, L.: 1970, *Astrophys. J.* **162**, L155.
Schmidt, M.: 1966; *Astrophys. J.* **146**, 7.
Schmidt, M.: 1969, *Ann. Rev. Astron. Astrophys.* **7**, 527.
Sérsic, J.L.: 1968, *Astron. J.* **73**, 892.
Sérsic, J.L. and Pastoriza, M.G.: 1967, *Publ. Astron. Soc. Pacific* **79**, 152.
Sérsic, J.L., Pastoriza, M.G., and Carranza, G.J.: 1972, *Astrophys. Space. Sci.* **19**, 469.
Sérsic, J.L., Bajaja, E., and Colomb, F.R.; 1977, *Astron. Astrophys.* **59**, 19.
Simkin, S.M.: 1971, *Bull. Am. Astron. Soc.* **3**, 352.
Strittmatter, P.A. and Williams, R.E.: 1976, *Ann. Rev. Astron. Astrophys.* **14**, 307.
Sulentic, J.W.: 1976, *Astron. J.* **81**, 582.
Tifft, W.G.: 1961, *Astron. J.* **66**, 390.
van den Bergh, S.: 1971, *Astron. Astrophys.* **12**, 474.
van der Laan, H.: 1977, *Ann. N.Y. Acad.* **302**, 637.
Weedman, D.W.: 1976, *Quart, J. Roy. Astron. Soc.* **17**, 227.
Wilson, A.S.: 1979, *Proc. Roy. Soc. London* **A366**, 461.
Wolfe, A.M.: 1974, *Astrophys. J.* **188**, 243.
Zwicky, F.: 1965, *Astrophys. J.* **142**, 1293.
Zwicky, F.: 1966, *Astrophys. J.* **143**, 192.

Bibliography

Axon, D. J. and Taylor, K.: 1978, *Nature* **274**, 37.
Bertola, F., Cuccin, F., and Nassi, E.: 1971, *Mem. Soc. Astron. Ital.* **42**, 4, 517.
Burbidge, G.R., Burbidge, E.M., and Sandage, A.R.: 1963, *Rev. Mod. Phys.* **35**, 947.
Burbidge, E.M., Burbidge, G.R., and Rubin, V.C.: 1964, *Astrophys. J.* **140**, 942.
Calderón, J.H. and Sérsic, J.L.: 1979, *Observatory* **99**, 215.
Casini, C. and Heidmann, J.: 1976, *Astron. Astrophys.* **47**, 371.

Cottrell, G.A.: 1978, *Nature* **274**, 13.

Chamaraux, P., Heidmann, J., and Lauqué, R.: 1970, *Astron. Astrophys.* **8**, 424.

Dupuy, D.: 1968, *Publ. Astron. Soc. Pacific* **80**, 29.

Kellerman, K.I.: 1972, in D. S. Evans (ed.), 'External Galaxies and Quasi-Stellar Objects', *IAU Symp.* **44**, 190.

Kiang, T.: 1967, *Astrophys. J.* **150**, L31.

Peimbert, M. and Spinrad, H.: 1970, *Astrophys. J.* **160**, 429.

Perola, O.C.: 1976, *Mem. Soc. Astron. Ital.* **47**, 387.

Rieke, G.H. and Lebofsky, M.J.: 1979, *Ann. Rev. Astron. Astrophys.* **17**, 477.

Sakka, K., Oka, S., and Wakamatsu, K.: 1973, *Publ. Astron. Soc. Japan* **25**, 153.

Sargent, W.L.: 1970, *Astrophys. J.* **160**, 405.

Sargent, W.L.: 1971, *Pontificiae Academiae Scientiarum Scripta Varia* **35**, 1.

Searle, L. and Sargent, W.L.: 1972, *Astrophys. J.* **173**, 25.

Sérsic, J.L.: 1973, *Publ. Astron. Soc. Pacific* **85**, 108.

Sérsic, J.L., Pastoriza, M.G., and Carranza, G.J.: 1968, *Astrophys. Letters* **2**, 45.

Stein, W.A., O'Dell, S.L., and Stritmatter, P.A.: 1976, *Ann. Rev.* **14**, 173.

Tifft, W.G.: 1969, *Astron. J.* **74**, 354.

Ulrich, M.H.: 1974, in J.R. Shakeshaft (ed.), 'The Formation and Dynamics of Galaxies', *IAU Symp.* **58**, 279.

Wakamatsu, K., Oka, S., and Sakka, K.: 1974, *Memoirs of the Fac. of Science, Kyoto, Univ.* **XXXIV**, No. 3.

Zwicky, F.: 1964, *Acta Astronómica* **14**, 151.

GALAXIES AND THEIR ENVIRONMENT

Most galaxies are isolated systems because of their large mutual distances and the quasi-stationary character of their evolution (Section I.1). This is not always true; a high degree of clustering is evident in the space distribution of galaxies and there are volumes where close encounters are frequent, producing exchanges of energy and mass between them. As a consequence, matter is lost to the surrounding medium.

The intergalactic medium exerts a ram-pressure on the gaseous component of galaxies moving through it, stripping them and stopping their subsequent star formation. Reciprocally, gas clouds in the intergalactic medium can be captured, producing sudden out-bursts of star formation in otherwise normal galaxies (Section III.4). Also, the release of internal energy in active galaxies pollutes the intergalactic space with high energy particles, diffuse matter and probably compact objects (Section III.9).

Galaxies in groups and clusters provide a way to study their properties as a class, such as luminosity and mass functions. Whilst the first can be reasonably established, the second is marred with the so called problem of the 'missing mass'. There is not yet any clear cut solution to it, although the margin of disagreement has been reduced in the last few years.

The mean mass-density of the Universe is a fundamental parameter to define the kind of Universe we live in; it is through the study of the statistical properties of galaxies that we can arrive at an order-of-magnitude estimate.

IV.1. Tidal Interactions

The nature of galaxies which have gone through gravitational interactions due to near passages or collisions with another galaxy was conjectural until a recent past. The reason for this is to be found in the insurmountable difficulties posed by the analytical treatement of the N-body problem. Although interaction between those bodies is the well known Newtonian attraction, its long range character makes the problem very difficult to solve.

Since the early studies by Chandrasekhar (1942) sensible progress has been made in the field, due principally to the advent to high speed computers and the development of elaborated computational techniques.

Interacting Galaxies (IG)

These objects have been the subject of extensive optical studies by Zwicky (1959, and references therein), while the catalogues of Vorontsov–Velyaminov (1959) and Arp (1966) provide a wealth of examples for these systems. Early lists of them in the Southern Hemisphere were published by Klemola (1969), Agüero (1971), and Sérsic (1968, 1974). The Uppsala Catalogue now covers the whole hemisphere (Section VIII).

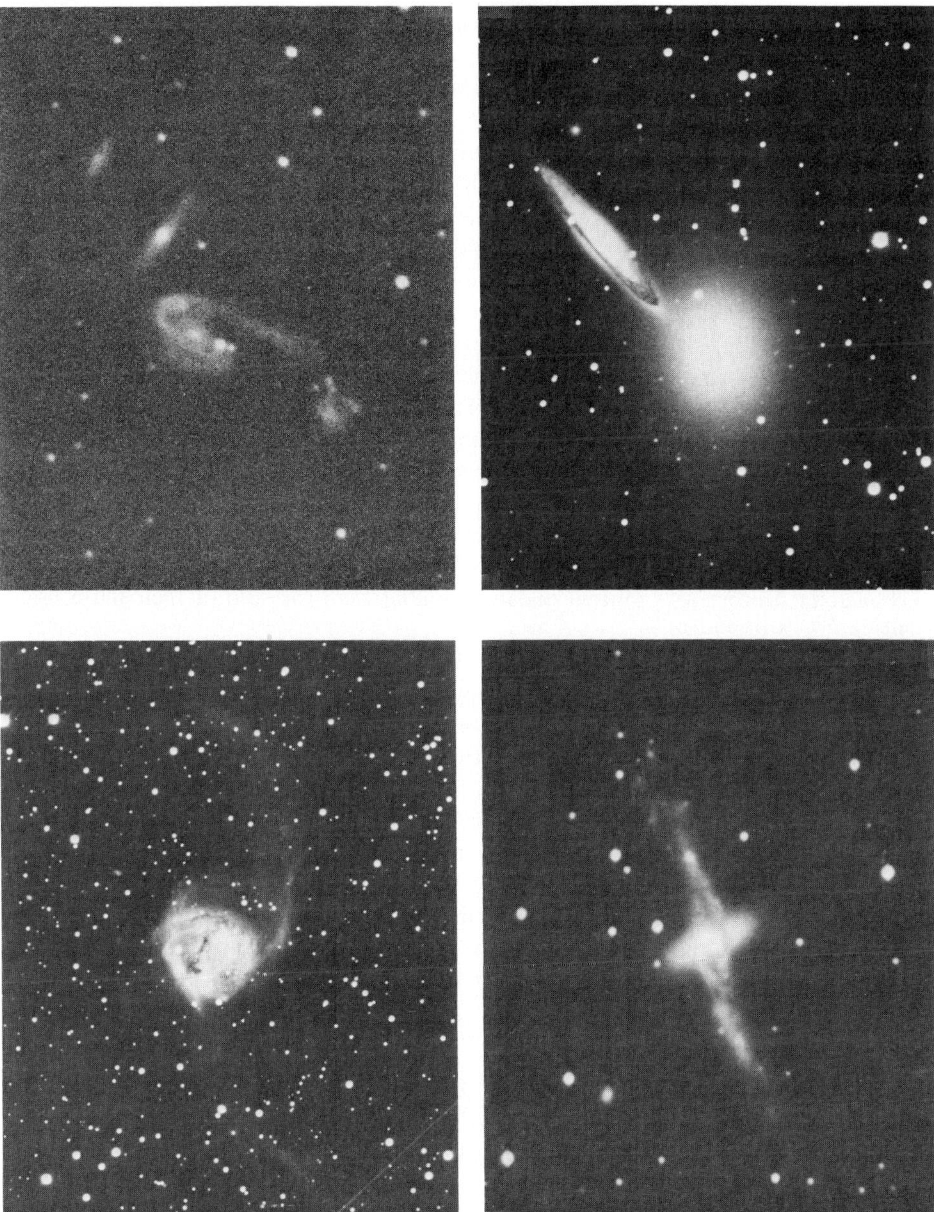

Fig. IV.1. Interacting galaxies. *Above left*: The Klemola 30 Group. (*Courtesy of Cordoba Observatory*.) Right the NGC 5090 group. *Below* left: NGC 3256 a probable merger. *Right*: A peculiar galaxy in Centaurus. (*The last three Courtesy of CTIO*.)

Spectral information has been obtained by several authors including Burbidge *et al.* (1963), Rubin *et al.* (1970), and Pedreros (1978). Wright (1974), Allen *et al.* (1973), and Kaufmann *et al.* (1975) have surveyed interacting galaxies in radio frequencies. Their spectral indices between 3 and 5 GHz are large and the spectra positively curved, as are many normal objects. Therefore no evidence for abnormal properties in radio spectrum seems to be present.

The 'tails' which are observed in certain close galaxy pairs are sometimes not accompanied by 'bridges' joining both members of the pair. This contradicts the expected bimodal character of tides. Given the circumstances, a certain degree of asymmetry would be expected, but just the opposite happens; the 'tails' on the far side are more developed and prominent, whilst the 'bridges', if any, are weaker notwithstanding that they are affected by the gravitational action. This (now) apparent contradiction led some authors to postulate non-gravitational interactions to interpret these phenomena (Vorontsov–Velyaminov, 1958, 1961; Burbidge *et al.*, 1963; Arp, 1966; Sérsic, 1973).

Numerical experiments, particularly those made by Pfleiderer (1963), Yabushita (1971), Wright (1972), and Toomre and Toomre (1972) have clarified the problem, demonstrating that gravitation alone is able to produce the most unexpected configurations in a near passage or collision between galaxies, the bridges and tails being their tidal relics (Figure IV.1).

In order to have substantial structural alterations in a passage or collision between galaxies, both objects must have comparable masses and be in a bound system and approach each other at a mutual distance smaller than the sum of their main body radii.

On the basis of statistical arguments, Ambartsumian (1957) demonstrated that the frequency of galaxy pairs requires that they had formed together and not through a capture process. Hyperbolic encounters between field galaxies are difficult to find, most of the IGs are pairs in high ellipticity orbits with perigalactic distances smaller than the sum of their respective radii.

Close passages in the direct sense (that is to say, in the sense of rotation of the 'victim') with parabolic speed or less, favors the accretion of matter on the nearside of the 'victim' at the expense of the perturbator. The capture of matter from the nearside tidal arm produces asymmetric galaxies, with a tail or plume on the opposite side which increases and reinforces with time.

Examples of this process are NGC 4038/9 and NGC 4676, described next.

NGC 4038/9: Also known as 'the antennae'. (Figure IV.2) illustrates this spectacular object and the results of the numerical experiences by A. and J. Toomre. The H I observations made by several independent groups found a fairly narrow range in radial velocities such as that necessary to numerically reproduce the observed configuration, although some difficulties in the interpretation of the velocity field still remain. The southern extension seems to have a radial velocity in the opposite sense to that in the model.

On the optical side, Schweitzer (1978) has found that both extensions or 'antennae' are much wider on deep-exposed plates, in agreement with the figures following from the Toomres model. Schweitzer also found proof that active star formation is going on in a cloud (also detected in H I) on the tip of the southern 'antennae'. In this spot H II regions and OB supergiant stars are observed.

NGC 4676: Also known as 'the Mice', which Vorontsov–Velyaminov gave the number 44 in his pioneer catalogue (1959) of peculiar and interacting galaxies.

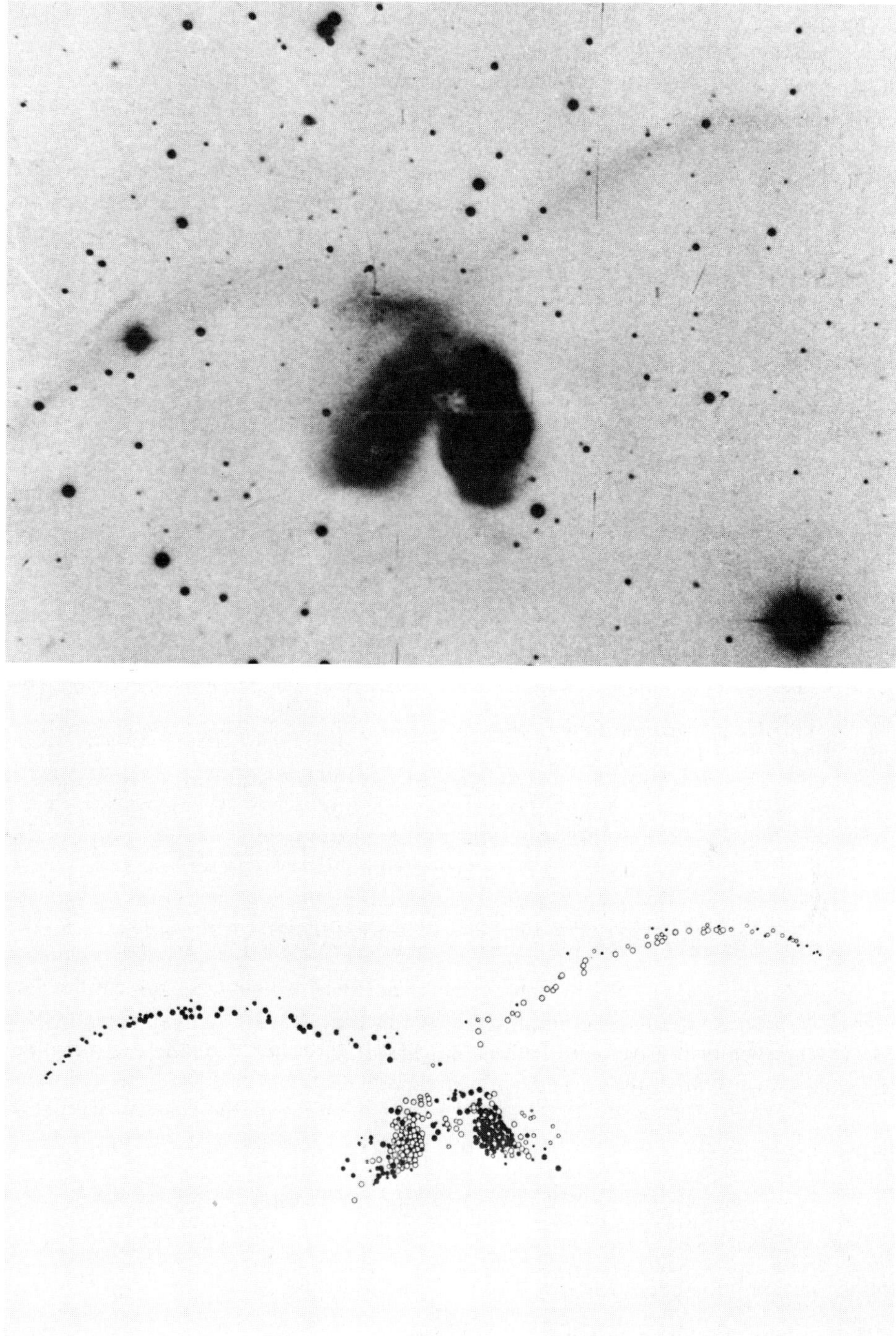

Fig. IV.2.   The antennae. *Above*: Negative copy of the Palomar Sky Survey of NGC 4038–9. *Below*:
Predicted structure by Toomre (1974).

The Mice is a system of two galaxies linked by a bridge. The one with irregular appearance has a long, almost linear tail in which some knots can be observed. This galaxy shows no nucleus in blue plates, the brightest part of it being the point from where the 'tail' joins it. The other galaxy shows some spiral structure (Burbidge and Burbidge, 1959, 1961). The Toomres (1972) computed a model consisting of the two equal disk galaxies which go through an elliptical encounter. Both disks are assumed flat and circular at the epoch of the last apocenter, although with different orientations. The resulting composite object, when seen projected onto the orbital plane simulates the morphology of the Mice, although several disagreements appear in connection with the observed velocity field.

Ring-like Peculiar Galaxies

The peculiar galaxies showing ring-like structure share several common characteristics abstracted by Freeman *et al.* (1974): (i) An early type system. E or SO galaxy, is associated with a late-type ring-shaped or chaotic object. (ii) The angular separation S of the two components is of the same order of the diameter D of the ring-shaped or chaotic component. The appearance often suggests tidal interaction, and the difference in radial velocity between the components is usually small (10 to 200 km s^{-1}). The linear separation is also small, of the order of 15 kpc. (iii) The shape of the ring varies from nearly circular (IC 298) to highly elliptical (NGC 4474). This means the ring is probably a flat circular annulus or torus. (iv) The chaotic component is frequently pear-shaped, almost triangular and often with one or more condensations near its center.

Several atempts to provide an interpretation of these weird structures have been given in the literature. Burbidge and Burbidge (1959) suggested they are the aftermath of a close collision between an elliptical and a spiral. Sérsic (1973) considered the formation of transient annular structures or 'lagrangian rings' in the ejecta of exploding galaxies and Canon *et al.* (1970) proposed they may represent the disk material of galaxies stripped of their nucleus by interaction with another galaxy. Freeman *et al.* (1974) instead proposed encounters of normal spirals with intergalactic H I clouds.

Toomre and Lynds (1976) has succeeded in showing that ring galaxies result from galactic collisions. An E galaxy falling perpendicularly to the symmetry plane of a disk galaxy, or two disk galaxies approaching flat-on, produce structures with striking resemblance with the observed ones (Figure IV.3).

Dynamical Friction

Close encounters between galaxies are strongly influenced by the two-body interaction. Under certain conditions the cumulative effect of the encounters can be described by a friction force resisting the motion of a massive body (Chandrasekhar, 1943). In fact, such a body moving through a previously statistically uniform star-field tends to produce a density enhancement behind it, caused by its attraction on the stars. Reciprocally, the same enhancement behaves as a statistically significant departure from uniformity in the field, producing a negative acceleration on the massive body with respect to the reference frame of the stars.

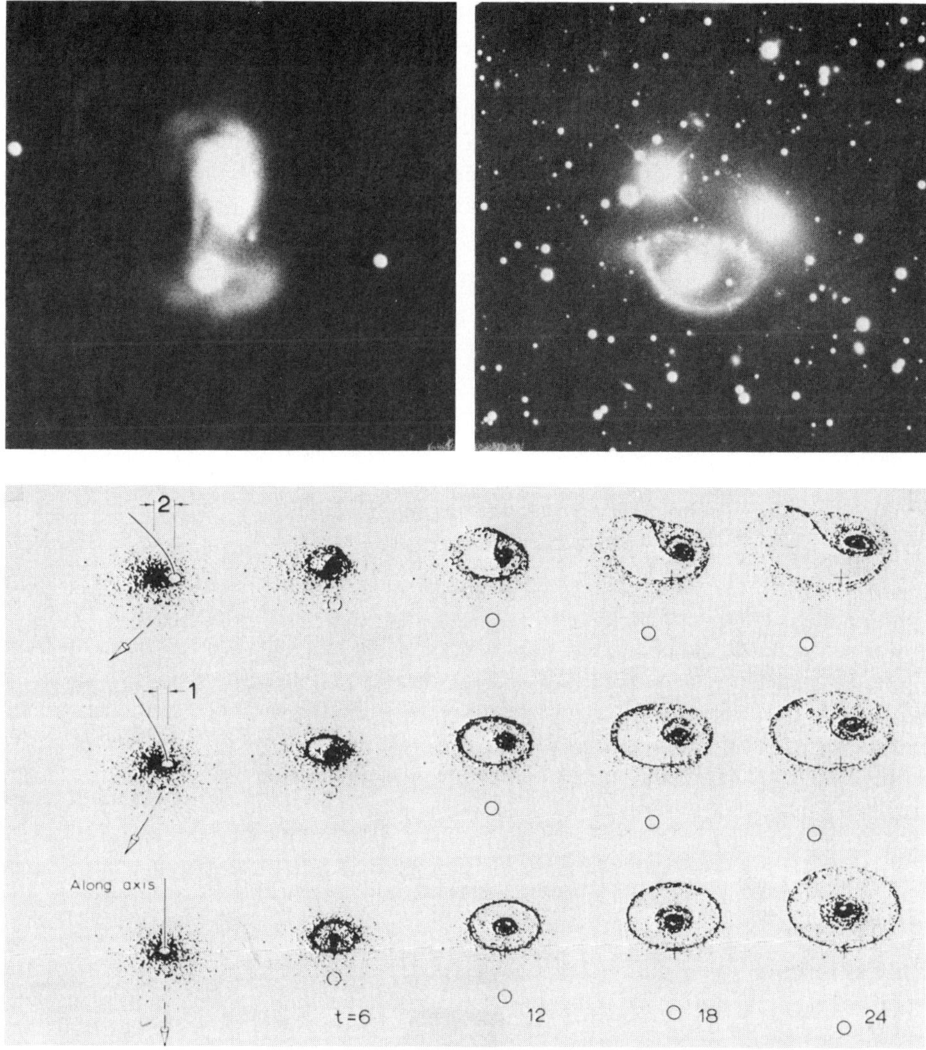

Fig. IV.3.   Ring galaxies. *Above*: Anonimous objects A $2^h23^m{-}40°$ (left) and $10^h07^m{-}38°$ (right), Agüero (1967). (*Courtesy of CTIO.*) *Below*: Calculations of an encounter leading to ring structures reproduced from Toomre (1978).

The energy balance shows that the kinetic energy lost by the body goes to increase the random motions of the star field at expenses of the motion of the massive body, which tries to reach equipartition with them.

In the general case, dynamical friction causes the degradation of the energy associated with ordered collective motions into energy of random motions through two-body interactions.

The possibility transforming translational energy into internal motions when two galaxies come close to each other was first considered by Holmberg (1941) and Chandrasekhar (1943), and later by Zwicky (1959). Alladin (1965) computed the energy exchange between two spherical politropes in the impulsive approximation: in this case the translational velocity is so large that it can be considered constant in the encounter, although internal motions become disturbed at the expense of the change in the mutual potential energy. The results of Alladin (1965), Sastry and Alladin (1970), and Rood (1965) demonstrate that a considerable fraction of the orbital energy is transferred to internal motions in the case of a head-on collision (Table IV.1).

TABLE IV.1

Energy transfer in head-on collision of two politropes of index N^*

N	$2\,\Delta T/\,W$
2	0.451
3	0.456
4	0.463
5	0.480

* Adapted from Sastry and Alladin (1970).

These results can the seen in a crude model consisting of two identical spherically symmetric galaxies of mass M approaching from a great distance with velocity V_i towards a head-on collision. The total energy of the system when they are at a large mutual distance is $E = MV_i^2 + 2(T + W)$, where T and W respectively are the internal kinetic and potential energies in each galaxy. At the moment of the maximum interpenetration, the total energy is written $E = MV_i^2 + 2(T + \Delta T) + W'$, where W' is now the potential energy of the composite configuration and ΔT the change in the internal kinetic energy. Conservation of the total energy E requires $2\Delta T + W' = 2W$ so that we have

$$\frac{2\,\Delta T}{|'W'|} = 1 - 2\,\frac{W}{W'}$$

for the ratio between the internal energy increase and the potential energy of the composite configuration. If spherical symmetry is roughly conserved at that time, we simply have $W' \lesssim 4W$ and obtain

$$\frac{2\,\Delta T}{|W'|} \lesssim 0.5$$

in rough agreement with the Sastry *et al.* results.

After the encounter, the energy of the system is $MV_f^2 + 2(T + \Delta T + W)$ at a large mutual distance, so that the kinetic energy of the translational motion reduces to $MV_f^2 = MV_i^2 - 2\Delta T \approx MV_i^2 + (W'/2) \approx MV_i^2 + 2W$. Rood (1965) has defined a 'capture velocity' V_c such that the resulting V_f be zero, in our case $V_c^2 = 2\Delta T/M = (-2W/M)$. If the initial velocity V_i were close to V_c the possibility of capture and formation of a losely bound system is given. Once this loose binary is formed, repeated perigalactic encounters may further decrease the energy of the relative motion, and both galaxies merge in a common system.

Detailed computations by Alladin *et al.* (1975) have shown that double galaxies without disruption can be formed provided the more massive one is less concentrated than the other and both are of comparable masses.

Mergers

The tidal interaction and dynamical friction suffered by two galaxies during a low-velocity encounter can sometimes lead to their coalescense through violent relaxation to form a relatively amorphous remnant or merger.

Dynamical friction is an effective mechanism in clusters of galaxies where intergalactic matter is known to exist (cf. Section I.2). The galaxies try to reach equipartition with the stars but only slow down because of their large mass, which means they fall into the gravitational field of the cluster towards the central regions. The relative size of a cluster member with respect to the distance of its nearest neighbor is only $\frac{1}{5}$ in the center of rich clusters of galaxies (Ostriker, 1977). The probability a galaxy has to be accreted by another in such an environment is rather high and calculations by several authors (Hausman and Ostriker 1978; Aarseth *et al.*, 1979, etc.) based on numerical simulations have shown that the effect of several successive mergers give definite properties to the resulting object.

From a given initial mass and space density of their neighbors, which are assumed to be encountered at random, the size and luminosity of the accreting galaxy is computed after successive mergers. The resulting object asymptotically approaches a stationary structure which develops in the central region. That structure is mainly determined by the accretion process, so that the memory of the initial conditions, as well as that of the particular encounters, is lost with time the final state is assimilated to a cD galaxy (Section I.2, III.6, and IV.2.1). This result also explain why the giant early-type galaxies in clusters have statistically stable values of their luminosities (Section V.32), as seen in Figure IV.4.

With time a tendency towards complete removal of halos is established after many encounters. As a result of tidal encounters, the outer parts tend to expand, so that

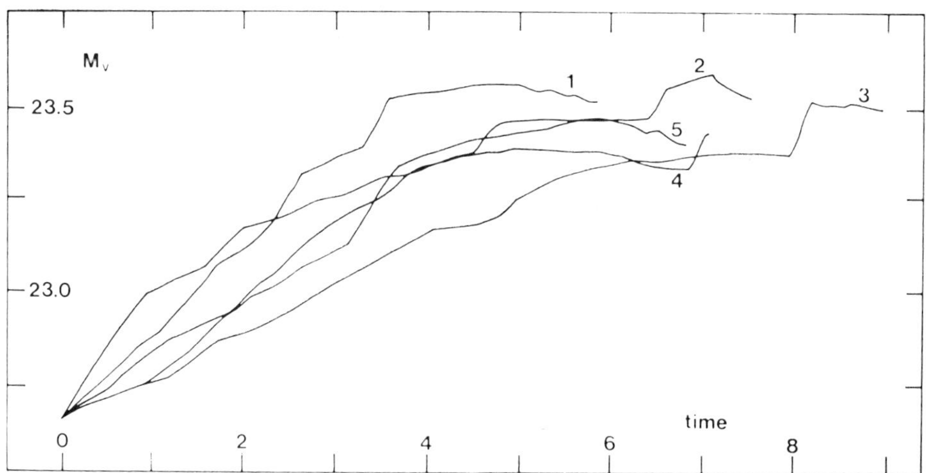

Fig. IV.4.   Evolution of the absolute magnitude of a galaxy after sucessive mergers. (*After Hausman and Ostriker* (1978).)

most of them become less bound, an easy target for further stripping (see below). Galaxies in rich clusters, therefore, are expected to have lost most of their halos due to successive encounters during their clustering process, while galaxies observed to have halos could not have undergone too many encounters. The intracluster medium, on the other hand, results enriched by a loose stellar population pervading the central regions of the cluster.

IV.1.2. Non-tidal Interactions

Active galaxies (Chapter III) are also a source of interaction with their environment. We have already seen that the liberation of internal energy in some of these objects leads to the ejection of matter and the radiation of vast amounts of energy. Reciprocally, the fall-in of intergalactic gas clouds in the deep potential well of an early-type galaxy results in bursts of stellar formation with subsequent ejection of gas jets back to the surrounding medium (Section III.4).

SRGs as Environmental Probes

The interaction of SRGs with their surrounding medium gives immediate information about it, provided we know about the nature of the radio source and the physics of the interaction.

The observations of extended SRGs are interpreted in terms of their evolution through sucessive crises leading to multiple ejections which originated in the confinement pressure of plasma clouds, whose instabilities produce phenomena such as hotspots and particle acceleration. The evolution of the plasma clouds in a multiple explosion of a SRG can be described through the following stages (Pacholczyk *et al.*, 1977):

(i) The plasma clouds created by the last explosion adiabatically expand through the channel previously formed by a former explosion. The surface brightness is low.

(ii) The plasma clouds rapidly decelerate when encountering the remnants of a previous explosion and become adiabatically compressed and brighten.

(iii) Turbulence due to Rayleigh–Taylor instability appears in front of the cloud, and propagates to the rest of the cloud. Particles become accelerated through the Fermi mechanism.

(iv) Deceleration reduces the velocity to zero and the clouds expand. Instabilities and particle aceleration decay and cease.

If the channel through which the plasma clouds propagate is not straight, due to anisotropic presure caused by the surrounding medium, the cloud becomes decelerated after colliding with the transverse walls of the channel. This is the reason for the brightnenings in RS components which present kinks and 'tails'. Such structures are mostly observed in galaxy clusters, where the IGM have large enough density and pressure gradients.

A classical example is the nearest SRG, Cen A = NGC 5128, which lies 4 Mpc away (Sérsic, 1960; de Vaucouleurs, 1980; Figure IV.5). The structure of the optical object shows two components: An elliptical system with color and luminosity similar to a gE2 galaxy and an almost edge-on disk containing gas, dust, and a young metal-rich stellar population (van den Bergh, 1978). The disk has a diameter of 9 kpc, lies in the equatorial plane of the system, and is responsible for the strong dark-line

Fig. IV.5. NGC5128 = Centaurus A. The nearest SRG. (*Courtesy of CTIO.*)

cutting across the smooth brightness distribution of the E-component. Detailed spectroscopic work by Graham (1979) improves earlier findings (Burbidge and Burbidge, 1959; Sérsic, 1969) regarding the velocity field. The disk system rotates

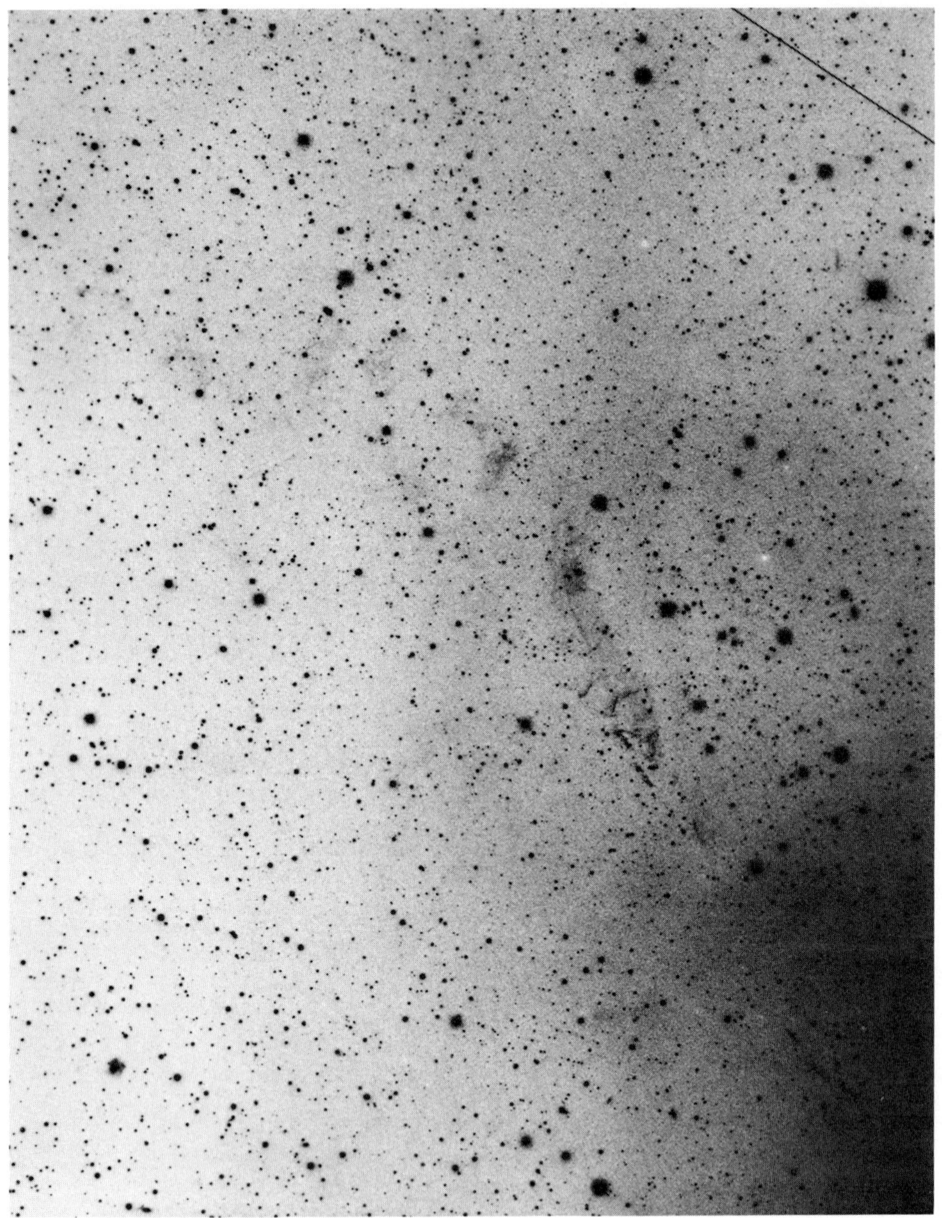

Fig. IV.6. Ejecta from NGC 5128 (see text). (*Courtesy of CTIO.*)

around the E-galaxy with a period of about 5×10^7 yr. The latter system also rotates, but at a much slower rate roughly around the same axis. H I-line observations (M.C.H. Wright, 1974) give a mass of neutral hydrogen of the order of 10^8 to 10^9 solar masses. According to Dufour and van den Bergh (1978), star formation proceeds in the disk, particularly on the periphery, as follows from the colors there. The distribution and size of the H II regions (Sérsic, 1960) agree with that interpretation, which is consistent with a stellar formation outburst which took place 4×10^7 yr ago and still continues.

The total power of the radio source and its energy content (Section III.6, Table III.5) suggests Cen A may once have been a powerful SRG like Cyg A. The non-thermal radiation comes predominantly from two giant lobes placed 0.7 Mpc away from each other, symmetrically with respect to the galaxy.

The central source, with dimensions comparable to those of the optical object, is itself resolved into two other lobes and a compact nuclear source. Both theses inner and outer lobes are aligned along the rotation axis of the system. The duplication of the lobes suggests a reiteration of explosive events.

Blanco et $al.$ (1975) and Peterson et $al.$ (1975) (Figure IV. 6) have optically observed jet-like chains of emission regions, blue knots and filaments in the N E inner radio lobes which correspond to the outskirts of the optical E-main body. Graham (1975) and Osmer (1978) have reported that the radial velocities of these features are smaller than that of the galaxy. From surface photometry Dufour et $al.$ (1978) have shown the jet structure to be moving radially outward from the nucleous of NGC 5128.

Kunkel and Bradt (1971) have identified the optical nucleus of the E component under the heavy obscuration of the dark-lane. This object has been shown to be a variable radio and X-ray source (Kellermarn, 1974; Kaufmann and Rafaelli 1979; Davidson et $al.$, 1975) in time-scales ranging from days to years.

The small diameter of the radio-source components in SRGs (Section III.6) in relation to its separation suggests that an external medium with density ρ_E impedes their free expansion. If they move at supersonic velocity V through this medium, they will be subjected to a dynamical pressure $\rho_E V^2$ independently of the external temperature. An estimate of the density ρ_E can be made if we assume the dynamical pressure cannot be smaller than the energy density ε stored in the radio-lobe, for which minimum equipartition energies can be computed (Section III.6). The energy densities $\varepsilon_{\min} = E^* R^{-3}$ are about 10^{-11} to 10^{-12} erg cm^{-3} (De Young, 1977) for most of the SRGs. The value of V probably does not exceed $0.1c$ (Section III.6) so that, from $\rho_E(0.1c)^2 \approx \varepsilon_{\min}$ follows $\rho_E \approx 10^{-28}$ to 10^{-30} g cm^{-3} the required minimum densities of the local IGM to have the observed confinements. These values are a factor of one hundred higher than the average density of matter in the Universe (Section IV.3.). Such high densities can exist in clusters of galaxies (Section IV.2.1), as follows from the interpretation of X-ray fluxes observed in them, and also are plausible values around any massive galaxy as a consequence of steady accretion (see below).

The deceleration of the sources by the external gas proves that most of their mass is also in a gaseous form. This deceleration produces a density enhancement and a radio brightnening close to the leading edge of the sources. As deceleration proceeds, the dynamical pressure $\rho_E V^2$ drops and the source expands in the transverse direction.

The Head-tail Radio Sources

The morphology of these objects is very suggestive of a SRG moving through a local intergalactic medium such that the ejected material becomes decelerated and blown back by dynamical pressure. Since the galaxy itself is not affected by the ram-pressure, to create such associated radio sources continual ejection is needed along the same axis of the compact object whose relativistic clouds are stripped away by the dynamic

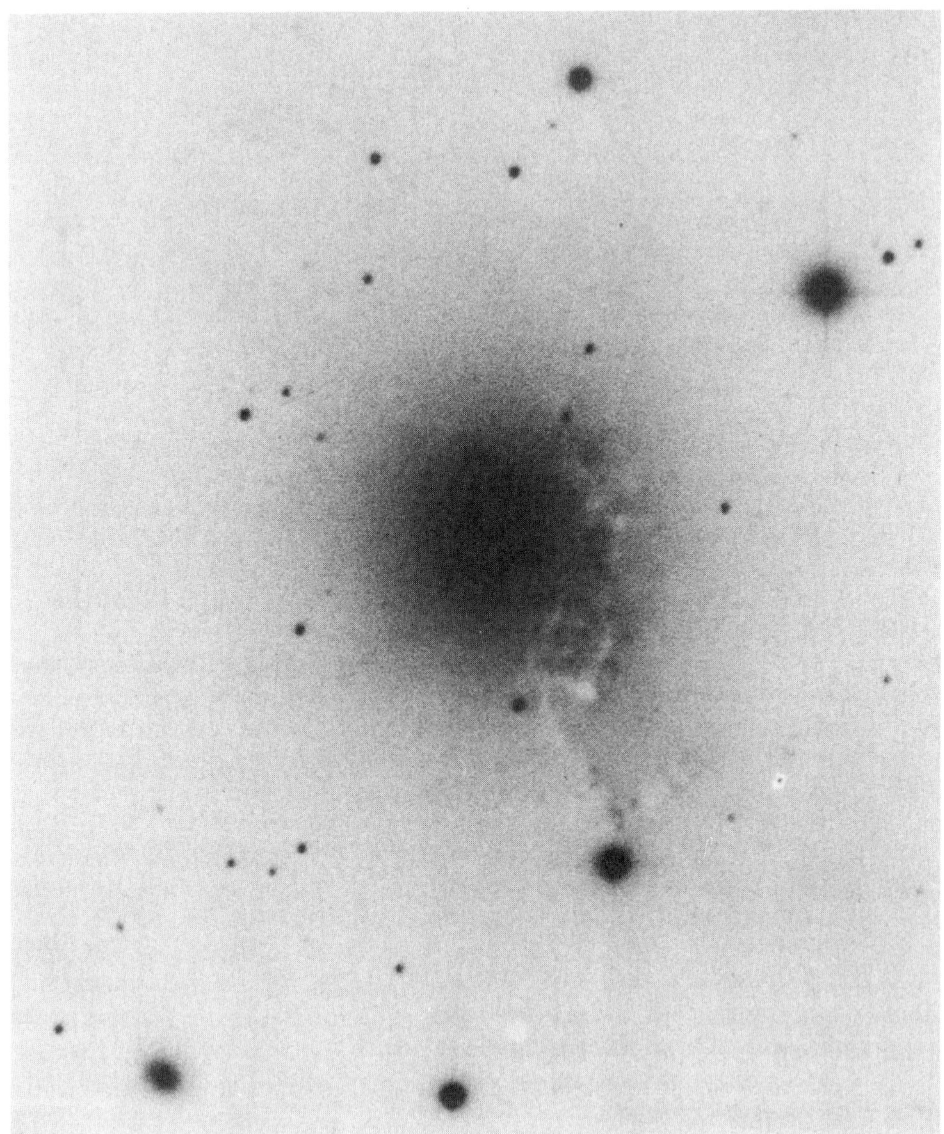

Fig. IV.7.   NGC 1275 = Perseus A. A SRG in the center of the Perseus Cluster. (*Courtesy of Halo Observatories.*)

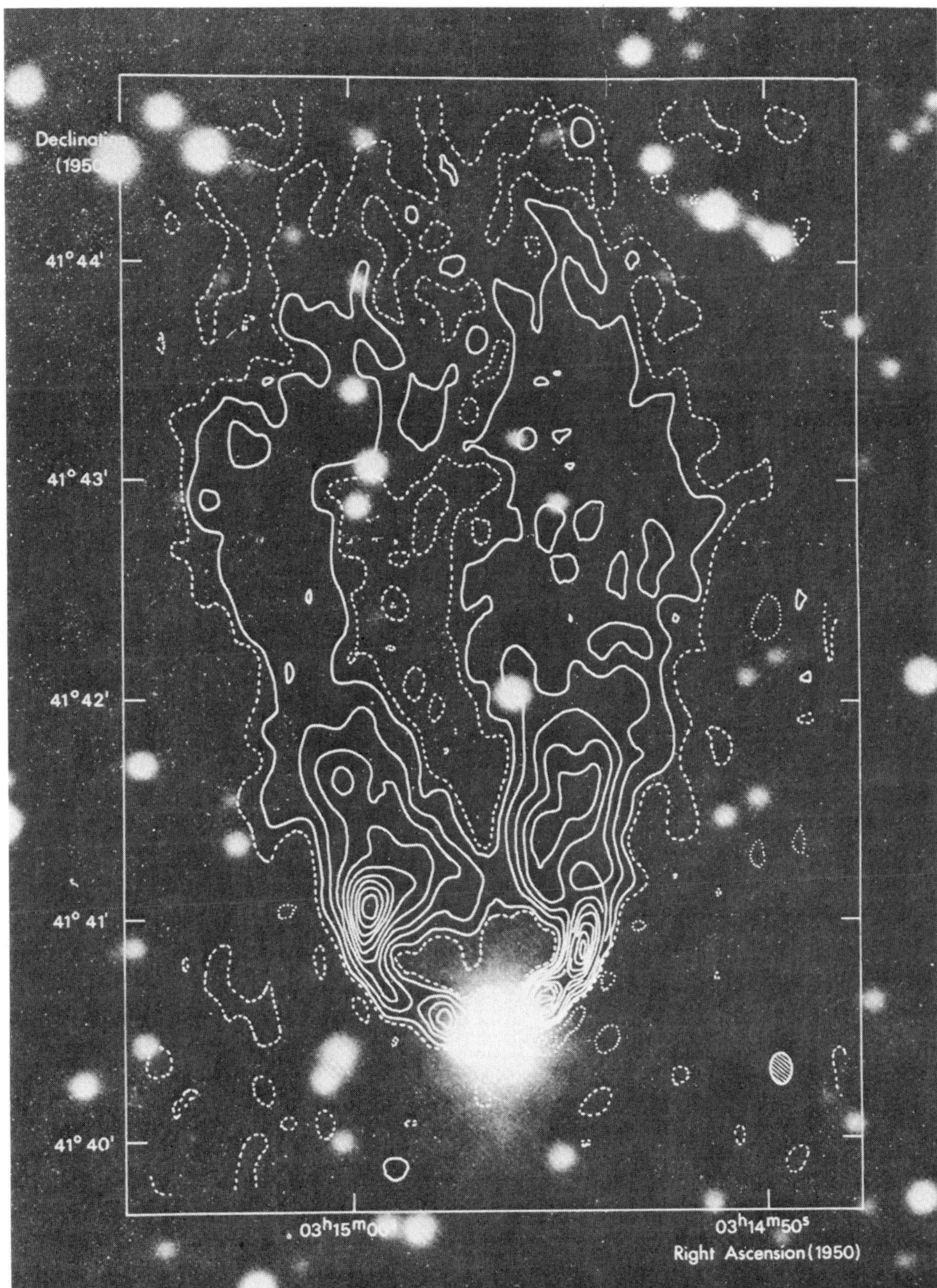

Fig. IV.8. NGC 1265 and the head-tail associated radio source. (*Reproduced from Miley et al.* (1974).)

pressure (Section III.9). Because of the symmetry between both 'tails', the ejected clouds must be of roughly equal energies and blown away at about the same distance from the central galaxy. The 'head-tail' radio sources are only found in clusters and the velocities required to explain them are consistent with the velocity dispersion of galaxies in clusters.

The 'head-tail' phenomenon depends on the relative motion between the SRG and the surrounding medium. In certain cases the medium is being blown away by a violent event and the 'head-tail' structures of nearby radio galaxies all point to the place from where the gas comes. For instance:

NGC 1275 = Per A = 3C84, (Figure IV.7) is a SRG in the center of the Perseus cluster. The optical object looks like an early-type galaxy, possibly E, with a complex filamentary structure of excited gases. As spectroscopically has been shown, they are ejected from the central source with at least two sets of velocities differing in several thousands km s^{-1}. From the distance at which the gas is observed, the outburst must have taken place 10^6 yr ago. The nuclear structure of NGC 1275 is complex (Section III.6) and the outer radio morphology is quite different from typical SRGs. Ryle *et al.* (1968) found two other nearby galaxies, NGC 1265 and IC 310, which are weaker radio sources with 'head-tail' morphologies. Both radio structures have extended tails trailing away up to 300 kpc from the parent galaxy and pointing to NGC 1275 (Figure IV.8).

Burbidge (1974) suggests that a high energy flux of particles ejected from the NGC 1275 event induced radio-emission in the nearby objects and conditioned their radio structures.

In the SRGs with the highest energies, the active galaxies emit relativistic jets and particle streams which are issued through very collimated channels with a diameter hardly superior to that of the central compact source (Section III.9). The orientation of these jets seems to be stationary in space during its lifetime for most cases.

Example: 3C449 is associated with a cD galaxy belonging to a group catalogued as VV 6–49–29 in the center of the open cluster of galaxies Zw 2231.1 + 3732. The common redshift to the SRG, the group and the cluster is cz = 5400 km s^{-1} and a distance of 54 Mpc follows. High resolution observation of 3C449 show the presence of two highly collimated symmetric jets on each of an unresolved core (Figure IV.9). The radiojets have a somewhat lumpy and turbulent structure and are not seen within 10^{11} (2 kpc) of the central source. Each jet widens steadily beyond the turn-on point, and gradually diminishes in brightness until each makes a sharp bend to the west and terminates in large, diffuse blobs with a very relaxed appearance. The emission is detected up to 8 arc min (120 kpc) from the center (Perley *et al.*, 1979).

It is assumed that the surprising collimation these jets takes place in the neighborhood of a rotating collapsed central object, the jets being issued along the rotation axis. In this way the orientation of the streams is kept invariable by the strong gravitational field, irrespective of the continuity of the ejections, which depends on the matter supply to the accreting central body (Section III.9). Although there are no definite models yet for these phenomena, it is agreed that the initial ejection velocities must be close to that of light. Some observed asymmetries are probably due to relativistic effects, such as the jet in M87.

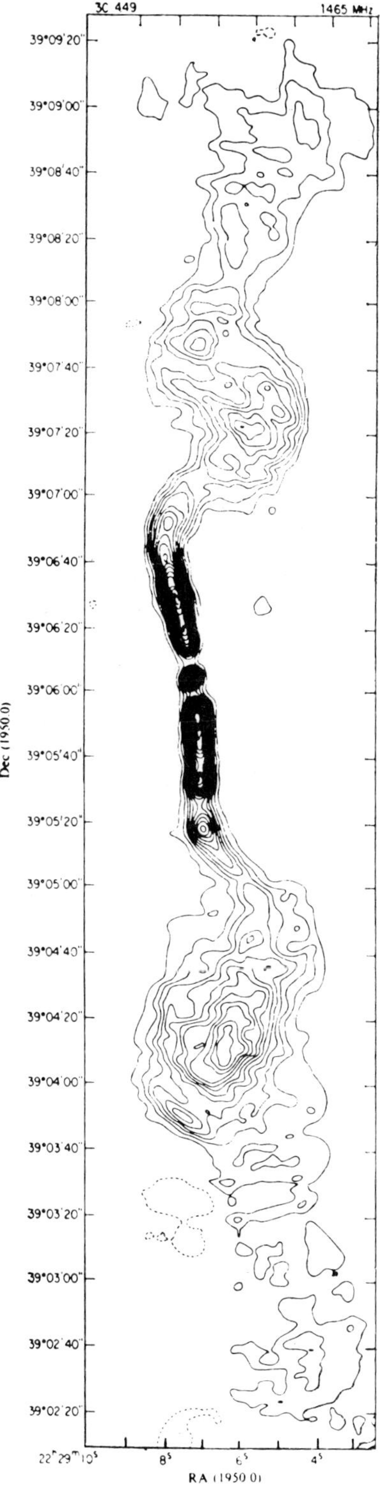

Fig. IV.9. The SRG 3C449 and its collimated beams. *(Reproduced from Perley et al. (1979).)*

Some instances of SRGs (e.g. NGC 326 = 4C26.03) with collimated jets can only be understood if the axis of ejection had changed with time. Rees (1978) proposes to explain them through a collision or a merger of the SRG with another galaxy, so that their interaction causes the rotating source to precess when it accretes matter from the intruder galaxy, which has a different orientation of angular momentum.

Stripping Process

Spitzer and Baade (1951) first suggested that galactic collisions in clusters could be responsible for the SO formation, when disk gas is removed from spirals in such events. Other ways to removed gas from spirals are stripping by dynamical pressure, proposed by Gum and Gott (1972), and gas evaporation. The stripping mechanism strongly depends on the ICM density, and also on temperature in the case of evaporation. In consequence, the most dense ICM regions are favorable environments for such processes. Galaxies moving through the central regions would be most affected by this process which would produce there a higher abundance of SO-type objects. Dressler (1980) has contested this argumentation on observational grounds. He finds a slow function correlating density and morphological types in clusters, so that a significant percentage of SO galaxies exists in environments unfavorable to the stripping process. Moreover, the relation between density and morphological type is virtually identical in rich and poor clusters despite the expected excess in former (Section IV.2.1). As an alternative, Dressler proposes the density vs morphological type relation to be a consequence of the long-time scale associated with disk formation in galaxies. In dense regions, most of the matter would be used up to form ellipticals in a shorter time-scale.

Accretion

The intergalactic medium modifies the structure of normal galaxies through accretion. A body of mass M moving with velocity V through the surrounding medium with density ρ_E accretes mass according to the law

$$\frac{\mathrm{d}M}{\mathrm{d}t} = kM^2 \quad \text{where} \quad k \approx 2\pi G^2 \rho_E / |V|^3$$

(Hoyle and Littleton 1940; Bondi and Hoyle, 1944). If the body is at rest with respect to the medium, the same law applies, but the velocity V has to be substituted by the speed of sound c_s in the medium (Bondi, 1952).

The motion of a test particle in the gravitational field of a variable mass was discussed by Sérsic (1970, 1973, 1980). Under certain conditions his results can be used to find consequences of steady accretion for the structure of a galaxy. In the spherically symmetric case, the model predicts a shell around the accreting galaxy, like those observed by Maline and Carter (1980) around elliptical galaxies. Particularly noticeable are those around NGC 3923, which extend up to a distance of 100 kpc from the central elliptical object. The observations suggest the shells to be constituted by stars. Malin and Carter, however, think they resulted from shock-induced star formation, caused by some explosive event.

Ring structures around bars in SB galaxies also result from an accretion process

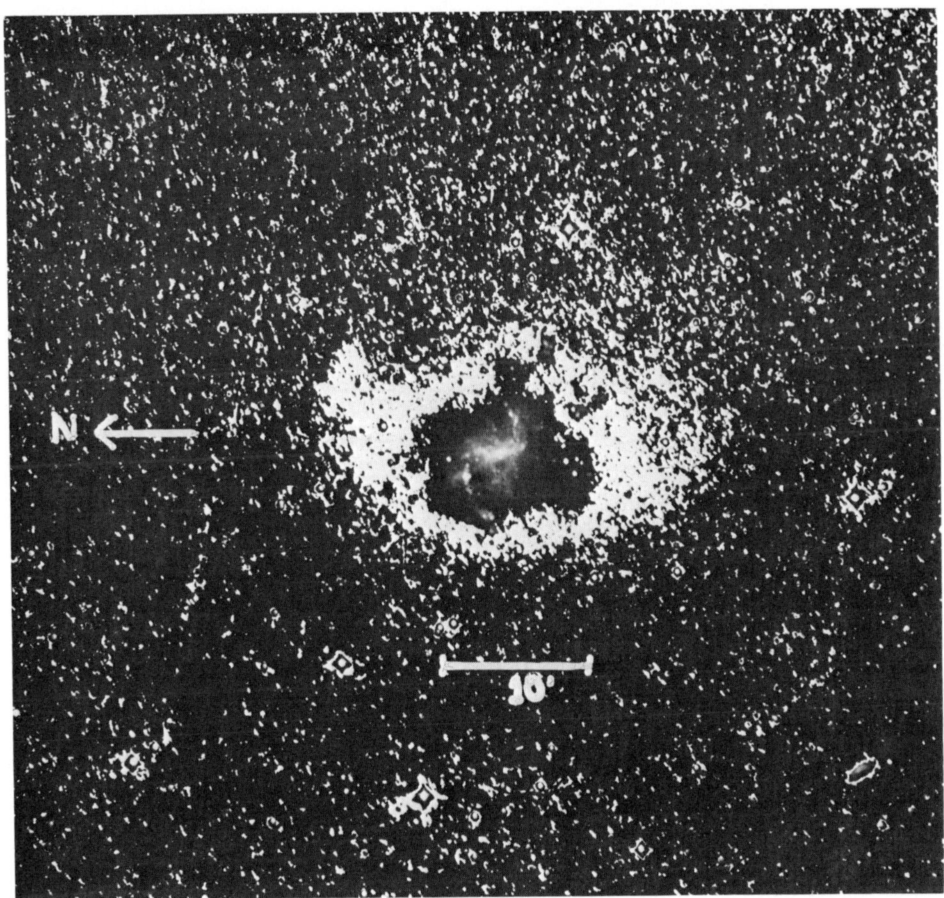

Fig. IV.10.   Composite picture of NGC 1313. The inner part is from Córdoba Observatory and shows the ring structure around the bar. The outer part is an isodensity of the outer envelope from a Maktsutov plate of the Observatorio Nacional, Chile.

(Sérsic *et al.*, 1979, 1980). The motion of a star in the field of a rotating and accreting bar has singular points on a circular ring around the bar whose plane is normal to the plane of rotation and is tilted with respect to the bar. If the mass variation is due to accretion, the ring should be in the inter-arm region as observed. Ring-like structures observed around NGC 1313, NGC 1672 and NGC 2685 suggest accretion is going on in these objects, which is in agreement with the presence of faint luminous matter in the outer fringes of these galaxies, most spectacularly in the NGC 1313 case (Figure IV.10).

IV.2. Agregates of Galaxies

Galaxies occur in a wide variety of systems which go from binaries through small groups and clouds to clusters of various degrees of richness and to superclusters.

These systems in turn cover a wide range of densities and sizes. The existence of isolated galaxies, that is galaxies which cannot be assigned to any of these structures, is still under debate: while Turner and Gott (1975) concluded that about 40% of the galaxies belong to the general field, de Vaucouleurs (1975) and also Soneira and Peebles (1977) agree on only about 15%, and Tully and Fisher (1978) claim that there is no evidence for a significant number of field galaxies.

IV.2.1. PAIRS AND MULTIPLETS

At scales smaller than 0.1 Mpc we find galaxies in pairs, triplets and multiplets which in turn show a tendency to form groups. This was first discussed by Holmberg (1941, 1969). His analysis led to a statistical separation of physical and optical companions: in circular areas around assumed central galaxies, physical and optical companions were chosen. If the radius of the surveyed areas is large enough, the statistical distribution of the observed separations can be divided into two parts: one distribution corresponding to the field galaxies and another to physical companions. Holmberg found in this way that the mean separations of physical pairs ranged between 140 and 380 kpc, the relative number of companions with separations less than 60 kpc being 30%.

A correlation between luminosity and separation in binary galaxy systems has been found by Ostriker and Turner (1979). The number of pairs which are both bright and close is smaller than would be expected statistically from independent distributions of the mutual distances and luminosity. The proposed interpretation is that dynamical friction causes the bright close pairs to spiral together, although a genetic origin cannot be discarded.

Compact groups, as well as linear associations of galaxies, have received much attention as possible places where discrepant redshifts are found. Arp believes that at least some redshifts cannot be attributed to the Hubble flow (Section III.8) whilst other astronomers strongly support the conventional point of view. A comprehensive discussion of the pros and cons in this debate can be found in (Field, 1973). The catalogue of Vorontsov-Velyaminov (1958) and Arp's Atlas of Peculiar Galaxies (1966), as well as the Uppsala Catalogue (Chapter VIII) for the southern hemisphere are valuable sources of information (Figures IV.11 and IV.12).

Satellites of spiral galaxies are found in 50% of the cases (Swakina, 1966). Statistically they have a tendency to avoid the symmetry plane of the central parent galaxy (Holmberg, 1969).

From a discussion of 600 isolated pairs from which 260 had measured relative radial velocities, Karachentsev (1978) found more than 50% have $\mathfrak{M}/\mathfrak{L}$ ratios smaller than 10, but 10% show $\mathfrak{M}/\mathfrak{L}$ ratios larger than 100, which apparently are related with accidental optical pairs.

Sargent (1974) has pointed out the high proportion of groups with very small physical separations like Stephan's Quintet or Seyfert's Sextet, which have discrepant redshifts associated to one of their presumable members. Rose (1979) has simulated loose groups of 7 members and followed its evolution to examine whether the ocassional compact configurations of 4 galaxies are formed within a looser system. From his calculations Rose concludes that the most plausible explanation for compact groups is that they are temporarily unbound configurations within loose groups

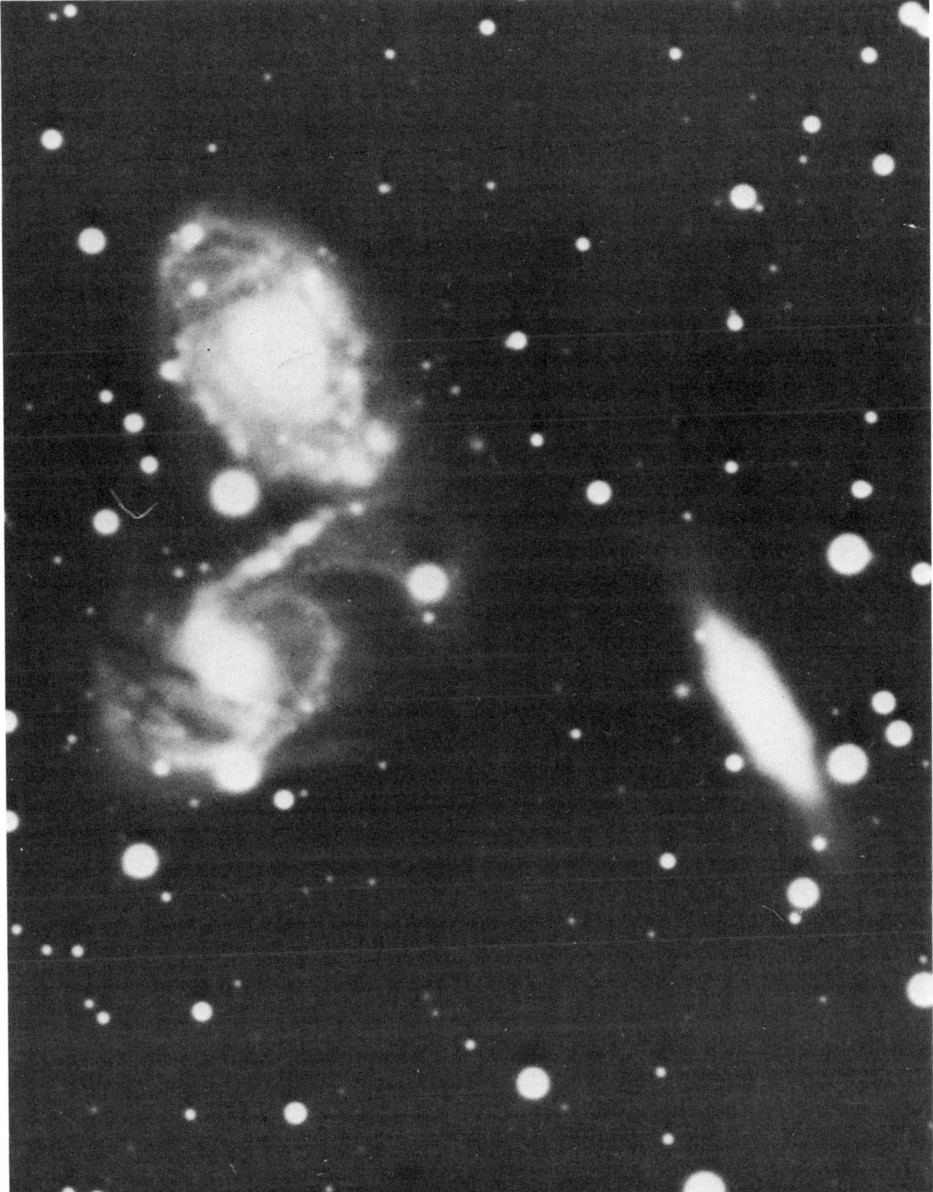

Fig. IV.11.   Triplet of Galaxies in Pavus = NGC 6769, 6770, 6771. (*Courtesy of CTIO.*)

almost as intrinsically compact as they appear to be and which disperse in a time of the order of the group crossing-time. According to this view, the face value application of the virial theorem is not warranted.

Example: The Seyfert Sextet is a very compact group of galaxies for which Burbidge

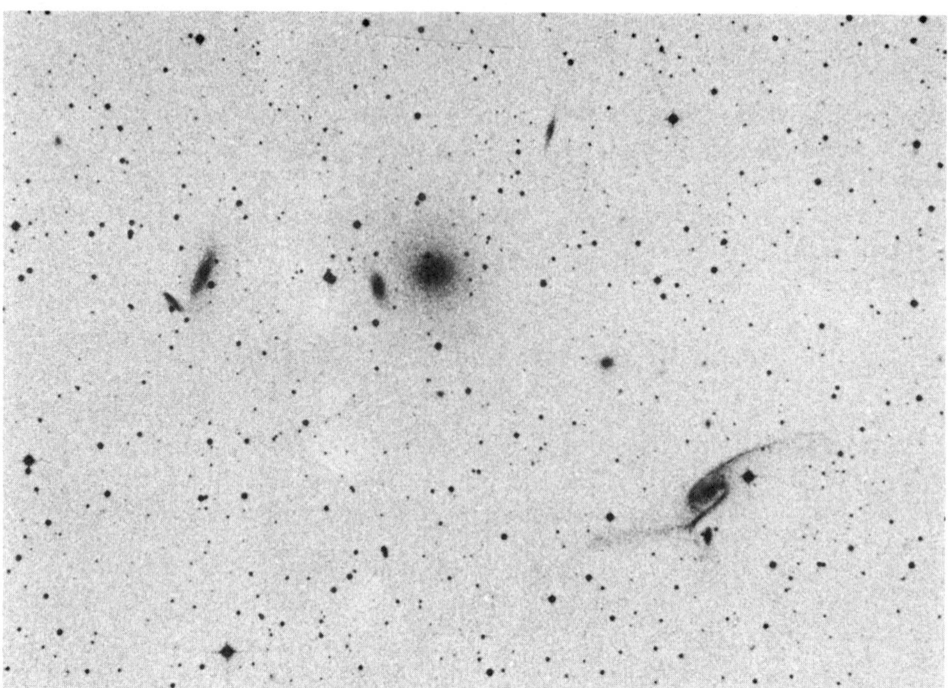

Fig. IV.12.   Linear group of NGC 6876, 6877, and 6880 in the deep south. (*Courtesy of the Observatorio Nacional, Chile.*)

and Burbidge (Sargent, 1974) estimated at less than a few hundredth the probability of the group being a chance coincidence. All members with one exception, show obvious features of gravitational interaction. These objects have radial velocities close to 4400 km s^{-1} while the remaining one not partaking in the interaction has a discrepant redshift close to 20 000 km s^{-1}, and probably is a background galaxy. If this were the case, it would become difficult to conciliate the absolute magnitude with its morphological type (Sargent, 1974).

The size of the Seyfert's Sextet is only 8 kpc, which explains the interaction between its members. The application of the virial theorem leads to a mass luminosity ratio $\mathfrak{M}/\mathfrak{L} \approx 100$, close to that of E galaxies, provided we exclude the galaxy with the discrepant redshift.

IV.2.2. GROUPS OF GALAXIES

At scales ranging between 1 and 2 Mpc we find loose groups of galaxies like the one around M81 (Figure III.5) and the Local one. An extensive study of the space distribution of nearby (< 20 Mpc) galaxies was made by de Vaucouleurs (1975) who found that most of the surveyed galaxies (85%) belonged to loose groups. About 50–60 of such systems are within 16 Mpc, which leaves 300 Mpc^3 for their average volume and 7 Mpc for the average distance between their centers. The mean diameter for these groups was found to be 1.5 Mpc, with a range between 0.3 to 2.5 Mpc.

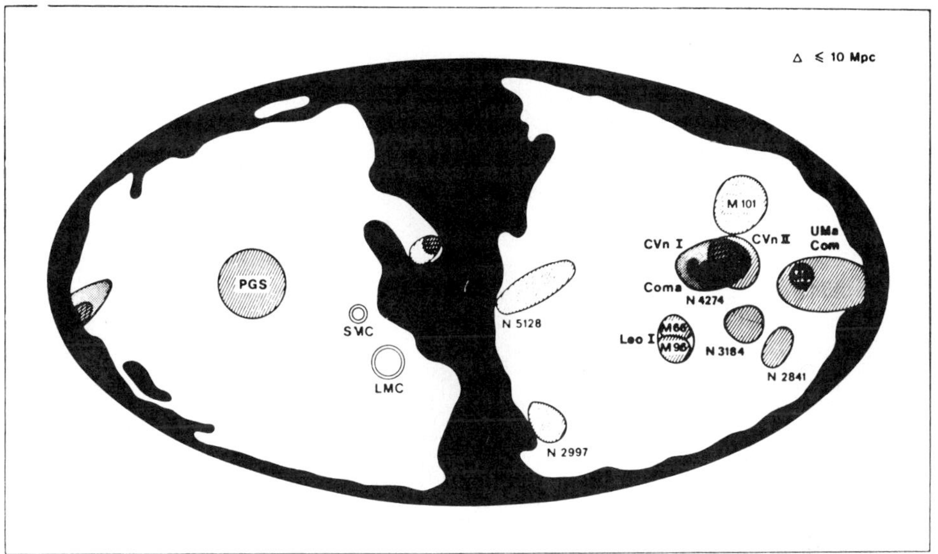

Fig. IV.13.   Distribution of nearby groups ($\Delta \leqslant 10$ Mpc) of galaxies. Notice the asymmetry between N and S galactic hemispheres. (*Adapted from de Vaucouleurs* (1975).)

The shapes of the loose groups are rather irregular, like low density clouds with much sub-clustering. Spiral and irregulars, as well as some dwarf-ellipticals, are the dominant population. Giant ellipticals and SO galaxies are conspicuously absent.

The space distribution of the nearby groups is far from random (Figure IV.13 and IV.14), they show a marked tendency to concentrate towards a great circle in the sky (Section (V.2.4)).

A composite luminosity function $\phi(L)$ for groups of galaxies was introduced by Turner (1978), where $\phi(L)$ gives the probability of having galaxies in the luminosity range $L, L + \mathrm{d}L$. Schechter's (1976) form of the luminosity function is

$$\phi(L)\, \mathrm{d}L \propto (L/L^*)^{-1} e^{-(L/L^*)}\, \mathrm{d}L,$$

where $L^* = 1.7 \times 10^{10}$ solar units, corresponding to an absolute magnitude $M_p^* = -20.1 \pm 0.1$. There is no sensible difference between $\phi(L)$ for early and late-type galaxies in group.

Rood *et al.* (1970) analyzed the 54 groups of de Vaucouleurs work (1975) with the virial theorem technique described in Section II.2.2. Assuming $\mathfrak{M}/\mathfrak{L} = 7$ for spirals and irregulars and $\mathfrak{M}/\mathfrak{L} = 50$ for ellipticals, they found most of the groups have log $(\mathfrak{M}_{VT}/\mathfrak{M}_L) > 0$, with a median log $(\mathfrak{M}_{VT}/\mathfrak{M}_L) \approx 1.6$. Among several other works on the same material and various techniques, that of Turner and Sargent (1974) is decisive in showing that many of the groups of galaxies in de Vaucouleurs' list are not bound systems. From a critical analysis, they have demonstrated that more than one half of the groups have estimated crossing times for the member galaxies in excess of the Hubble time H_0^{-1} (Section V.2.2) and consequently are not bound, and also that those with shorter crossing times have smaller virial discrepancies.

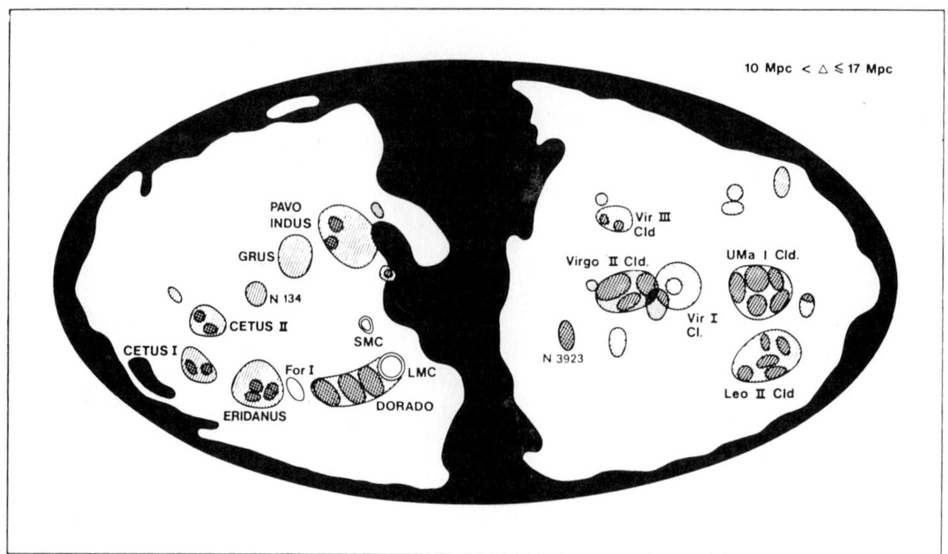

Fig. IV.14.   Distribution of nearby groups (10 < Δ ⩽ 17 Mpc) of galaxies. (*Adapted from de Vaucouleurs* (1975).)

IV.2.3. CLUSTERS

Galactic concentrations on a larger scale ($\geqslant 10$ Mpc) have been known for many years. The general properties of the projected density distribution and luminosity function of galaxies in these systems have been studied extensively. Several reviews on the subject of clusters of galaxies have been given in recent years (Abell, 1976; Bahcall, 1977b; van den Bergh, 1977; Quintana, 1979).

Classification

From the five classification systems currently used in the literature, two of them (Zwicky's and Abell's) are based on the morphology and structure of the clusters. A third one (Rood-Sastry) uses the morphology and distribution of the ten brightest members, while that of Bautz and Morgan (BM) relies on the brightness contrast of the brighter galaxies with respect to the remaining ones. Oemler's classification is based on the galaxy content of the clusters.

According to their morphology, Zwicky (1967) distinguishes three types of clusters:

(i) *Compact*: containing a single conspicuous concentration of the brightest galaxies, with at least 10 members in apparent contact.

(ii) *Medium Compact*: Either a single concentration of at least 10 galaxies separated by several diameters, or several distinct concentrations.

(iii) *Open*: There are no obvious concentrations.

The degree of symmetry and the central concentration are the parameters considered by Abell (1958, 1976) in his classification, which reduces to:

(i) *Regular clusters*: High central concentration, spherical symmetry and more than 10^3 members on the first 5 magnitudes of the luminosity function.

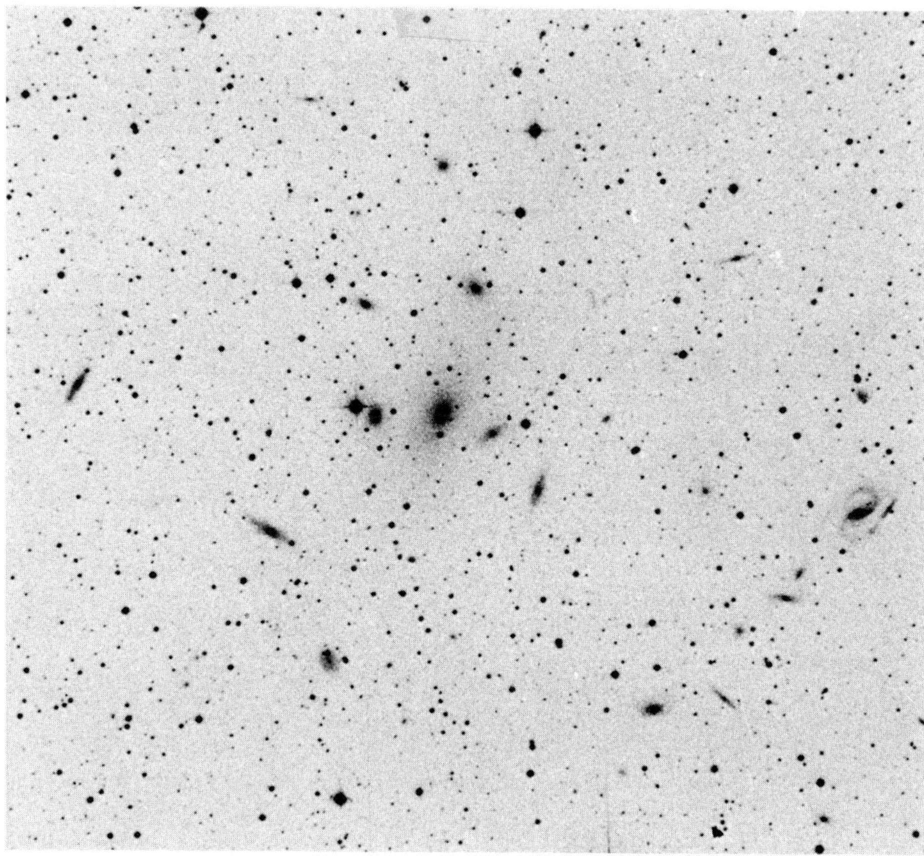

Fig. IV.15. Rich cluster of galaxies in 18^h43^m–$63°$ with a dominant cD galaxy in the center. (*Courtesy of the Observatorio Nacional, Chile.*)

(ii) *Irregular clusters*: No spherical symmetry, no single definite but possibly several central concentrations.

Since they have some overlap because of similitude of criteria, Zwicky's and Abell's systems are not independent.

Further refinements were introduced by Rood and Sastry (1971), whose classification is based in the morphology and distribution of the 10 brightest galaxies. They introduce the scheme

$$cD - B \Big\langle \begin{matrix} L - F \\ C - I \end{matrix}$$

where the sucessive types are described as follows:

cD: The cluster is dominated by a single galaxy of Morgan's cD type (Section III.6).
 Examples: IC 2082.

B: The cluster has two supergiant galaxies separated by no more than ten diameters of the larger one.
Example: Coma.

L: At least three of the brightest galaxies are aligned.

C: At least four of the brightest members are in the center, its separations being all of the same order.

I: There is no definite center. The members are irregularly distributed.

F: A flattened configuration is suggested by the space distribution of the 10 brightest members.

The frequency of types between more than one hundred nearby Abell clusters is given in Table IV.2.

TABLE IV.2

Frequency of types in Rood and Sastry classification

Type	%
cD	21
B	9
L	9
C	14
F	18
I	29

Bautz and Morgan (BM) (1970) introduced three principal classes exclusively defined by the relative contrast between the few brightest with the other galaxies in the cluster. The BM scheme is equivalent to a rough description of the luminosity function of the cluster. They used the following criteria to define the BM classes:

BM I: Clusters containing a centrally placed cD galaxy.
Example: IC 2082. (Figure IV.15).

BM II: Clusters where the brightest galaxy or galaxies are intermediate in appearance between class cD and gE′ in the Virgo Cluster.
Example: Coma cluster.

BM III: No dominant galaxy in the cluster.
Example: Virgo cluster.

The authors also consider intermediate classes I–II and II–III and give a list of clusters intended to be used as standards for classification. The classification pro-

TABLE IV.3

Mean absolute magnitude for the first three-ranked galaxies in rich clusters as function of BM-type*

BM Type	$-M_V(1)$	$-M_V(2)$	$-M_V(3)$
I	22.9	22.0	21.7
II	22.6	21.9	21.6
III	22.3	21.5	21.1

* Adapted from Sandage (1976), with $H = 100$ km s^{-1} M^{-1}.

cedure is not free from subjectivity although the comparisons of results from several authors are in general agreement. Anyway, it is convenient to have standard clusters in the same plate material. BM standards for the Southern Hemisphere are being established by the Quintana and White (1978) work.

van den Bergh (1977) as well as Bahcall (1977a) have discussed distance and contamination by foreground galaxies effects in the Sastry and BM Classifications.

Table IV.3 gives the calibration of the absolute magnitudes for the mean of the three first-ranked cluster galaxies as a function of BM type for rich clusters.

A classification based on the galaxy content of clusters was first proposed by Morgan (1962, also Section I.4) and later developed by Oemler (1974), who introduced three types of clusters:

cD: Sphericals in shape. Rich in early-type galaxies (E, SO). One or more cD systems may be found at the center.

S-rich: Rich in spirals. Irregular morphology, low mean density and no definite central concentration.

S-poor: Intermediate between the two preceding classes. SO galaxies dominate the scene.

From all the described classification it is clear that clusters of galaxies can be arranged in a roughly continuous sequence between two extreme situations: At one end there are the regular, spheroidal, highly concentrated and rich clusters. Some of them are dominated by cD or D galaxies placed at the center. At the other end of the sequence there are the irregular shaped, spiral-dominated, low density clusters without symmetry nor concentration (Table IV.4).

TABLE IV.4

Classification and properties of clusters of galaxies*

Classif./ Property	Regular	Intermediate	Irregular
Zwicky	Compact	Medium-compact	Open
B–M	I, I–II, II	(II), II–III	(II–III), III
Rood–Sastry	cD, B. (L, C)	(L), (F), (C)	(F), I
Content	E-rich	S-poor	S-rich
Symmetry	Spherical	Intermediate	Irregular
Degree of concentration	High	Moderate	Very little
Profile	Steep	Intermediate	Flat

* From Longair (1978).

The role of cD galaxies must be emphasized: Observation shows that they are not simple extrapolations of bright ellipticals but a special class of galaxies with particular characteristics already recognized by Morgan (Section III.6). In fact, recent observations have ratified and enlarged this point of view, showing that cD galaxies seem to be also abnormally active at X-rays and radio frequencies as compared with normal Es (Kriss *et al.*, 1980). Moreover, cD galaxies have the same X-ray, optical and radio properties irrespective of the cluster richness, they are placed at the center of the deep potential well of the cluster (White, 1978), something not necessarily shared by the gEs. This fact became a key clue in our understanding of those systems (Section IV.1.1).

Dynamics and Evolution

The structural regularity of rich clusters of galaxies such as that in Coma, whose space density distribution approaches that of an isothermal distribution (Zwicky, 1957), has suggested the virial theorem could be applied with confidence to estimate its total mass (Section II.2.2). Early workers such as Zwicky (1933) and Smith (1936) soon noticed the ratio of these systems greatly exceeded the mean individual values for field galaxies. This problem still persists without a definite answer. Page (1965) and also (Karachentsev) (1966) have shown that this discrepancy increases with the size of the cluster (Table IV.5).

Taking adventage of the regular structure of the richest clusters, Bahcall (1977a) followed Zwicky's steps and fitted isothermal distributions to 15 clusters finding a mean core radius $R_c = 0.13 \pm 0.02$ Mpc. Other authors also computed R_c and found small dispersions, although different mean values. Knowing the velocity dispersion σ_R for member galaxies, the mass of the cluster can be calculated through a model (an isothermal distribution, or that with a projected number density of galaxies following the $\rho^{1/4}$-law for spheroidal systems, Section II.1.1, as it was done for E galaxies in Section II.2.2). The agreement between different calculations for the same clusters is encouraging, although the values of $\mathfrak{M}/\mathcal{L}_V$ are still large. The way the scale parameter R_c has been derived points to excessive values for σ_R to explain the discrepancy.

TABLE IV.5

Mass-lumionisty ratios for agregate of galaxies*

Sample	$\mathfrak{M}/\mathcal{L}_V$
6 rich clusters	1700 ± 300
9 poor clusters	890 ± 200
29 groups	660 ± 200
11 triplets	170 ± 60

* Adapted from Karachentsev (1966) for $H = 100$ km s^{-1} Mpc^{-1}.

Holmberg (1961) has noticed that subclustering in form of pairs or small groups, if unaccounted for will tend to exaggerate σ_R; the effect however is small because the binaries and small groups (Section IV.2.1) have velocity dispersions far smaller than those in clusters of galaxies. Contamination by field galaxies is another source of bias for σ_R, but little has been done in this sense.

Oemler (1974) has demonstrated that the large clusters of galaxies are still collapsing and consequently are far from virial equilibrium. He determined detailed density distributions for each cluster type: cDs, S-rich, S-poor of a given sample, and compared them with evolutive models computed by Aarseth. Through this procedure, Oemler suceeded in uncovering two important facts. The first is that clusters achieve their virialization only after several characteristic times (crossing-time) and secondly, the models yield improved projection factors for radial velocities (cf. Section II.2.2). Through fitting the observed mass distributions and velocities to the computed ones, final values for 8 clusters were obtained giving an average $\mathfrak{M}/\mathcal{L}_V \approx 340$ independently of the cluster type. For the Perseus Cluster, dominated by the SRG NGC 1275 (Section IV.2.), the value obtained for $\mathfrak{M}/\mathfrak{L}_V$ was as large as 1400, a figure independently confirmed by Bahcall (1973). Even in the well studied nearby spherical cluster in Coma, the discrepancy is as large as a factor of 20 ($\mathfrak{M}/\mathfrak{L}_V \approx 500$), and has not been understood through 'missing masses' such as the existing ICM (Section IV.2.1), the overlapping of massive galaxy halos, or a common cloud of intergalactic stars. This situation seems characteristic for all large clusters (Table IV.5).

According to Butcher and Oemler (1980), irregular and regular clusters are instances of pre- and post-virialization states of cluster evolution. Relaxed clusters should have a wide range of densities, while the unrelaxed ones, contracting at a roughly uniform density, should have a narrow range. Dressler (1980) draws interesting conclusions from this interpretation. Most clusters are of the low concentration type, thus they must be in the contracting phase, although because of this relatively high density, they are very close to virialization. This is not surprising, because there is a strong selection which favors higher density clusters to be recognized as such, while those of lower densities are only extended density enhancements of a factor 2 or 3 over the general field, their identification being difficult.

The presence of X-ray emission in the most dense clusters (see below) leads Dressler to identify these objects with clusters closest to virialization, at their maximum densities. In this stage the gravitational potential is at a minimum, galactic and gas-density is high, and the heating of the gas cluster is maximized.

Intracluster Medium (ICM)

The morphology of radio sources in clusters is clear evidence for the existence of an ICM (Section IV.1.2). Most of the RS in clusters have significant distorsions and misalignements, as compared with the typical field SRGs (Section III.6) according to Rudnick and Owen (1977).

Spiral galaxies would be stripped of their gas content if the ICM were gaseous and dense enough. Evidence that this is the case, however, is contradictory. Haynes and Roberts (1978), who observed individual spirals for H I in most of Abell's nearer clusters, found the ratio $\mathfrak{M}_{H\,I}/\mathfrak{M}$ for the spirals in E dominated clusters is significantly smaller than that prevailing in S-rich clusters, which equals that in field S-galaxies. Melnick and Sargent (1977) and Bahcall (1977a) discovered that the radial distribution of spirals often differs from that of the ellipticals on X-ray clusters. Also Tytler *et al.* (1978) found that the percentage of (E + SO) galaxies correlates well with the X-ray luminosity L_x. The simplest interpretation of this would be that spirals moving through the X-ray emitting ICM are stripped of their gas and thus become SO systems. The empirical relationship is $P(E + SO) \propto \log L_x$, where $P(E + SO)$ is the percentage of E + SO galaxies in a cluster with X-ray luminosity L_x. If thermal bremsthralung is the mechanism for the X-ray emission, we have $L_x \propto \rho^2 V$. Assuming a roughly constant emitting volume V, from the two preceding relations follows $P(E + SO) \propto \log \rho$, a semi-empirical relation between the density of the ICM and the percentage of early type galaxies (Tytler *et al.*, 1978). This shows that the (E + SO) production would be a weak function of the ICM density, which is in agreement with Dressler's (1980) findings of a correlation between $P(E + SO)$ and density from cluster to

cluster but not inside them (Section IV.1.2) which would be against the stripping of
spirals by dynamical pressure as a main source of SO galaxies.

The X-ray observations of several clusters, particularly those in Coma, Perseus
and Virgo have detected an emission feature originated in highly ionized Fe (Serlemitsos *et al.*, 1977) which gives support to the thermal origin for the X-ray emission.
Calculations by Lea (1976) have shown that such emission can be produced by a gas
with a temperature of 10^8 K and a density in the range of 10^{-26} to 10^{-28} g cm^{-3}, depending on the location in the cluster. This leads to a mass of the order of 10^{14} solar masses
for the ICM in rich clusters (Section IV.1).

Luminosity Function

The integrated luminosity function of galaxies in a cluster is the number $N(m)$ of
galaxies brighter than magnitude m. $N(m)$ is now known in a range of seven magnitudes, thanks mainly to the work of Abell (1962, 1976), Bautz and Abell (1973),
Oemler (1974), and others.

Zwicky suggested an exponential form $N(m) = K (10^{0.2M} - 1)$ with K being a constant, but this function over represents the bright end of the luminosity function.
Instead Abell used a broken linear representation of log $N(m)$, namely

$$\log N(m) = K_1 + S_1 m \quad \text{if} \quad m \leqslant m^*$$
$$\log N(m) = K_2 + S_2 m \quad \text{if} \quad m \geqslant m^*,$$

where $S_1 = 0.75$ and $S_2 = 0.25$ (in agreement with Zwicky's). The discontinuity
in m^* appears to be universal and corresponds to an absolute magnitude $M_V^* =
-20.41 \pm 0.15$, as follows from a dozen of clusters studied by Bautz and Abell (1973)
and Austin *et al.* (1975).

The differential form of the luminosity function $\phi(L)$ dL given by Schechter (1976)
(Section IV.2.2) with an exponential cut-off toward high luminosities

$$\phi(L) \, dL \propto (L/L^*)^\alpha \, e^{-(L/L_*)} \, dL$$

has $\alpha = -1.25$ (corresponding to Abell's $S_2 = 0.25$) and $L^* \approx 1.7 \times 10^{10}$ solar units.

Felten (1977) has shown that this representation of the luminosity function $\phi(L)$
gives an excellent fit for many clusters with the above values of α and L_*, including
field galaxies. This is not, however, to be taken as a suggestion for the universality of
$\phi(L)$; several and important departures have been found.

IV.2.4. Second Order Clustering

The subject of a second order clustering between galaxies was under debate between
its leading opponent Zwicky (1957) and the proponent Abell (1958), until sucessive
contribution by this latter author (1967), Kiang (1967) and Kiang and Saslaw (1969)
settled the question in favor of the existence of second order clustering.

The idea of a local super cluster of galaxies (LSC) was introduced by de Vaucouleurs
(1956, 1958) and through his own and other contributions its reality was established.
A list of probable superclusters (SC) was given by Abell (1961). All these investigations
suggested that matter is clustered in at least two different orders or, most probably in
a continuous way up to scales of 50 Mpc. The SCs involve then a whole hierarchy of
minor systems: clusters, groups and even individual galaxies.

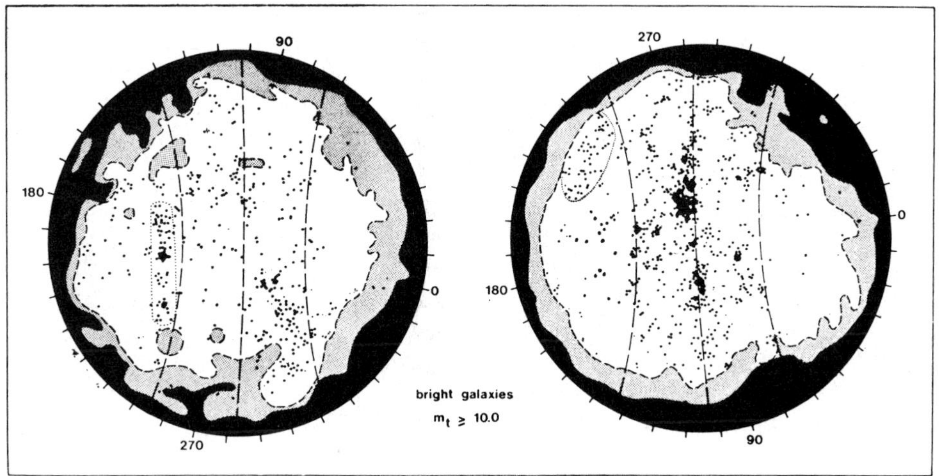

Fig. IV.16.   Distribution of the bright Shapley–Ames galaxies. (*Adapted from de Vaucouleurs* (1956).)

The Local Super Cluster (SC)

Evidence for a 'galaxy belt' were mentioned by early researchers, in particular Reynolds (1921). Figure IV.16 based on the Shapley–Ames catalogue of bright galaxies (1932) put in evidence what de Vaucouleurs in a series of papers in the fifties termed 'Local SC'. This system can be traced all around the sky in both hemispheres, although the southern galactic hemisphere is much less populated. Large clouds of galaxies tend to be flattened, their planes being parallel to the general distribution, and also nearby groups cluster preferentially around the same plane (Section IV.2.3). The distribution of intergalactic H I clouds has also been shown to be controlled by the LSC (de Vaucouleurs and Corwin 1975).

Intergalactic extinction has been detected; the absence of very faint galaxy clusters behind the foreground clusters (Zwicky, 1957), such as those in Coma and Virgo (VC), is a clear indication of that. Holmberg (1974) has estimated the intracluster extinction is 0.25 mag. in the VC case.

The direction of the center of the LSC is that of the VC, irrespective of whether that cluster is included in the galaxy counts. Behind the VC there is an empty region, which stresses the inappropiateness of a disk-model for the LSC. String-like structures such as those of the Perseus and also Fornax SCs are suggestive of a similar situation in our system.

The kinematic of the LSC produces departures from the ideal Hubble flow (Section V.2.3). The velocity field reveals conspicuous departures from linearity and also from isotropy in the velocity-distance relation for galaxies in the equatorial belt of the LSC. de Vaucouleurs (1972) has fitted a simple ellipsoidal model with differential rotation and expansion which gives an excellent representation of the data.

IV.2.5. CLUSTERING

Much effort has been given in recent years to develop a description of galactic clustering in probabilistic terms and to put it within a cosmological context. In this approach, mostly due to Peebles, clustering is described in terms of correlations in the distibution of galaxies. One of the best established results is the so called two-point correlation function for galaxies. Peebles (1974) has defined a function $\xi(r)$ for galaxies in terms of the probability dP of finding a galaxy within a volume dV at distance r from a randomly chosen galaxy

$$dP = n_G^2(1 + \xi(r))\, dV,$$

where n_G is the average number density of galaxies. If the distribution were uniformly random, the two-point correlation function $\xi(r)$ would be zero and P would depend on the average density n_G. The cosmological principle (Section VI.1) requires the probability to be isotropic, depending only on the relative distance. In principle, from a large sample of galaxy positions and redshifts we would be able measure $\xi(r)$ directly but in practice this is impossible. We are forced to measure the angular two-point correlation on the sky $w(\theta)$ and convert it afterwards into a spatial correlation

$$dP = n_G'(1 + w(\theta))\, d\Omega,$$

where n_G' is the mean number of galaxies per steradian and θ the angular distance. This is done solving Limber's integral equation (Fall, 1979) which requires introduction of the sample selection function. For a sample which is magnitude-limited, the selection function is expressed in terms of the integral luminosity function for galaxies $\phi(M)$.

Because $\phi(M)$ is a rather broad function (Sections IV.2.2 and IV.2.3), most of the structure in $\xi(r)$ is lost when convolved through $\phi(M)$. From Limber's equation follow two simple although important results:

(i) If the angular two-point correlation function admits a power-law representation $w(\theta) = A\theta^{-1-\gamma}$, then the space two-point correlation $\xi(r)$ assumes the form $\xi(r) = Br^\gamma$, where A, B, are connected through a definite expression in terms of $\phi(M)$.

(ii) A scaling relation exists between two samples of galaxies with different depths D. Since $\xi(r)$ depends on spatial scales and $w(\theta)$ on the angular ones, it is clear that $w(\theta)$. $D = F(\theta D)$, because $r \propto \theta D$. Through this scaling property it is possible to compare results from different samples.

The Observed and Computed Correlations

Since the first definite measurements by Totsuje and Kihara (1969), several authors (Peebles, 1974; Davis and Geller, 1976; Davis *et al.*, 1978) using various samples have all led to similar results which can be expressed by the function

$$\xi(r) = 28\, r^{-1.8}$$

in the range 0.1 Mpc $\leqslant r \leqslant 10$ Mpc, with $H = 100$ km s^{-1} Mpc^{-1}.

The $\xi(r)$ correlation obtained from numerical N-body calculations is in essential agreement with the observed one. The experiments by Miyos and Kihara (1978), Aarseth *et al.* (1979), Gott *et al.* (1979), and Efstathiou (1979) following the evolution of N-points ($N \approx 10^3$ to 10^4) interacting through gravitational attraction on an expanding background, all lead to a persistence of the r^{-2} correlation. A wide variety of initial distribution (Poisson and no-Poisson) and velocities, end with the same two-point correlation. Moreover, once acquired, the correlation repeats itself with great stability.

The preceding results mean that the distribution of galaxies in space can be interpreted as a hierarchy of clustering in scales going from 0.1 Mpc to 10 Mpc. Moreover, estimates of the three- and four-point correlation functions indicate that the distribution is indeed hierarchical (Soneira and Peebles, 1978). The absence of a preferred clustering scale suggests the hierarchy is self-similar, a most natural consequence of gravitational instability, because Newtonian attraction lacks a preferred scale (Section VII.4.2).

Saslaw's Interpretation

An approach to the interpretation of the two-point correlation has been made by Saslaw (1980) through non-linear dynamics and thermodynamics. Saslaw demonstrated first that the time evolution of the exponent γ in the power-law $\xi(r) \propto r^{-\gamma}$ has a strong non-linear dynamical stability for $\gamma = 2$. This was done using the Layzer-Irvine cosmic energy equation, thus Saslaw argues that the reason for the stability is thermodynamical.

The rate of entropy production in gravitational systems depends on the structure correlations, besides the changes in the thermodynamical variables. Saslaw demonstrated that the rate of entropy production in clustering becomes larger with $\gamma \to 2$, which means that once clustering begins, it proceeds to the state of maximum entropy production. Saslaw (1979) has also noticed the relation between the tendency of gravitating systems to maximize their entropy production and their tendency to maximize their clustering efficiency and argues that this is due to the long-range unsaturated gravitational forces which promote continual formation of bound structures (Section VII.4.1).

IV.3. Mean Mass Density of Matter in the Universe

The density of matter in the form of galaxies has been computed by Oort (1958), Kiang (1961), van den Bergh (1961), Schapiro (1971), etc. All authors are coincident in the value

$$\rho_0 = 2 \times 10^{-31} \left(\frac{H_0}{100} \right)^2 \text{g cm}^{-3}.$$

The usual procedures to compute ρ_0 are:

(i) To measure the luminosity density Λ and to assign some mean mass luminosity ratio $\langle \mathfrak{M}/\mathfrak{L} \rangle$ to it, thus $\rho_0 = \Lambda \langle \mathfrak{M}/\mathfrak{L} \rangle$;

(ii) To obtain the average number of countable galaxies $\langle \nu \rangle$ in a representative volume and multiply it by the average mass $\langle \mathfrak{M} \rangle$ per countable galaxy independently derived through the methods described in Section II.2.

The luminosity density Λ is usually computed separately for E and S galaxies to account for their differences in mass-to-luminosity ratios (Section II.2.3) so that the mass density follows from

$$\rho_0 = \Lambda_E \langle \mathfrak{M}/\mathfrak{L} \rangle_E + \Lambda_S \langle \mathfrak{M}/\mathfrak{Z} \rangle_S.$$

TABLE IV.6

Mass, size and density of large astronomical systems*

Object	log $\mathfrak{M}$ (g)	log R (cm)	log ρ (g cm^{-3})	Examples
Nuclei of galaxies	40.3	19.3	-18.2	M32 Nucleus
	41.0	19.5:	-18.1	M32 Core
Compact dE	42.5	20.65	-20.0	M32
	43.2	21.1	-20.7	NGC 4486–B
Large spirals	43 5	21.8	-22.5	M33
	44.6	22.3	-22.9	M32
gE's	44.3	22.0	-22.35	N3378
	45.1	22.5	-22.95	N4486
Compact groups	45.5	22.6:	-23.1	Seyfert/Stephan
Dense E clusters	46.5	23.7	-25.2	(Virgo E core) Fornax I.
Loose groups of spirals	46.5	24.1	-26.4	Nearby Groups
Small clouds o spirals	47.0	24.3	-26.5	Virgo S, UMa Cloud
Small clusters of ellipticals	47.2	24.3	-26.3	Virgo E
Large clusters of ellipticals	47.9	24.5	-26.2	Coma I
Super clusters	48.7	25.5	-28.4	SCL

* After de Vaucouleurs (1974).

To compute Λ_E, Λ_S it is necessary to have the luminosity function for both galaxy types, which was done by van den Bergh (1961) and also Schapiro (1971) with good mutual agreement. Schapiro's results for a volume of 10^4 Mpc^3 which excludes the VC give

$$\rho_0 = 10^{-33}[1.33\langle\mathfrak{M}/\mathfrak{L}_B\rangle_E + 20\langle\mathfrak{M}/\mathfrak{L}_B\rangle_S](H_0/100)^2 \text{ g cm}^{-3}.$$

For $\langle\mathfrak{M}/\mathfrak{L}_B\rangle_E = 50$ and $\langle\mathfrak{M}/\mathfrak{L}_B\rangle_S = 5$ follows $\rho_0 = 1.7 \times 10^{-31} (H_0/100)^2$ g cm^{-3}.

All matter which is not associated with galaxies can be missed with this type of calculation, so that the proceding estimate is only a lower limit.

Burbidge (1972, 1974) has reviewed the arguments in favor and against the presence of a sensible contribution to ρ_0 originated in matter between galaxies. He considered intergalactic bridges (Section IV.1), intergalactic matter in galaxy clusters (Section IV.2.3), low luminosity galaxies (Section III.1) and QSO's (Section III.8) but despite all attempts to detect IGM, scarce evidence has been obtained, except in the central regions of galaxy clusters. A discussion on the global limits to ρ_0 will be given in Section VI.6.2.

A Hierarchical Universe?

The definition of an average density for the Universe has been contested by de Vaucouleurs on grounds that the presence of clustering on many scales makes it difficult, it not impossible, to define how large a volume must be to become a fair sample for measuring ρ_0. If there is no upper limit to the clustering scale, the mean-density can be specified only for a given volume, either in an integral form $\rho(V) = \mathfrak{M}(V)/V$ or as a differential density $\rho(\mathrm{d}V) = \mathrm{d}\mathfrak{M}/\mathrm{d}V$. For a homogeneous statistical distribution it is $\rho(V) = \langle\rho(\mathrm{d}V)\rangle$ but empirical evidence (Table IV.6), de Vaucouleurs argues, demonstrates that $\langle\rho(V)\rangle > \langle\rho(\mathrm{d}V)\rangle$ that is $\langle\rho(V)\rangle > \langle\rho(V + \mathrm{d}V)\rangle$.

References

Aarseth, S.J., Gott, J.R., and Turner, E.L.: 1979, *Astrophys. J.* **228**, 664.
Abell, G.O.: 1958, *Astrophys. J. Suppl.* **3**, 211.
Abell, G.O.: 1961, *Astron. J.* **66**, 607.
Abell, G.O.: 1962, in McVittie (ed.), 'Problems of Extragalactic Research', *IAU Symp.* **15**, 213.
Abell, G.O.: 1967, *Astrophys. J.* **72**, 288.
Abell, G.O.: 1976, in A. R. Sandage, M. Sandage, and J. Kristian (eds.), *Galaxies and the Universe*, Chicago. Univ. Press, p. 601.
Agüero, E.L.: 1971, *Publ. Astron. Soc. Pacific* **83**, 493.
Alladin, S.M.: 1965, *Astrophys. J.* **141**, 768.
Alladin, S.M., Potdar, A., and Sastry, K.S.: 1975, in A. Hayli (ed.), 'Dynamics of Stellar Systems', *IAU Symp.* **69**, 167.
Allen, R.J., Ekers, R.D., Burke, B.F., and Milley, G.R.: 1973, *Nature* **241**, 260.
Ambartsumian, V.A.: 1957, *Publ. Armenian Acad. Sci. Ser.* **11**.
Arp, H.: 1966, *Atlas of Peculiar Galaxies*, California Institute of Technology.
Arp, H.: 1969, *Astron. Astrophys.* **3**, 418.
Austin, T.B., Codwin, J.G., and Peach, J.V.: 1975, *Monthly Notices Roy. Astron. Soc.* **171**, 135.
Bautz, L.P. and Abell, G.O.: 1973, *Astrophys. J.* **184**, 709.
Bautz, L.P. and Morgan, W.W.: 1970, *Astrophys. J.* **162**, L149.
Bahcall, N.: 1973, *Astrophys. J.* **186**, 1179.
Bahcall, N.: 1977a, *Astrophys. J.* **217**, L19, L77.
Bahcall, N.: 1977b, *Ann. Rev. Astron. Astrophys.* **15**, 505.
Blanco, V.M., Graham, J., Lasker, B.M., and Osmer, P.: 1975, *Astrophys. J.* **198**, L67.
Bondi, H.: 1952, *Monthly Notices Roy. Astron. Soc.* **112**, 195.
Bondi, H. and Hoyle, F.: 1944, *Monthly Notices Roy. Astron. Soc.* **104**, 273.
Butcher, H.R. and Oemler, A.: 1980, *Astrophys. J.* **226**, 559.
Burbidge, E.M.: and Burbidge, G.R.: 1959, *Astrophys. J.* **130**, 23.
Burbidge, E.M. and Burbidge, G.R.: 1961, *Astrophys. J.* **133**, 726.
Burbidge, E.M., Burbidge, G.R., and Hoyle, F.: 1963, *Astrophys. J.* **138**, 873.
Burbidge, G.: 1972, in D.S. Evans (eds.), 'External Galaxies and QSO's', *IAU Symp.* **44**, 492.
Burbidge, G.R.: 1974, in K. Brecher and G. Setti (eds.), *High Energy Astrophysics*, MIT Press, p. 111.
Canon, R.D., Lloyd, C., and Penston, M.B.: 1970 *Observatory* **90**, 153.
Chandrasekhar, S.: 1942, *Principles of Stellar Dynamics*, Univ. of Chicago Press.
Chandrasekhar, S.: 1943, *Astrophys. J.* **97**, 255.
Davidson, P.J.N., Culhane, J.L., Mitchell, J.R., and Fabian, A.C.: 1975, *Astrophys. J.* **199**, L139.
Davis, M. and Geller, M.J.: 1976, *Astrophys. J.* **208**, 13.
Davis, M.: Geller, M.J., and Huchra, J.: 1978, *Astrophys. J.* **221**, 1.
Dressler, A.: 1980, *Astrophys. J.* **236**, 351.
de Vaucouleurs, G.: 1956, in A. Beer (ed.), *Vistas Astron.* **2**, 1584, Pergamon.
de Vaucouleurs, G.: 1958, *Astron. J.* **63**, 253
de Vaucouleurs, G.: 1972, in D.S. Evans (ed.), 'External Galaxies and QSO's', *IAU Symp.* **44**, 353.
de Vaucouleurs, G.: 1974, in J.R. Shakeshaft (ed.), 'Formation and Dynamics of Galaxies', *IAU Symp.* **58**, 1.
de Vaucouleurs, G.: 1975, in A.R. Sandage, M. Sandage, and J. Kristian, (eds.), *Stars and Stellar Systems*, Chapter IX, Chicago Univ. Press, p. 14.
de Vaucouleurs, G.: 1980, 'The Distance of M83 and the Centaurus Group', Preprint.
de Vaucouleurs, G. and Corwin, H.G.: 1975, *Astrophys. J.* **202**, 327.

de Young, D.S.: 1977, *Ann. N.Y. Acad. Sci.* **302**, 669.

Dufour, R.J. and van den Bergh, S.: 1978, *Astrophys. J.* **226**, L73.

Efstathiou, G.: 1979, *Monthly Notices Roy. Astron. Soc.* **187**, 117.

Fall, S.M.: 1979, *Rev. Mod. Phys.* **51**, 21.

Felten, J.E.: 1977, *Astrophys. J.* **82**, 861.

Field, G.: 1973, *The Redshift Controversy*, Freeman.

Freeman, K.C. and de Vaucouleurs, G.: 1974, *Astrophys. J.* **194**, 569.

Gisler, G.R.: 1980, *Astron. J.* **85**, 623.

Gott, J.R., Turner, E.L., and Aarseth, E.L.: 1979, *Astrophys. J.* **234**, 13.

Graham, J.: 1975, *Bull. Am. Astron. Soc.* **7**, 414.

Graham, J.: 1979, *Astrophys. J.* **232**, 60.

Gum, J.E. and Gott, J.R.: 1972, *Astrophys J.* **176**, 1.

Hausman, M.A. and Ostriker, J.P.: 1978, *Astrophys. J.* **224**, 320.

Haynes, M., Brown, R.L., and Roberts, M.S.: 1978, *Astrophys. J.* **221**, 414.

Holmberg, E.: 1941, *Astrophys. J.* **94**, 385.

Holmberg, E.: 1961, *Astron. J.* **66**, 620.

Holmberg, E.: 1969, Uppsala Medd. No. 166.

Holmberg, E.: 1974, *Astron. Astrophys.* **35**, 121.

Hoyle, F. and Littleton, R.A.: 1940, *Proc. Camb. Phil. Sur* **36**, 424.

Karachentsev, I.D.: 1966, *Astrofizika* **2**, 81.

Karachentsev, I.D.: 1978, in M.S. Longair and J. Einasto (eds.), 'The Large Scale Structure of the Universe', *IAU Symp.* **79**.

Kaufmann, P. and Rafaelli, J.C.: 1979, *Monthly Notices Roy. Astron. Soc.* **187**, 23p.

Kaufmann, P., Marques dos Santos, P., Braz, A., and Borges, R.M.: 1975, *Astrophys. Space Sci.* **32**, L25.

Kellerman, K.I.: 1974, *Astrophys. J.* **194**, L135.

Kiang, T.: 1961, *Monthly Notices Roy. Astron. Soc.* **122**, 263.

Kiang, T.: 1967, *Monthly Notices Roy. Astron. Soc.* **135**, 1.

Kiang, T., and Saslaw, W.C.: 1969, *Monthly Notices Roy. Astron. Soc.* **143**, 129.

Klemola, A.R.: 1969, *Astron. J.* **74**, 804.

Kriss, G.A., Canizares, C.R., McClintock, J.E., and Feigelson, E.D.: 1980, *Astrophys. J.* **235**, L61

Kunkel, W.E. and Bradt, H.V.: 1971, *Astrophys. J.* **170**, L7.

Lea, S.M.: 1976, *Astrophys. J.* **203**, 569.

Limber, D.N.: 1961, *Astron. J.* **66**, 572.

Longair, M.S.: 1978, in M.S. Longair and J. Einasto (eds.), 'The Large Scale Structure of the Universe', *IAU Symp.* **79**, 454.

Malin, D.F. and Carter, D.: 1980, *Nature* **285**, 643.

Melnick, J. and Sargent, W.L.W.: 1977, *Astrophys. J.* **215**, 601.

Miley, G.K., van der Laan, H., and Wellington, K.J.: 1974, in J.R. Shakeshaft (ed.), *IAU Symp.* **58**, 'Formation and Dynamics of Galaxies', 347.

Miyoshi, K. and Kihara, T.: 1978, *Publ. Astron. Soc. Japan* **27**, 333.

Morgan, W.W.: 1962, *Astrophys. J.* **135**, 1.

Oemler, A.: 1974, *Astrophys. J.* **194**, 1.

Oort, J.: 1958, *XI-Solvay Conference*, Bruxelles, p. 113.

Osmer, P.: 1978, *Astrophys. J.* **226**, L79.

Ostriker, J.: 1977, in B.M. Tinsley and R. Larson (eds.), *The Evolution of Galaxies and Stellar Populations*, Yale Univ. Observatory, p. 369.

Ostriker, J.P. and Turner, E.L.: 1979, *Astrophys. J.* **234**, 785.

Page, T.L.: 1965, Smithsonian Spec. Rept. No. 95.

Pacholczyk, A., Christiansen, W., and Scott, J.: 1977, *Nature* **266**, 593.

Pedreros, M.: 1978, *Publ. Astron. Soc. Pacific.* **90**, 14.

Peebles, P.J.: 1974, *Astrophys. J.* **189**, L51.

Perley, R.A., Willis, A.G., and Scott, J.S.: 1979, *Nature* **281**, 437.

Pfleiderer, J.: 1963, *Z. Astrophys.* **58**, 12.

Peterson, B.A., Dickens, R.J., and Cannon, R.D.: 1975, *Proc. Astron. Soc. Australia* **2**, 367.

Quintana, H.: 1979, in A. Gutierrez and H. Moreno (eds.), Vol. III, Dept. Astronomía Univ. Chile.
Quintana, H. and White, R.A.: 1978,
Reynolds, J.H.: 1921, *Monthly Notices Roy, Astron. Soc.* **81**, 129, 598.
Rees, M.J.: 1978, *Nature* **275**, 516.
Rood, H.J.: 1965, Thesis, Univ. of Michigan.
Rood, H.J. and Sastry, G.N.: 1971, *Publ. Astron. Soc. Pacific* **83**, 313.
Rood, H.J., Rothman, V.C., and Turnrose, B.E.; 1970, *Astrophys. J.* **162**, 411.
Rose, J.A.: 1979, *Astrophys. J.* **231**, 10.
Rubin, V.C., Ford, W.K., and D'Odorico, S.: 1970, *Astrophys. J.* **160**, 801.
Rudnick, L. and Owen, F.N.: 1977, *Astron. J.* **82**, 1.
Ryle, M. and Dindram, M.D.: 1968, *Monthly Notice Roy. Astron. Soc.* **138**, 1.
Sandage, A.R.: 1976, *Astrophys. J.* **205**, 6.
Sargent, W.L.W.: 1974, in K. Brecher and G. Setti (eds.), *High Energy Astrophysics*, The MIT Press, p. 453.
Schapiro, S.L.: 1971, *Astron. J.* **76**, 291.
Schechter, P.: 1976, *Astrophys. J.* **203**, 297.
Schweitzer, F.: 1978, in E.M. Berkhuijsen and R. Wielebinski (eds.), 'Structure and Properties of Nearby Galaxies', *IAU Symp.* **77**, 279.
Sastry, K.S. and Alladin, S.M.: 1970, *Astrophys. Space. Sci.* **7**, 261.
Saslaw, W.C.: 1979, *Astrophys. J.* **229**, 461.
Saslaw, W.C.: 1980, *Astrophys. J.* **235**, 299.
Sérsic, J.L.: 1960, *Z. Astrophys.* **51**, 64.
Sérsic, J.L.: 1968, *Atlas de Galaxias Australes*, Univ. Nacional de Cordoba.
Sérsic, J.L.: 1969, *Nature* **224**, 253.
Sérsic, J.L.: 1970, in G.E.O. Giacaglia (ed.), *Proc. Symp. Periodic Orbits, Stability and Resonances*, D. Reidel Publ. Co., Dordrecht, Holland, p. 314.
Sérsic, J.L.: 1973, *Bull. Astron. Inst. Czech.* **24**, 150.
Sérsic, J.L.: 1974, *Astrophys. Space Sci.* **28**, 365.
Sérsic, J.L. and Calderón, J.H.: 1979, *Astrophys. Space Sci.* **62**, 211.
Sérsic, J.L. and Calderón, J.H.: 1980, *Bull. Astron. Inst. Czech.* **31**, 253.
Serlemitsos, P., Smith, B., Bolt, E., Holt, S., and Swank, J.: 1977, *Astrophys. J.* **211**, L63.
Smith, S.: 1936, *Astrophys. J.* **83**, 23.
Soneira, R.M. and Peebles, P.J.E.: 1977, *Astrophys. J.* **211**, 1.
Soneira, R.M. and Peebles, P.J.E.: 1978, *Astron. J.* **83**, 845.
Spitzer, L. and Baade, W.: 1951, *Asthrophys. J.* **113**, 413.
Swakuna, E.W.: 1966, *Soviet Astron. J.* **43**, 34.
Toomre, A.: 1974, in J.R. Shakeshaft (ed.), 'Formation and Dynamics of Galaxies', *IAU Symp.* **58**, 347.
Toomre, A.: 1978, in E. M.S. Longair and J. Einasto (eds.), 'The Large Structure of the Universe', *IAU Symp.* **79**, 109.
Toomre, A. and Lyns, R.: 1976, *Astrophys. J.* **209**, 382.
Toomre, A. and Toomre, J.: 1972, *Astrophys. J.* **203**, 72.
Turner, E.L.: 1978, in M.S. Longair and J. Einasto (eds.), 'The Large Scale Structure of the Universe', *IAU Symp.* **79**, 21.
Turner, E.L. and Gott, J.R.: 1975, *Astrophys. J.* **197**, L89.
Turner, E.L. and Sargent, W.L.W.: 1974, *Astrophys. J.* **194**, 587.
Tully, R.B. and Fisher, J.R.: 1978, in M.S. Longair and J. Einasto (eds.), 'The Large Scale Structure of the Universe', *IAU Symp.* **79**, 31.
Totsuji, H. and Kihara, T.: 1969, *Publ. Astron. Soc. Japan* **21**, 221.
Tytler, D. and Vidal, N.V.: 1978, *Monthly Notices Roy. Astron. Soc.* **182**, 33p.
van den Bergh, S.: 1961, *Z. Astrophys.* **53**, 19.
van den Bergh, S.: 1977, *Vistas Astron.* **21**, 71.
van den Bergh, S.: 1978, *Vistas Astron.* **22**, 307.
Vorontsov-Velyaminov, B.A.: 1958, *Astron. J. URSS* **35**, 858.
Vorontsov-Velyaminov, B.A.: 1959, *Atlas and Catalogue of Interacting Galaxies*, Moscow Univ.

Vorontsov-Velyaminov, B.A.: 1961, in McVittie (ed.), 'Problems of Extragalactic Research', *IAU Symp.* **15**, 194.
White, R.A.: 1978, *Astrophys. J.* **226**, 591.
White, R.A.: 1980, *Astrophys. J.*, in press.
Wright, A.C.: 1972, *Monthly Notices Roy. Astron. Soc.* **157**, 309.
Wright, A.C.: 1974a, *Monthly Notices Roy. Astron. Soc.* **167**, 251.
Wright, A.C.: 1974b, *Astron. Astrophys.* **31**, 283.
Yabushita, S.: 1971, *Monthly Notices Roy. Astron. Soc.* **153**, 97.
Zwicky, F.: 1933, *Helv. Phys. Acta* **6**, 110.
Zwicky, F.: 1957, *Morphological Astronomy*, Springer.
Zwicky, F.: 1959, *Handbuch der Physik* **53**, 390.

CHAPTER V

MEASURING THE UNIVERSE

The lines of the optical spectra of galaxies show a shift towards the red side which increases with the decreasing apparent brightness (Hubble's law). The facts seem to be consistent with the idea that the galaxies recede with velocities proportional to their distances.

The interpretation of the redshift as a cosmological Doppler-shift was strengthened with the observation of the same phenomenon at other frequencies such as radio frequency. Now the compatibility of the optical and radio redshift is widely accepted (Roberts, 1972). A small percentage of cases where the redshift could arise from another source (i.e. intrinsic or caused by another physical effect) is considered as possible by some authors (Arp, 1976; Vigier and Pecker, 1976).

The extragalactic scale of distance depends on distances established in the Local Group and within our Galaxy through Cepheids and Main Sequence fitting. Ultimately the scale becomes tied to locally determined parallaxes, in particular to the distance of the galactic open cluster in Taurus known as the Hyades.

The distance of the Hyades is determined through the moving cluster method which lies at the basis of the whole scale of distances, both galactic and extragalactic. The Hyades cluster has a large apparent diameter on account of its proximity. The projected motions of their stars over the sky converge towards a point which lies at an angular separation η from the center of the cluster. Since the radial velocity of their stars can be measured, it is easy to see that the tangential velocity of a cluster star V_t is related to the radial velocity V_r by $V_t = V_r \tan \eta$ thus the distance $R = V_t/\mu$ follows from the corresponding proper motion μ. The analysis of the proper motions of the cluster stars provides the angle to the convergent point so that the distance can be established. Hanson (1975) has given the best estimate of the distance modulus of the Hyades $\mu_0 = 3\overset{m}{.}29 \pm 0\overset{m}{.}05$ although figures in the literature give a spread ranging from 3.03 to 3.48. Thus, already at its foundation the scale of distances has a 10% indetermination.

V.1. Distance Indicators

The fundamental postulate of isotropy and homogeneity (Section VI.1) is the basis for our knowledge of the Universe at large, because without it, it would be impossible to proceed with the basic tool of Physics: the comparation.

The important point is that the principle refers to samples of the Universe which are large enough, such that inhomogenities can be considered local fluctuations not interfering seriously with the elements of information which are really comparables. In other words, the samples must be large enough to define mean values free from systematic errors. The investigations by Morgan (Section I.4) or de Vaucouleurs (Section IV.2.4) indicate respectively that there are fluctuations in the stellar content of the galaxies and perturbations of the Hubble flow in volumes of the order of clusters and superclusters of galaxies. Up to the present it is not possible to state with certitude what the dimensions are of a minimum cell for the averaging process, it is only possible to admit that they are larger than the volume of a cluster of galaxies, some 10^3 to 10^4 Mpc^3.

The astronomical objects with at least one measurable characteristic (q) enabling us to define a mean value ($\bar{q}$) and its corresponding dispersion (σ_q) are called 'distance indicators' if they are comparable with those of our cosmic vicinity, in the sense that a permutation between them does not alterate the postulated homogeneity of the Universe.

After elimination of errors in measurements, an intrinsic dispersion (σ_q^*) remains, also called cosmic dispersion, which measures the quality of the object as a distance indicator. Naturally, the only objects useful to that end are those having small intrinsic dispersion and also a mean value $\bar{q}$ which makes them easily observable.

From the present state of the astronomical techniques, three categories of distance indicators can be distinguished: photometrics, geometrics and dynamics. The redshift is the ultimate distance indicator. Through it and knowing a definite model of the Universe which best represents the observations, we should in principle be able to obtain the corresponding distance. To this end, we are led to calibrate the Hubble law by means of the previous distance indicators through the slow, usually deceiving, process of building the scale of distances.

The fundamental cause of the difficulty has been the use of the observed luminosity as a distance indicator (Section V.1.1). In determining H the intrinsic luminosity of the various indicators must be inferred by indirect, statistical methods involving a piecewise calibration procedure to ever most distant objects (Section V.2). Also methods involving the ratio of the observed angular size to intrinsic linear diameters (Section V.1.3) suffer from similar inherent difficulties, as does any other method which must make assumptions about the intrinsic properties of the object which can never be strongly verified by other observations (Section V.1.4).

V.1.1. Primary Distance Indicators

The distance indicators which can be calibrated in our Galaxy through fundamental geometric methods are considered primaries. The distances of the Local Group galaxies as well as those of a few belonging to the nearest groups can be established by use of primary indicators, usually the Novae, the Cepheids, and the RR Lyrae stars.

Some authors have proposed additional primary indicators such as AB-supergiants and the eclipsing binaries (de Vaucouleurs, 1978a) or the W Virginis variables (van den Bergh, 1977), but these objects have neither been exploited nor calibrated enough.

Primary calibrators, as all photometric indicators, are affected by intersteller absorption both in the Galaxy and the parent galaxy, which becomes a source of large errors.

Interstellar Absorption

The comparison of the apparent magnitude (m) of a photometric indicator with its estimated absolute magnitude (M) does not necessarily suffice to produce a distance estimate: it is necessary to know the absorption (A_G) caused by the interstellar medium in our Galaxy and in some cases also the internal absorption (A_I) in the parent galaxy. In this way we obtain the true distance indicator μ^0 from the apparent one $\mu = m - M$ by means of

$$\mu^0 = \mu - A, \qquad A = A_G + A_I.$$

The galactic absorption A_G can be estimated with the so-called cosecant-law

$$A_G = a_0 \operatorname{cosec} |B|,$$

where $|B|$ is the absolute value of the galactic latitude. The preceding relation implies a continuous and uniform distribution of the interstellar absorbing matter in the galactic plane, a_0 being its mean optical half-depth. A usual procedure to find a_0 is through galaxy counts. Let N be the average number of galaxies per steradian up to magnitude m in a field of latitude B, then a fair representation is obtained through

$$\log N = P + Q \operatorname{cosec} |B|.$$

Since $d \log N/dm = 0.6$ for a uniform galaxy distribution in depth, the optical half-depth of the absorbing layer becomes $a_0 = -Q/0.6$. Through this technique Hubble (1934) found $a_0 = 0.25 \, (pg)$, a value essentially confirmed by Oort (1938) and Mineur (1938), who also found an asymmetry for the values of a_0 between both galactic hemispheres.

The properties of the spectral distribution of the interstellar absorption (Whitford, 1958) in the range between 3000 and 11 000 Å show an increase of absorption and also reddening of the spectral energy distribution towards the shorter wavelengths. If the two indices k, k' refer to two photometric windows with effective wavelengths $\lambda_k < \lambda_{k'}$, the reddening $E(k, k') = A(k) - A(k')$, also called color excess, is related to $A(k)$ through

$$A(k) = E(k, k') \, R,$$

where the ratio R can be determined observationally. In the case of $k = pg$ and $k' = pv$ the value of R is 4.0. Holmberg (1958) used this property to find $A(pg)$ in our Galaxy from the analysis of the reddening of a homogeneous sample of galaxies at different latitudes,

$$E = e_0 \operatorname{cosec} |B|$$

finding $e_0 = 0\overset{m}{.}062 \pm 0\overset{m}{.}007$, from where he deduced $a_0(pg) = 0\overset{m}{.}248 \pm 0\overset{m}{.}018$, with a N–S asymmetry of $0\overset{m}{.}048 \pm 0\overset{m}{.}024$ in good agreement with earlier findings.

de Vaucouleurs and Malik (1969) considered local variations in longitude on the basis of counts by Shane and Wirtanen on the Lick astrograph plates. The absorption is given as a function of both galactic latitude (B) and longitude (L) in the form

$$A_G(LB) = a_0 z(L, B)$$

tabulating the values of $z(L, B)$. The mean half-thickness a_0 is left free, on account of possible changes in the value of a_0, In fact, Shane and Wirtanen (1967) derived $a_0 = 0\overset{m}{.}51$ from their galaxy counts. From the reddening technique van den Bergh (1967) derived instead $E_0 = 0\overset{m}{.}06$ in good agreement with Holmberg (1958). Later observations by Peterson (1970a) using a larger photometric range and galactic clusters led again to $E_0 = 0\overset{m}{.}06$, in serious conflict with the large values from Lick counts. Hence, the simple expressions

$$A_G(pg) = 0\overset{m}{.}25 \operatorname{cosec} |B|; \qquad A_G(pv) = 0\overset{m}{.}18 \operatorname{cosec} |B|$$

are generally accepted (Humason *et al.*, 1956). Since local deviations in low galactic latitudes are sometimes very large, specific absorption estimates to improve these average values are usually made.

Within our Galaxy the absorption is determined in terms of the color excess E_{B-V} through separate observing in each case. If the absorption is large, higher order terms for R are to be considered.

TABLE V.1

A_{pg}/E Ratios in galaxies

Object	$A(pg)/E$	Source
NGC 5194–5	4.3	Holmberg (1950)
224	4.0	Stebbins (1950.)
3031	3.8	Holmberg (1950.)
5128	4.2	Sersic (1958)
Galaxy	4.0	Holmberg (1958.)

The intrinsic absorption A_I in the parent galaxy is a more difficult problem to solve. The normal dispersion of the period-luminosity (P–L) relationship for Cepheids in M31 was interpreted by Baade and Swope (1963) as produced by the internal absorption in the spiral arms where they are found. These authors found it reasonable to use the bright envelope to define the P–L relation (see below).

For non-variable stars it is necessary to know the color and the reddening law which is normal in external systems. Table V.1 gives the values for A_{pg}/E found in several galaxies.

Because of the internal extinction in other galaxies, indicators placed on the far side of the dust layer tend to be under represented or even absent in magnitude-limited samples so that the statistics have to be corrected for this effect.

Novae

These object were historically the first to be used for distance estimates and to be called distance indicators by Lundmark (1926). The galactic calibration of the brightest novae is due to Schmidt-Kaler (1957). A survey of the novae in M31 was carried out

TABLE V.2

Absolute magnitudes of the galactic novae at maximum*

Group	I	II	I + II
M_{pg}	−8.61	−6.82	−7.89
σ_M	0.36	0.53	1.00
n	9	6	15

* Adapted from Schmidt-Kaler (1957).

by Arp (1956) and later by Rossino (1964). To the several novae already known in the Magellanic Clouds we must add those found in the survey by Graham (Smith, 1977).

Arp found the maxima of the M31 novae clustered around two values. The same phenomenon is found in Schmidt-Kaler values for the galactic novae (Table V.2). The general mean of his data is in fair agreement with Cechini and Gratton's pioneer work on the subject.

Through comparison of the light-curves of Novae in the Galaxy and the nearby Local Group galaxies, Mc Laughlin (1945) has shown that the time t_3 during which the novae decreases its brightness 3 magnitudes from the maximum, is related to the peak absolute magnitude M_B through

$$M_B^{\max} = 1.80 \log t_3 - 10.7$$
$$\pm.20 \qquad\qquad \pm.30$$

(Pfau, 1976). The observed relation between $M_B^{\max}$ and t_2 has also been used as a distance estimator (van den Bergh, 1977) by fitting it to the same relation for the galactic novae.

Buscombe and de Vaucouleurs (1955) noticed instead that the absolute magnitudes of novae 15 days after maximum have closely the same value $\langle M_B \rangle_{15} = -5^m5 \pm 0^m18$ (m.e.) as follows from calibration with the galactic novae (de Vaucouleurs, 1978a).

Population I Cepheids

The discovery of the period luminosity relation by Leavitt in 1908 and the determination of its zero-point by Hertzprung gave astronomers the best tool for distance determinations within a distance modulus $\mu^0 \leqslant 27$.

The understanding of the P–L relation begins with the fundamental formula for stellar pulsations

$$p \sqrt{\bar\rho} = Q, \tag{V.1}$$

where P is the pulsation period. $\bar\rho$ the mean density and Q the so called pulsation constant. The virial equation

$$\frac{1}{2} \frac{d^2 J}{dt^2} = 2E - W, \tag{V.2}$$

where $J = AMR^2$ is the polar momentum of inertia, $W = -BM^2/R$ the potential and E the mechanical energy of a star with mass M and radius R, is satisfied for homologous oscillations if A, B are constants. Let us assume small-amplitude oscillations δR around the equilibrium radius $R_0 = BM^2/(-2E)$. We have $J = J_0 + \delta J$ and $W = W_0 + \delta W$, where

$$J = 2J_0(\delta W/W_0), \qquad W = -W_0(\delta R/R_0)$$

so that Equation (V.1) results

$$(\delta R)'' = -(B/A)(M/R_0^3)\, \delta R$$

and δR will oscillate with a period

$$P = 2\pi (A/B)^{1/2} (R_0^3/M)^{1/2} = Q/\sqrt{\bar{\rho}},$$

where $Q = (3\pi A/B)^{1/2}$ shows the dependence on the internal structure of the star through the coefficients A, B.

The lines of constant density (and hence period) in the theoretical H–R diagram follow from considering the relation between luminosity L, effective temperature T_e and radius R_0 of a star $L = 4\pi\sigma T_e^4 R_0^2$, and the mass-luminosity law $L \propto M^\alpha$ (Section II.3.2), which give

$$\frac{d \log T_e}{d \log L} = \frac{1}{6}\left(\frac{3}{2} - \frac{1}{\alpha}\right).$$

Combining this result with the period-density law (V.1), we obtain

$$\log L = \frac{4\alpha}{3\alpha - 2}(\log P + 3 \log T_e) + \text{const.} \tag{V.3}$$

a period-luminosity relation, provided the considered stars be in a narrow strip $T_e = f(L)$ in the theoretical H–R diagram. Stellar pulsations naturally happen in several regions of it, but the so-called Cepheids strip is the best suited for a P–L relation, because the instability arising in the second He ionization takes place on a narrow line in the H–R diagram.

To transform Equation (V.3) to the usual magnitude-color diagram, we use the UBV photometric system for which the relations

$$M_{BOL} = M_V - 0.32(B - V) + \text{const.}, \qquad \log T_e = -0.18(B - V) + \text{const.}$$

are valid in the Cepheid strip. Since $M_{BOL} = -2.5 \log L + c$ we obtain

$$M_V = -\frac{10\alpha}{3\alpha - 2} \log P + \left(\frac{5.4\alpha}{3\alpha - 2} + 0.32\right)(B - V) + \text{const.}$$

a period-luminosity-color relation first derived by Sandage (1968). If we take $\alpha = 4$, which is a representative exponent for the mass-luminosity law in the range of masses of the Cepheids (1 to 15 solar units) we finally have

$$M_V = -4 \log P + 2.4 (B - V) + C. \tag{V.4}$$

Since the theoretical approach to the P–L relation is marred by uncertainties due to differences in chemical composition of the stars, by an inappropriate theory for the convective transport, etc., astronomers rely on the empirically derived P–L relation.

The slope of the lines of constant period in the H–R diagram has been found to be

$$dM_V/d(B - V) = 2.65$$

on the basis of semi-empirical considerations by Sandage (1972). Since the ratio $A(V)/E_{B-V} \approx 2.5$ to 3 (Section V.1.1), both the reddening and the constant period lines are nearly parallel. Thus, the quantity

$$M_V - 2.65 (B - V) = M_W$$

is independent of reddening and the intrinsic color effect (van den Bergh, 1975a).

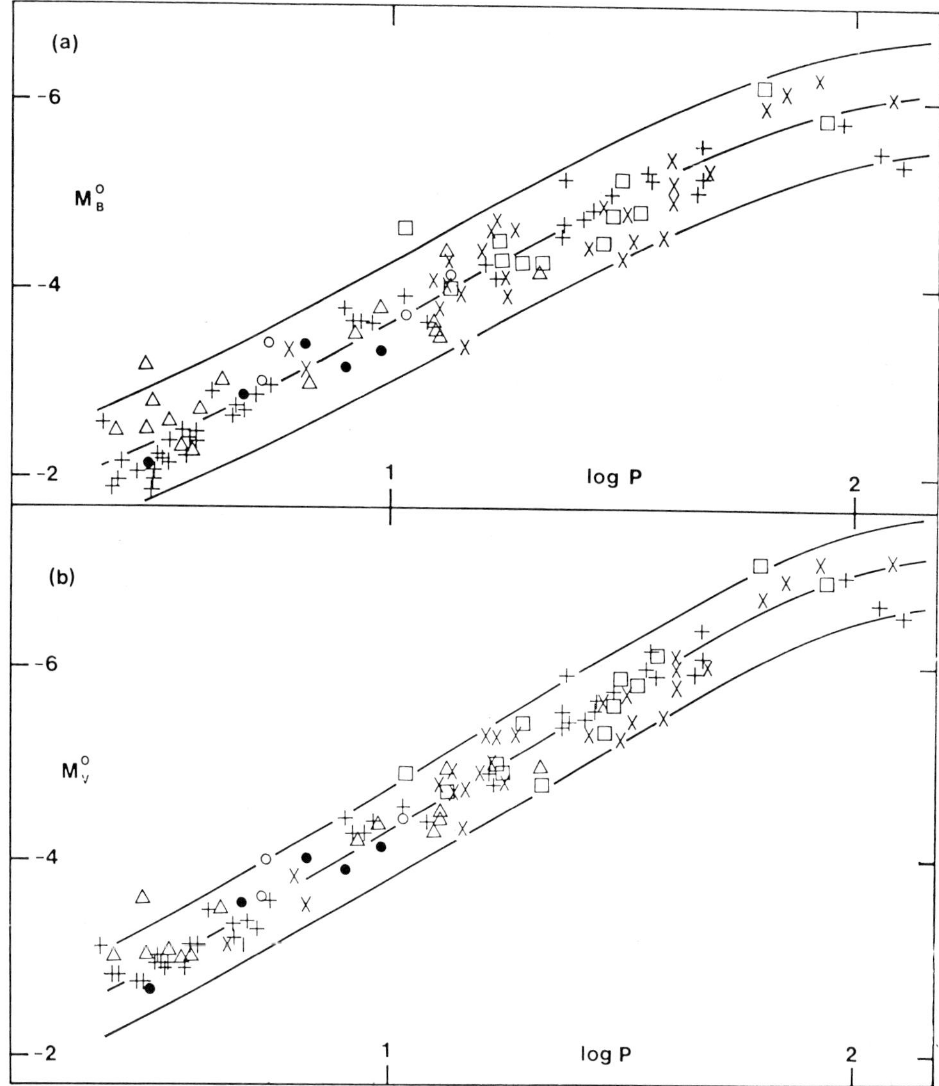

Fig. V.1.   The composite P–L relation for Cepheids. (*Adapted from Sandage and Tammann* (1968).)

The shape of the P–L relation was found by Sandage (1972) on the basis of the Cepheids of the Local Group galaxies. He produced a composite P–L diagram (Figure V.1) through vertical shifts to minimize the scatter.

Some departure from linearity was found, as well as an appreciable scatter at a given period, but no evidence for different slopes in different galaxies.

To calibrate the P–L relation, Sandage (1972) used 13 galactic Cepheids belonging to clusters and associations (Figure V.1). From a rediscussion of these data, Tammann

(1970) found

$$M_V = -2.83 \log P - 1.49 \quad \text{and} \quad (B - V)^0 = 0.37 + 0.26 \log P$$

which together gives

$$M_V = -3.53 \log P + 2.65(B - V)^0 - 2.47$$

with a mean error of $\pm 0^m14$ between periods of 2 and 42 days.

van den Bergh (1977) used more galactic Cepheids to the same end but after discussing membership to clusters, and discarding those associated with reflection nebulae and pertaining to binaries, his calibration was based on 14 galactic Cepheids which give

$$M_V = -2.90 \log P - 1.18 \quad \text{and} \quad (B - V)^0 = 0.27 + 0.48 \log P$$

which together lead to

$$M_V = -4.18 \log P + 2.65 (B - V)^0 - 1.90.$$

A comparison of both calibrations shows the degree of uncertainty implied in their derivation.

de Vaucouleurs (1978a) prefers instead to define the mean absolute magnitude of Cepheids at a given period $P = 0.8$ day, $\langle M_V \rangle_{0.8} = -3^m60 \pm 0^m15$, with $\langle (B-V)_0 \rangle = +0^m65 \pm 0^m15$.

Several sources of systematic errors weaken these procedures, the most important being the uncertainty of the absorption derived from reddening, used to estimate the distances of the galactic Cepheids. When applied, the P–L relation can also be biased since galactic Cepheids are of solar composition, and differences with other galaxies can produce systematic errors.

The Cepheids have as a useful range of $\mu_0 = 27^m$ as indicators, which means they are only applicable for the Local, M81 and Sculptor (SGP) groups which account for less than twenty late-type galaxies.

RR Lyrae

The first statistical researches by Woolley and independently by Pavloskaya suggested in the fifties that the absolute visual magnitude of RR Lyr stars was $+0^m5$ against the convention which assigned them the value 0^m0. From the distances of globular clusters Sandage (1964) derived a value $\langle M_V \rangle_{RR} = +0^m41$ whilst van Herk (1965) obtained $\langle M_V \rangle_{RR} = +0^m68$ from the analysis of the proper motions of two hundred field RR Lyr stars. Van den Bergh (1976a) uses as a compromise $\langle M_V \rangle_{RR} = 0^m5 \pm 0^m2$ and $\langle M_B \rangle_{RR} = 0^m8 \pm 0^m2$ whilst de Vaucouleurs (1977) instead prefers $\langle M_V \rangle_{RR} = 0^m86 \pm 0^m15$ and $\langle B - V \rangle_0 = 0.20 \pm 0.05$ derived by Kukarkin (1974) for the mean of RR Lyr in 89 globular clusters.

The low luminosity of this indicator makes it useful only for the nearest Local Group galaxies, particularly the dwarf ellipticals and the Magellanic Clouds.

V.1.2. OTHER PHOTOMETRIC INDICATORS

Because the primary indicators have a limited range for reliable distance estimates, secondary and tertiary indicators must be used. The secondary indicators are calibrated with galaxies whose distance have been derived with the primaries and so, through a step-by-step procedure, the scale is extended farther and farther away. Although the precision of the indicators becomes lower, more galaxies are available to calibrate the following ones as the explored volume of space enlarges. This permits the use of indicators whose absolute or intrinsic properties are a function of the integral parameters of the parent system, as for instance the morphological type or the luminosity class of galaxies in case the indicators were features of the galaxies.

The rank of an indicator varies in some cases according to the calibration criteria followed in the construction of the scale (Section V.2), but there is general agreement in considering the brightest stars, the globular clusters and the largest H II regions as secondary, while the absolute luminosities (whether luminosity class or absolute magnitude) and isophotal diameters of galaxies as tertiary.

The Brightest Stars

Although it is possible, in principle, to measure 22nd magnitude stars and to reach up to distances of the order of 20 Mpc, there is no way to tell a single bright star from a cluster at these distances. The considerable change in the post-war scale of distances is due greatly to an improved technique which disclosed earlier misunderstandings (Sandage, 1958). A more realistic estimate for the distance up to which we are able to recognize individual bright stars in galaxies would perhaps be 10 Mpc. Since the bright end of the luminosity function is sparsely populated, confusion arises with faint foreground galactic stars, particularly for the LG galaxies. Separation through proper motions and radial velocities are, of course, desirable but the amount of work implied makes progress in this sense much slower than with other indicators.

TABLE V.3

Mean absolute magnitudes of the three brightest stars in the Galaxy*

Type	M_V^*
OB	-8.4 ± 0.2
A	-8.9 ± 0.5
FGK	-9.4 ± 0.1
M	-7.7 ± 0.2

* After Humpreys (1978).

The most recent contribution to the knowledge of the brightest stars in the Galaxy comes from a complete survey by Humphreys (1978) carried out within 2 to 3 kpc from the Sun. For the supergiants in clusters and associations she derived the absolute magnitudes given in Table V.3.

Blue Bright Stars

The luminosity of the brightest stars in the galaxies depends on their total population content. This sample-size effect was considered by Holmberg (1950) on grounds of a common luminosity function for these objects. A luminosity function $\varphi(L) \propto L^{-\alpha}$ leads to a total luminosity of the system $L_T \propto \int^{L_*} \varphi(L)L \, dL \propto L_*^{2-\alpha}$ and thus $M_T = (2 - \alpha)M_* + $ const. Sandage and Tammann (1974b) derived such a relation from total corrected face-on magnitudes M_T^0 and the absolute magnitudes for the mean of the three brightest stars in galaxies

$$\langle M_B^* \rangle = -9.6 + 0.33\,(M_T^0 + 20)$$

from a discussion of the late-type galaxies in the Local and M81 groups. de Vaucouleurs (1978c) arrived at a similar result:

$$\langle M_B^* \rangle = -9.9 + 0.38(M_T^0 + 20)$$

in reasonable agreement with the former. Sandage and Tammann however, prefer the use of luminosity classes L_c instead M_T^0 (see below).

The Hubble-Sandage Stars

The brightest blue variables of Population I in the Galaxy were suggested by Hubble and Sandage (1953) as potential distance indicators. Recent work by Sandage and Tammann (1968) suggests, however, that these objects may not be suitable as distance indicators because they seem to stay below its maximum magnitude for long periods, requiring long series of observations and they can also can be mistaken for objects like η Carinae which are much brighter at maximum. de Vaucouleurs (1978c) argues that the observation of several of these indicators in the same galaxy should lead to an establized mean absolute magnitude and proposes $\langle M_B \rangle_{HS} = -9\overset{m}{.}2 \pm 0\overset{m}{.}2$.

Red Giants

The brightest Population II indicators are the red giant stars. The calibration of the absolute magnitude of these stars in globular clusters was made in relation with those of the RR Lyr stars (Arp, 1960). Although their potential value as indicators is high, their absolute magnitudes are strongly composition-dependent.

Sandage and Tammann (1974b) have first suggested the Population I red supergiants might be good distance indicators. Humpreys and Sandage (1980) have succeeded in identifying and measuring apparent magnitudes of these objects in nearby galaxies (Table V.3).

Globular Clusters

The globular clusters (Section II.1.4) have simple properties which vary smoothly with these of their parent galaxies, making them suitable for distances estimates. They have, moreover, the advantage of avoiding the obscured regions, thus leading to distance moduli which are free from the intrinsic absorption of the parent galaxy (Figure V.2).

The luminosity function of the globular clusters seems to be universal (Section II.1.4) and adopts a gaussian form with mean $\langle M_V \rangle_{GC} = -7\overset{m}{.}3$ and dispersion $\sigma_V = 1\overset{m}{.}2$. The total population N_T depends on the absolute magnitude of the parent galaxy. For B-magnitudes, de Vaucouleurs (1977) gives $\log N_T = -0.3(M_B(G)+11.0)$. The knowledge of the shape of the luminosity function and the total population permits the calculation of a statistically expected value for the absolute magnitude of the brightest globular cluster. de Vaucouleurs (1977) used this property to devise a procedure to estimate the absolute magnitude of the parent galaxy and then the distance modulus. He computed the difference $M_B(1) - M_B(G)$ between the absolute magnitude of the brightest globular cluster and the total absolute magnitude of the parent galaxy and plotted it as a function of $M_B(G)$. Within the needed approximation

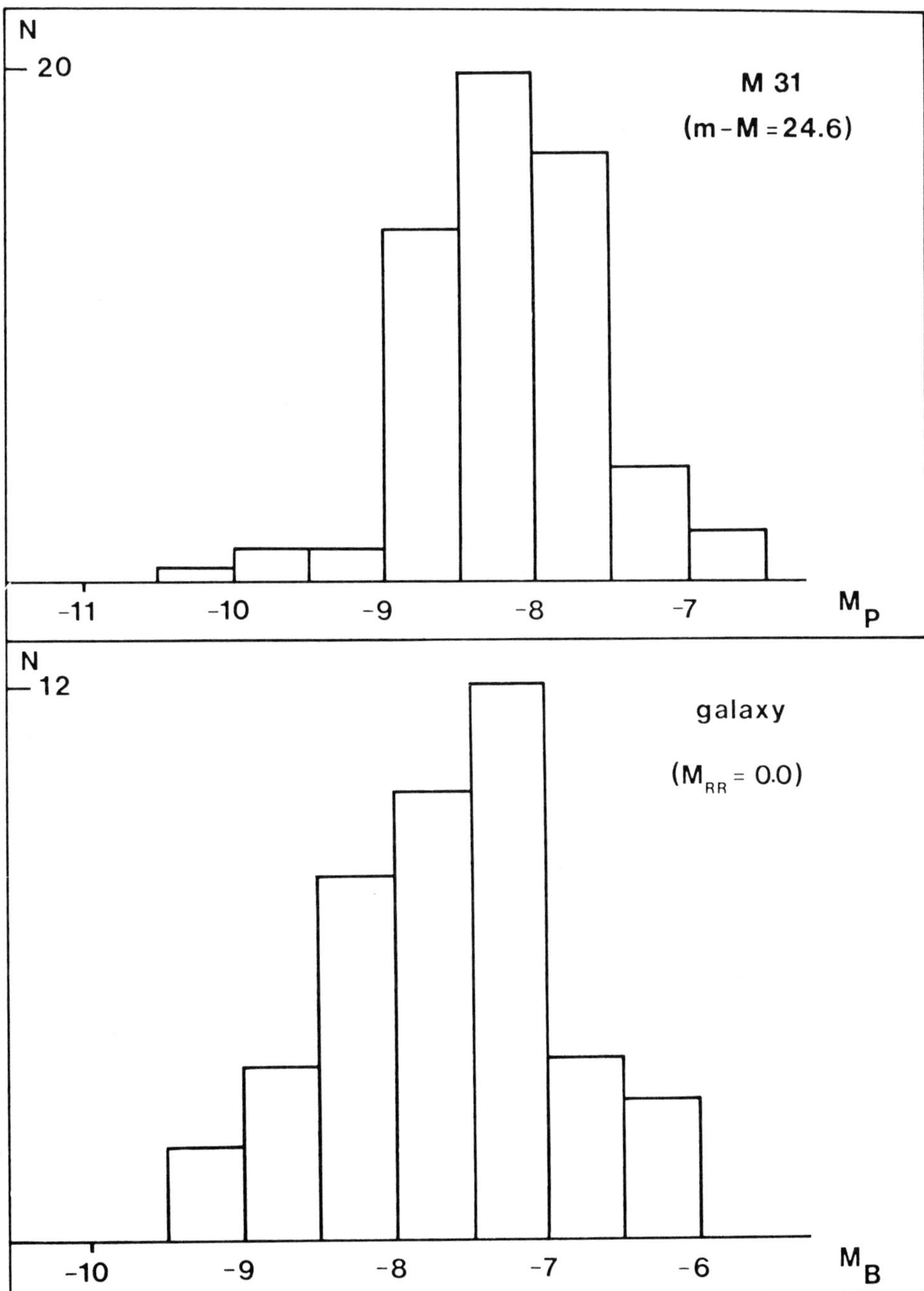

Fig. V.2. Frequency distribution of globular cluster luminosities in M31 and the Galaxy. (*After Kron and Mayall* (1960).)

the relation is linear and can be written

$$M_B(G) = -16.0 - 1.20(\Delta m_1 - 7)$$

where $\Delta m_1 = M_B(1) - M_B(G)$.

$B(1) - B_T(G)$ is the observed blue magnitude difference between the brightest globular cluster and the total magnitude of the parent galaxy. The same author observes that because $\delta M_B(1)/\delta M_B(G) = 0.2$ is small, the upper luminosity end of the globular cluster luminosity function was assumed earlier to be constant (Sandage, 1968).

The Color-Magnitude (CM) Effect

Early-type galaxies (E–SO) show a color-luminosity effect (Section II.1.1) which has been exploited by Sandage (1972) and Visvanathan and Sandage (1977) as a distance indicator. To this end they have selected a wider color base which maximizes the effect through the use of u ($\lambda \approx 3500$ Å) and $V(\lambda = 5500$ Å) magnitudes.

The origin of the CM effect is the gradual weakening of the Mg I b-band from giant to dwarf galaxies (Section II.1) indicating a change in the metalicity with absolute magnitude.

The $(u - V)_0$ corrected colors were found negatively correlated with the total corrected magnitudes of galaxies in the VC. With a distance modulus $\mu_{VC} = 30.47$ the CM relation for early-type galaxies becomes

$$M_V^0(\text{E–SO}) = +10.327(u - V)_0 + 2^m46$$

with an intrinsic dispersion $\sigma(M) = \pm 0^m5$ for a single galaxy. An extension of the CM-relation was made by Visvanathan and Sandage (1977) to early-type spirals (SO/a, Sa, and Sab) because the bulges in these objects are similar to ellipticals in surface brightness, continuum energy, and color-distributions. From the early-type spirals in the VC and identical distance modulus as above, it was found

$$M_V^0(S) = -7.73(u - V_0) - 6.34$$

with $\sigma(M) = \pm 0^m7$, a much higher dispersion than that for E–SO galaxies.

Luminosity Classes

The classification system due to van den Bergh (Section I.4) provides a way to estimate the absolute magnitude of a galaxy through the luminosity class L_c which is assigned after morphological inspection. The original calibration was performed by its author with the few galaxies of the Local Group to which a L_c can be assigned.

TABLE V.4

Callibration of luminosity classes*

L_c	$\langle - M_B^0 \rangle$
I	20.6
II	19.4
III	18.6
IV	17.7
V	14.7

* Adapted from Bottinelli and Gougueheim (1976), Mould *et al.* (1980) and van den Bergh (1980). Compare with Table I.2.

Sandage and Tammann (1974c) extended the procedure to the Sc spirals in the VC and used all Sc spirals with known Cepheids as calibrations. Bottinelli and Gougueheim (1976) found a bias in the Sandage and Tammann calibration, in the sense that the most luminous classes are over-evaluated, and traced the effect to several causes, the principal being related to an unwarranted extrapolation towards M101 (Section V.1.3). Table V.4 gives the luminosity calibration adapted from Mould *et al.* (1980) and van den Bergh (1980).

Absolute Magnitudes of Galaxies

Table I.1 defined the stage t along the sequence of galaxies. A luminosity index $\Lambda = t + 2L_c - 1$ was introduced by de Vaucouleurs, which correlates with the absolute magnitude M_T^0 of spirals. M_T^0 is in the same photometric system as the corrected B_T^0 magnitudes of the Second Reference Catalogue (de Vaucauleurs *et al.*, 1976). The regression line $M_T^0(\Lambda)$ was independently derived from 18 spirals in nearby groups, 18 spirals in the VC and 341 spirals with $cz > 100$ km s^{-1}. The calibration is based on 25 galaxies with distances obtained by means of primary and secondary indicators. Thus

$$M_V^0 = -22^m15 \pm 0^m12 + 0.3\Lambda$$

according to de Vaucouleurs and Bollinger (1979b). The moduli derived from M_T^0 are supposed to have an error of $\pm 0^m4$.

V.1.3. Geometric Indicators

A geometric distance indicator is an measurable linear feature in the galaxy image for which an angular diameter can be assigned. Because of its nature, geometrical indicators are absorption-free in a first approximation, leading directly to true moduli μ_0, provided we had a statistically valid estimate D of their absolute size, thus

$$\mu_0 = 17.26 + 5\log(D/\theta)$$

if D is given in parsec and θ in arc sec. To assimilate the foregoing expression with the apparent modulus $\mu = M - m$ obtained through the use of photometric indicators, Sérsic (1960, 1980) has used the quantities

$$q = 5\log\theta \quad \text{and} \quad Q = 21.57 + 5\log D$$

such that

$$\mu_0 = Q - q$$

is the true modulus. Obviously $\mu = Q - q + A$.

The distance up to which a geometrical indicator can be used is determined by seeing and particularly plate-diffusion. These effects produce stellar images with a finite size θ_s and spread the brightness distribution of the measured indicator producing an observed diameter θ which is larger than the true one θ_0. Because of the subjectivity implied in visual measurements, isophotal diameters derived from surface photometry should be preferred.

Diameters of H II *Rings*

Gum and de Vaucouleurs (1953) proved the constancy of the diameters of the largest H II rings in galaxies. The mean diameter D_1 of these features was found later weakly dependent on the luminosity of the parent galaxy (de Vaucouleurs, 1978b) through

$$\langle \log D_1 \rangle_{HR} \approx 1.85 - 0.05(M_T^0 + 10.0)$$

or

$$\langle Q \rangle_{HR} \approx 30.82 - 0.25(M_T^0 + 10)$$

as follows from the calibrators given in Table V.5.

H II rings leave little room for subjectivity in measurements, which are independent of exposure time, foreground extinction, and even wavelength (de Vaucouleurs, 1978b)

Diameters of the largest H II *Regions*

The H II complexes in galaxies were proposed by Sérsic (1959) as distance indicators and used to establish a distance scale by the same author (1960) and also by Sandage and Tammann (1974a). Sérsic measured the H II diameters in blue plates whilst Sandage and Tammann used Hα plates (Figure V.3).

The difficulties implicit in the method have been later stressed by Hodge (1976) who pointed out the ambiguities in defining a H II region when hierarchic multiplicity is present, as already noticed by Carranza (1968) and also Smith (1976). Sandage and Tammann were confronted, moreover, with ill defined borders for the H II regions, due in part to the greater specific sensibility of the Hα plates. Therefore they measured diameters at two density levels (halo and core diameters) but their results are subject to systematic effects due to the nonlinearity of the plate response to different densities as shown by de Vaucouleurs (1976).

Isophotal diameters measured by Kennicutt (1979a, b) seem to be free of these difficulties but their measurement is naturally a more laborious process. Corrections for instrumental dispersion and seeing are a common handicap and limiting factor for the application of the method, whatever measuring technique is used. The usual

TABLE V.5

Diameters of H II rings in nearby galaxies*

Object	$\langle \theta_1 \rangle$
LMC	840 arc sec
M31	78
M33	63
N6822	52
I1613	37
D155	15
I342	15
N2403	15
Sextans A	13
N6946	6.5

* Adapted from de Vaucouleurs (1978b).

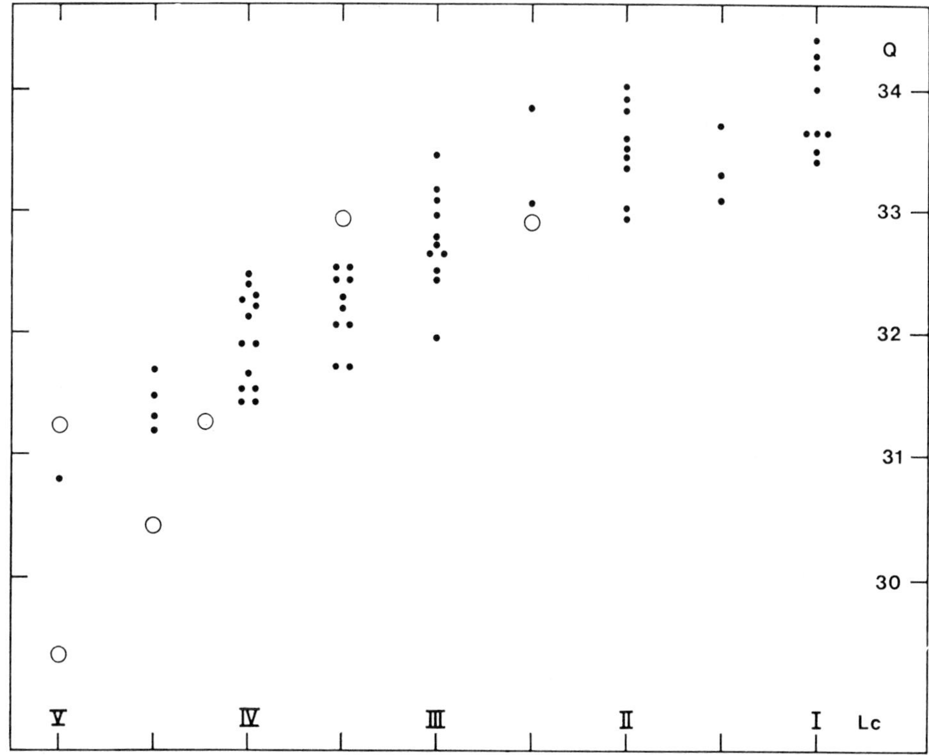

Fig. V.3. Correlation between H II-diameters with luminosity classes. (*After Sérsic* (1980).)

procedure is to assume the spreading function to be gaussian to deconvolve the observed profile.

The calibration of the absolute diameters of the three largest H II complexes in galaxies was carried out (Sandage and Tammann, 1974a) through direct use of galaxies with distances already obtained with Cepheids. Their sample was then reduced to nearby galaxies belonging on the Local and M81 group, which resulted in the absence of high luminosity Sc spirals. A linear correlation was then sought from the available data to relate $\langle D_3 \rangle_{\mathrm{H\,II}}$ with the luminosity of the parent galaxy

$$\log \langle Q_3 \rangle_{\mathrm{H\,II}} = -0.140 M_P^0 - 0.20$$

or

$$\langle Q_3 \rangle_{\mathrm{H\,II}} = 32.46 - 0.70(M_P^0 + 17)$$

(V.5)

which they assumed to be valid up to the brightest luminosities. There is some doubt, however, on the soundness of this procedure. From a rediscussion of H II diameters Sérsic (1978) derived the correlation $Q_3(L_c)$ from 68 galaxies in 19 clusters, a procedure which permits a considerable representation of the high-luminosity spirals. in the sample (Figure V.3). Two slopes can be defined for $Q_3(L_c)$ according to whether $L_c \gtrless 3$, namely

$$\langle \Delta Q_3 / \Delta L_c \rangle \approx 1.3 \quad \text{and} \quad \langle \Delta Q_3 / L_c \rangle \approx 0.33.$$

From Bottinelli and Gougueheim (1976) or Mould *et al.* (1980) we find

$$\langle \Delta M_{Bl}\Delta L_C\rangle \approx -1.96 \quad \text{if} \quad L_c > 3 \quad \text{and}$$
$$\langle \Delta M_{B}/\Delta V_c\rangle \approx -1.05 \quad \text{if } L_c < 3$$

and thus follows

$$\langle \Delta Q_3/\Delta M_B\rangle \simeq -0.65 \quad \text{and} \quad \langle \Delta Q_3/\Delta M_B\rangle \approx -0.33$$

in the same intervals. For the lower luminosity galaxies the slope is in essential agreement with that of Sandage and Tammann $\langle \Delta Q_3/\Delta M_P\rangle = 0.7$ as well as with that derived by van den Bergh (1980):

$$\langle Q_3\rangle_{\text{H\,II}} = 32.45 \pm 0.20 - (0.67 \pm 0.12)(M_B^0 + 17) \tag{V.6}$$

on the basis of Kennicutt's isophotal diameters and galaxies with $L_c > 2.5$. Thus the H II diameters used by Sandage and Tammann for $L_c > 3$, because of their extrapolated regression, are over evaluated. This fact also reflects on their calibration of the luminosity classes L_c (Section V.1.2. and also Mould *et al.* (1980)).

The calibration of the mean diameters of the three largest H II complexes in galaxies in the whole range of luminosities can hence be written

$$\langle Q_3\rangle_{\text{H\,II}} = 33.5 - \binom{0.67}{0.33}(M_B^0 + 18.6) \tag{V.7}$$

according to whether $M_B^0 \gtrless -18.6$ or $L_c \gtrless 3$. The foregoing expression is based on van den Bergh's (V.6) relation for lower luminosity galaxies and Sérsic's slope for higher luminosity objects. The absolute magnitude $M_B^0 = -18.6$ for $L_c = 3$ is a compromise value between van den Bergh's (1980) and Mould *et al.*'s (1980).

Diameters of Spiral Galaxies

Because of the good correlation between apparent diameters and total magnitudes of galaxies, it is possible to use them as tertiary distance indicators. As a consequence of the dispersion of intrinsic diameters for a given morphological type, another parameter with a better correlation is needed. de Vaucouleurs and Bollinger (1979a) relates the linear diameters D_0 corresponding to the isophotal apparent ones given in the Reference Catalogue with the luminosity index $\Lambda = r + 2 L_c - 1$. The correlation was established independently for 18 spirals in nearby groups, 18 spirals in the VC and 34 spirals with $cz > 100$ km s^{-1}, and calibrated with 25 galaxies whose distances were previously derived from primary and secondary indicators. Thus he found

$$\log D_0 = 4.78 \pm 0.04 - 0.06\Lambda,$$

where D_0 is given in pc and $3 < \Lambda < 20$.

V.1.4. OTHER INDICATORS

Beside the classical ones, other distance indicators have been proposed and used, some of them such as super associations and also the peak brightness of supernovae partake the general characteristics of the classic indicators but have not yet been elaborated enough to become reliable. In others, radio frequency observations are combined with optical data, with the usual sequel of corrections and optical frequency dependence.

Radically different are those approaches tried with radio source super-luminal and also with supernovae expansion.

Surface Brightness of Galaxies

The surface brightness of spiral galaxies was found by Holmberg (1958) to correlate with the absolute magnitude. Since the surface brightness is independent of distance, the absolute magnitude can be obtained once the relationship is calibrated. This was done by Holmberg with the spirals of the VC. The dispersion in the correlation arises on the uncertain corrections needed to allow for intrinsic absorption, which requires the knowledge of the inclination of the galaxy. Because of the small slope of the relationship, a little error in surface brightness produces a large error in the estimate of the absolute magnitude.

Super Associations (SAS)

The concept of super association is due to Ambartsumian (1963) and refers to the largest complexes of young OB stars and H II regions observed in late spirals and irregular galaxies. Wray and de Vaucouleurs (1980) have used a composite UBV-color technique (Wray, 1978) to identify them in color plates as blue knots, noticing that their intrinsic brightness surpasses the brightest blue stars by 3 to 5 magnitudes. Typical examples of super associations are those of 30 Doradus in the LMC and NGC 604 in M33. Previous work on these features was done by Seyfert (1940). Sérsic (1959) and Bok and Bok (1962) noticed the possibilities of these large complexes as distance indicators.

Wray and de Vaucouleurs (1980) have surveyed 78 galaxies for these blue knots and estimated their apparent B-magnitudes. From distance moduli of these objects derived from other distance indicators the following bivariate correlation was found (de Vaucouleurs and Bollinger, 1979c):

$$M^0_{SAS} = -13^m0 + 0.67(M^0_T + 19.75) + 2.5(B - V)^0_T - 0.50 \qquad \text{(V.8)}$$

between the corrected absolute magnitude M^0_{SAS} of the SAS with the corrected absolute magnitude M^0_T and color $(B - V)^0_T$ of the parent galaxy.

It is easy to show (Section VIII.6) that the former correlation is tied to that already found for the diameters of the largest H II complexes (Section V.1.2) through

$$M^0_{SAS} + \langle Q_3 \rangle_{H II} = 19.6 + \left(\begin{matrix} 2.5 \\ -0.54 \end{matrix} \right) [(B - V)^0_T - 0.45]. \qquad \text{(V.9)}$$

This relation is analogous to the one found by Sérsic (1959) which allowed him to estimate the absolute magnitudes M^0_{SAS} as shown in Table II.2. The upper coefficient has to be taken for $t > 6$ or $L_c > 3$ and the lower for $t \leqslant 6$ or $L_c \leqslant 3$ (Section VIII.6). Since $M^0_{SAS} + \langle Q_3 \rangle_{H II} = B^0_{SAS} + 5 \log \theta^c_s$ is an indicative surface brightness of the largest SASs in galaxies, the foregoing expression shows its dependence on galaxy color.

Supernovae

The distant supernovae (SN) can be compared directly with their galactic counterparts which avoids intermediate distance indicators. However, little is known about the intrinsic dispersion of absolute magnitudes at maximum in the different types of SN. On the other hand, only fragmentary information is available on most of the six SN which appeared in the Galaxy in the last thousand years.

The SN have been classified in types I and II according to their spectra, whose principal difference arises on the chemical composition of their respective envelopes. That of the SN II is solar-type while the SN I have hydrogen-poor envelopes. Because SN types are defined by their spectra, we ignore the types of the historic galactic SNs.

The light curves of SNs differ according to type, although some scatter exists particularly in the SN II. The principal difference is a bump some 50 days after maximum in SN II. There is strong indication that the maximum luminosity of SN I depends on the morphological type of the parent galaxy (Tammann, 1977). The brighter ones are found between the E–SO galaxies whilst the fainter ones are observed between the late spirals and the irregulars. No significant correlation with the morphology of the parent galaxy has been noticed in the case of SN II.

TABLE V.6

Frequency of SN*

Galaxy type	Galaxy frequency (%)	SN I	SN II
E + SO	22.9	5.5	0
Sa	7.7	0	0
Sb	27.5	7	5
Sc	27.3	14	11
Irr	2.1	1	0
Other	12.5	1.5	0

* From Maza and van den Bergh (1976).

The frequency of SN with respect to the type of the parent galaxy has been studied by Maza and van den Bergh (1976) and also by Tammann (1977). Table V.6 gives the pertaining information. SN I are found along the whole sequence of galaxies, from ellipticals to irregulars, but SN II are not observed in early-type objects from E to Sa or also in the Irr I, although in this last case could be a sample size effect (Table V.6).

The historical SN in the Galaxy were observed at the time of their explosion, with the only exception of Cass A. Information regarding these past events is so scarce that in the case of the Lupus SN (1006 AD) neither the apparent magnitude nor the distance is known. The Crab SN is the best known case but it is a non-typical object whose extragalactic counterpart has not yet been observed (Minkowski, 1966). Only Tycho's 1572 SN is well known. The work of Baade (1945) suggests a visual apparent magnitude $V = -4^m_{.}0 \pm 0^m_{.}3$. The distance was estimated to be between 3 and 7.5 kpc from H I absorption lines and rotation models for the Galaxy, which give $M_V \approx -19^m_{.}4 \pm 0^m_{.}6$, if an absorption of $M_V = 1.6$ is adopted (van den Bergh, 1975a).

Kowal (1968, 1969) derived the relation $M_V = -18^m_{.}65 + 5 \log (H/100)$ for SN I from their apparent magnitudes and redshifts of their parent galaxies.

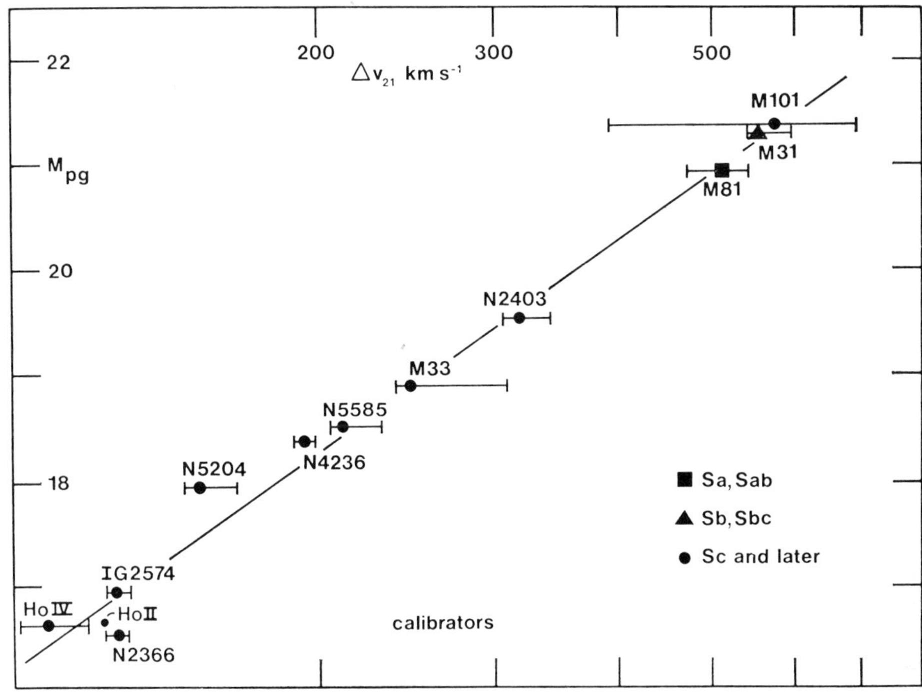

Fig. V.4.   The Tully–Fisher correlation (see text).

Tamman (1977) instead has obtained the following relations for the maximum blue luminosities

$$M_B = -17^m7 \pm 7^m1 + 5 \log(H/100) \qquad \text{(SN I)}$$

$$M_B = -15^m8 \pm 0^m8 + 5 \log(H/100) \qquad \text{(SN II)}$$

from a distance limited sample.

The 21-cm Line Width

Tully and Fisher (1977) have proposed the use of the 21-cm H I-line width as a distance indicator through a correlation with the luminosity of the galaxy. The galaxies with apparent diameters smaller than the beam of a radio telescope produce a Doppler-widened H I-profile due to internal motions. In spirals and irregular galaxies most of the width is due to rotation and the line HW is proportional to $2V(R_M)$, where $V(R_M)$ is the maximum rotational velocity. The method has to be applied to galaxies with well defined H I-profiles, i.e. late spirals and irregulars. The line width is measured at 20% of the peak intensity and the observed amplitude $2V_R$ has to be corrected to allow for the inclination of the plane of the galaxy $V(R_M) = 2V_R/\sin i$. Errors arising in the estimation of i are thus introduced.

In very rough terms, a correlation between the apparent luminosity $\mathcal{L}_K$ through a given photometric window and $V(R_M)$ can be expected, such as $M_K^0 = a(K) \log(\Delta V) + b(K)$. In fact, the mass of a galaxy is $\mathfrak{M} \propto V^2(R_M)R_M$ (Section II.2.2) and is expected to be roughly constant for late type galaxies, so that $\mathcal{L}_K \propto V^2(R_M)R_M$.

For a given photometric system a relation such as $\mathfrak{L}_K \propto R^{2\beta(K)+1}$ holds between luminosity and a isophotal radius R_K. Assuming moreover that $R_K \propto R_M$, we obtain $\mathfrak{L}_K \propto [V(R_M)]^{3\beta_k}$ and $R_K \propto [V(R_M)]^{3\beta_k/2\beta_k+1}$ and thus

$$M_K^0 = -7.5\beta_K \log [V(R_M)] + b(K),$$

$$\log R_K = [3\beta_K/2\beta_K + 1] \log[V(R_M)] + C(K),$$

as shown in Figure V.4.

Bottinelli *et al.* (1980) have made a detailed study of the empirical correlations and conclude: (i) the slope $a(K)$ in the Tully–Fisher relation is a continuous function of the photometric window where the magnitudes and diameters of galaxies are defined, increasing from $a(B) \approx 5$ in the B-band to $a(H) \approx 10$ in the H-band at the infrared. (ii) The main source of error is in log $V(R_M)$ rather than in the magnitudes, consequently the error in the estimate of M_K^0 is proportional to $a(K)$, becoming larger for longer wavelengths.

From de Vaucouleurs (1978a, b) primary calibrators Bottinelli *et al.* obtained the following correlations for the *B*-window,

$$M_T^0(B) = -19.4 \pm 0.15 - (5.0 \pm 0.4) [\log V(R_M) - 2.2]$$

and

$$\log R_B = 4.20 \pm 0.07 + (1.00 \pm 0.07) [\log V(R_M) - 2.2]$$

with $V(R_M)$ expressed in km s^{-1} and R_B in pc.

Supernovae Expansion

The idea of Baade (1926) to determine the distance of variable stars has been applied by several authors to SN (Kirschner and Kwan, 1974; Wagoner, 1979; Schurmann *et al.*, 1979).

During the early phases of the SN outburst the shell radiates as a black body so that its angular diameter θ and expansion rate θ' can be determined from photometry on a wide spectral range. Through the use of the expansion velocities inferred from spectral scans, the photospheric radius can be estimated and thus the distance.

Enough precision has not yet been achieved but, further theoretical improvements might give better estimates. The importance of this technique lies in the ad-hoc determination of the intrinsic properties of the indicator, which makes SN a self-calibrated distance indicator.

Super-Luminal Expansion

The components of some SRGs are observed to separate with an apparent large angular velocity. At the cosmological distances of these sources, the corresponding apparent linear velocities seem considerably larger than the speed of light.

Rees (1967) has proposed, however, that if the source expands with relativistic velocity, the difference in arrival time of the signals emitted from different parts can cause the appearance of expansion with a velocity $v > c$.

The observed characteristics of super-luminal expansion can be summarized as follows (Kellermann and Shaffer, 1977): (i) The component separation is independent of wavelength; (ii) only expansion motion is observed; (iii) the motion occurs in a straight line; (iv) changes in the component separation up to a factor of 10 are observed, with apparent linear velocities up to $10c$; (v) the epoch of null separation can be obtained through back extrapolation and often agrees with the time of a major outburst in the central source.

A simple 'light-echo' model for these phenomena was suggested by Lynden–Bell (1977), based on the interpretation that the apparent super-luminal motion is due to the differential time-travel of a signal emitted from a source which is moving towards the observer at relativistic velocity. Since ejections are generally symmetrical (Section III.6) the apparent transverse expansion would be $2c$ if the ejection is perpendicular to the line of sight, provided the observer and the central source were in relative rest. If the ejection is along a line forming an angle θ with the line of sight, the apparent transverse velocity is $2c/\sin \theta$.

In fact, a flash from the central source will illuminate and excite the material at, say, source S_1 distant ct from it. This event will be seen by the observer at time $t - (ct \cos\theta)/c = t(1 - \cos \theta)$ because S_1 is nearer than the central source and the observed transverse velocity of S_1 will be $V_1 = c \sin \theta/(1 - \cos \theta)$ while that of the source S_2 moving in the opposite direction is $V_2 = c \sin \theta/(1 + \cos \theta)$ so that apparent mutual velocity becomes

$$V_m = 2c/\sin \theta \geqslant 2c.$$

If the central and the sources S_1, S_2 are seen, the separation ratio

$$\frac{R(S_1)}{R(S_2)} = \frac{1 + \cos \theta}{1 - \cos \theta}$$

provides us with θ. Moreover, the apparent angular separation rate η' is related to the distance D through $\eta' = Vm/D$ and can be determined by VLBI (Section III.6). Thus the distance is given by $D = Vm/\eta' = 2c/\eta' \sin \theta$.

From the observed redshift cz, the Hubble constant H_0 is obtained through

$$H_0 = \tfrac{1}{2}z\eta' \sin \theta$$

provided $z \ll 1$. If the SRG has a large redshift Cohen *et al.* (1976) give the modified expression for H_0 in a Friedmann Universe (Section VI.2) with a deacceleration parameter q_0,

$$H_0 = \tfrac{1}{2}\eta' \sin\theta \ \frac{z}{1 + z} \ [1 + \tfrac{1}{2}(1 - q_0)z + \cdots]$$

provided $2q_0z \ll 1$.

V.2. The Scale of Distances

Hubble (1936) first estimated the expansion parameter H_0 at 536 km s^{-1} Mpc^{-1} but this value was very uncertain because the farthest galaxies used had redshifts hardly superior to their peculiar velocities. The estimation was based on Cepheids in the Local Group and the brightest stars in galaxies outside the Local Group calibrated with it.

Both indicators were later revised. Baade (1952) made an important change in the zero point of the P–L relation for Cepheids: he succeeded in discriminating Cepheids according to stellar populations and the change resulted in an equivalent decrease of

H_0 by a factor of 2. The brightest stars used by Hubble were actually composite compact objects, mostly clusters and compact H II regions intrinsically brighter than the 'bonafide' brightest stars, and Sandage (1958) revised down the Hubble constant to somewhere between 180 and 75 km s^{-1} Mpc. Three independent estimates were made in the meantime: Holmberg (1958) obtained $H_0 = 136$ km s^{-1} Mpc, van den Bergh (1960) gave $H_0 = 120$ km s^{-1} Mpc through callibration of his luminosity classes and Sérsic (1960) got $H_0 = 125$ km s^{-1} Mpc^{-1} with H II-diameters. A rediscussion of these estimates plus his own and the consideration of a revised zero point in the P–L relation led Sandage (1961) to propose $H_0 = 100$ km s^{-1} Mpc^{-1}. Before the next specific determination more than a decade later, a conventional value of 75 km s^{-1} Mpc^{-1} was used.

V.2.1. THE HUBBLE FLOW

The density fluctuations of the mass-distribution in the Universe together with their characteristic sizes show that the intergalactic gravitational fields are weak and slowly varying (Sections IV.3 and VII.2). Thus the velocity field $V(z)$ of the galaxies is given by the sum of two terms

$$V(r) = H_0 r + v(r), \tag{V.10}$$

where the first represents a uniform and isotropic expanding background or Hubble Flow, while the second term is the peculiar velocity field resulting from either large primordial peculiar velocities or from the accelerations produced by the lower gravitational potentials in and around the inhomogeneities.

An upper bound to motions primordial in origin can be set at 300 km s^{-1} in order to make the velocities of the perturbations at recombination ($z \approx 1000$) smaller than the speed of light, thus the density inhomogeneities become the major source of peculiar motions.

From Equation (V.10) we derive the Hubble ratio

$$H(r) = \frac{V(r)}{r} = H_0 + \frac{v(r)}{r}$$

which is a field function that asymptotically converges to the Hubble constant H_0 when we consider galaxies very distant from the origin.

de Vaucouleurs has found evidence for a considerable anisotropy of the expansion velocities $V(r)$ among galaxies which he atributes to rotation and expansion of the LSC. Shearing motions have also been proposed on scales much larger than the inhomogeneities of the spatial distribution but this can be ruled out on the grounds of the interpretation of the background radiation as primordial (Section VI.5). According to de Vaucouleurs the ratio $\langle H \rangle / H_0 \approx 1.35$ for the LSC, self-gravitation within it counteracting the general expansion to a certain extent. Sandage believes instead in a much smaller, if any, velocity anisotropy, and proposed eliminating the effect in fitting the Hubble diagram by going directly to distant galaxies outside the region of the anisotropy, for $cz > 4000$ km s^{-1}.

Corwin (1971) has compared the distance modulus of the VC with that of its closely similar Hercules Cluster which has redshifts about ten times as large, and hence far out of the LSC. From comparisons of luminosity classes of their galaxies as well as the diameters and magnitudes of the five brightest E and S galaxies in both clusters, resulted a difference of moduli $\Delta\mu° = 4^m3 \pm 0^m5$, from where follows $\langle H \rangle / H_0 \approx 1.19 \pm 0.20$.

Tully and Fisher (1977, 1978) confirmed the center-anti-center asymmetry in the observed radial velocities with extensive H I measurements of $V = cz$ for spirals and irregular galaxies and de Vaucouleurs and Bollinger (1979c) have confirmed his (de Vaucouleurs, 1958) early explorations regarding the velocity field of the LSC.

Rubin *et al.* (1973) have discussed a possible non-isotropic distribution in the velocities and magnitudes of faint Sc I galaxies. This much debated subject has now been settled as being a consequence of the Galaxy motion in space (Rubin *et al.*, 1976). When it is removed ($V_G = 454$ km s^{-1}) the Hubble flow becomes isotropic at a 4% level.

V.2.2. THE QUEST FOR H_0

Nowadays three main independent lines can be recognized in the quest for the Hubble parameter which are outlined next.

Seven Steps Toward the Hubble Constant

In an impressive series of seven papers Sandage and Tammann (1974a, b, c, 1975a, b, c, 1976) have founded a scale of distances based fundamentally on Cepheids as the sole primary distance indicator. The galaxies of the Local and M81 groups, for which distances can be found from Cepheids, are then used as calibrators for the diameters of the H II regions and the absolute magnitudes of the brightest stars in galaxies. These two secondary indicators were found to depend on the luminosity and morphological type of parent galaxies and the corresponding dependences established (Sections V.1.2 and V.1.3). Through these two secondary distance indicators the distance modulus of the M101 group was established, which then provided the nearest Sc I calibrator for van den Bergh's luminosity class I galaxies. In a fourth paper the same two indicators are used to derive the distances of 39 late-type field galaxies (Sc to Irr I) distant 3 to 20 Mpc which together with those belonging to the Local and M81 groups makes 48 galaxies to complete the calibration of the luminosity classes. In this way Sandage and Tammann escape the realm of Cepheids as valid indicators, calibrate the luminosity classes, and use them to determine the modulus of the Virgo Cluster whose peculiar velocity relative to the Hubble flow is found to be negligible. Since the VC contains gE galaxies and is the first of 84 other clusters defining the Hubble law for large redshifts (Section V.2.3) a first link with the Hubble flow is established.

The distances to groups of galaxies within 30 Mpc are determined in the fifth paper. To this end they use again H II region diameters and the L_c's. The same procedure is also applied to remote Sc galaxies with known L_c reaching distances up to 50 Mpc. The analysis of the peculiar velocities and the various estimates of H_0 from different samples led to the conclusion that the local velocity field is regular, linear and isotropic, and also that the absence of observable velocity perturbations suggests the gravitational potential to be weak in spite of observed mass in-homogeneities, which means $\Omega_0 < 1$ (Section VI.6.2).

The sixth paper uses again the Sc I galaxies now up to distances as large as 200 Mpc and also explores southern groups and clusters to test the isotropy of the flow.

Through a more detailed discussion of the upper bound for the disturbances of the Hubble flow, an estimate of the present value of the density parameter (Section VI.6.2) is made and $2q_0 = \Omega_0 \approx 0.06$ is found.

The last paper by Sandage and Tammann uses Tully and Fisher's correlation (Section V.1.4), which is recalibrated and applied to 19 spirals in the central part of the VC for which they give $\mu_c^\circ = 31^m50 \pm 0^m08$. This modulus provides the absolute magnitude of the first ranked E galaxies in clusters and consequently the calibration for the Hubble law derived from the remote great clusters.

The resulting value for $H_0 = 50 \pm 4.4$ km s^{-1} Mpc^{-1}. Since $H_0^{-1} = 19.4 \times 10^9$ yr, its comparison with the age of globular clusters and that of the Galaxy permits them to exclude all Friedmann models with $\Omega_0 > 1$, and to conclude that the Universe has happened only once (Section VI.6.1).

Seven Other Steps Toward the Hubble Ratio

de Vaucouleurs has produced another impressive series of papers devoted to the establishment of a distance scale (de Vaucouleurs, 1978a, b, c, d; de Vaucouleurs and Bollinger, 1979a, b, c). The scale is based on three main primary indicators (Novae, Cepheids, RR Lyrae) and two subsidiary ones (AB Supergiants and eclipsing binaries). The operational philosophy, stated from the outset, is a preference to rely in the use of every distance indicator instead of relying on a selected few. To support this point of view, de Vaucouleurs argues that unaccounted effects of age or chemical composition can affect any indicator, while it is unlikely that all indicators would be affected in the same sense and amount in different galaxies.

Primary indicators are restricted to galaxies in the Local Group. There are six secondary onces: brightest Population II red giants, globular clusters, brightest Population I stars, brightest red variables, the Hubble–Sandage blue-variables, and the largest H II rings. The distance moduli of some Local Group galaxies are discussed next: LMC, SMC, M31, M33, NGC 6822, IC 1613 through the primary indicators. A comparison with the moduli given by Sandage and Tammann (1974a) produces a mean difference $\Delta\mu = -0^m31 \pm 0^m08$ while with van den Bergh's moduli, he finds $\Delta\mu = -0^m05 \pm 0^m10$ (Table V.7).

The secondary indicators are calibrated now with the six Local Group galaxies which define the scale and then the distance moduli of 22 nearby S, and I galaxies. Among these 22 galaxies, a dozen belong to nearby groups (M81, Sculptor, M101).

TABLE V.7

Moduli of local group galaxies*

	μ°(ST)	μ°(dV)	μ°(vB)
LMC	18.6	18.3	18.4
SMC	19.3	18.6	18.8
M31	24.1	24.1	24.0
M33	24.6	24.3	24.1
NGC 6822	24.0	23.7	23.6
IC 1613	24.4	24.0	24.4

* Adapted from Sandage and Tammann (1974b, c); van den Bergh (1977), and de Vaucouleurs (1978b).

Twenty five galaxies define now the scale of distance and they are used to calibrate the tertiary indicators: isophotal diameters and corrected absolute B-magnitudes. Since this is done within the homogeneous photometric system of the Second Reference Catalogue, the statistical possibilities of these indicators are considerable, and lead to the distance determinations of 458 spirals from where distance moduli follow for 31 nearby groups within 35 Mpc. Tests on the linearity of the scale are conducted through independent indicators: super associations, metric (effective) diameters and SN I and the linearity of the scale is warranted in a range of 15 magnitudes.

The seventh paper of de Vaucouleurs series deals with the velocity-distance relations in different directions and the Hubble ratio $\langle H \rangle$ within the local Super-cluster and without it. The distances of 322 spirals selected from the 458 galaxies previously studied are used to analyze the velocity-distance relation in sky areas (seven on the equatorial belt of the local super cluster and four in the polar caps), and it is found that $\langle H \rangle$ varies from a minimum of 75 to a maximum of 110 km s^{-1} Mpc^{-1}. Neither $\langle H \rangle$ nor the slope of the relation are constant, both tend to be lower than average toward the center of the LSC and higher in the anticenter region, and significant departures from the Hubble flow are evident. If the sector dominated by the LSC is excluded, an approximation to the asymptotic value of H for empty space is found to be $\langle H_0 \rangle = 100 \pm 10$ km s^{-1} Mpc^{-1}.

Two leaps to H_0

van den Bergh (1970, 1975a, b, 1977) has consistently propiciated a two-step procedure to estimate H_0: First the distances of nearby Local Group galaxies have to be established. Next these galaxies are compared with others at distances large enough as to overcome the noise in the Hubble flow caused by peculiar and systematic motions.

The adopted distance moduli for the LG galaxies given in Table V.7 are weighted means from several separated estimations based in several other indicators besides those described in Section V.1.1 such as W Virginis stars, Population II giants, globular clusters, MK classification of the brightest stars (only the LMC) and the H-line strengths from Hβ and Hγ photometry (LMC, SMC, M31). A weight of 3 has been assigned to Cepheids and of 2 for Novae and RR Lyr, while weight one was given to the other remaining indicators.

The second and last step assumes that there exist distance indicators in distant galaxies. Although a particular indicator may be wrong because of differences of age, evolutionary history or composition, van den Bergh assumes that the systematic errors due to such differences will minimize through the use of the largest possible number of different indicators.

From the discussion of twelve distance indicators (some of them with very low weight) van den Bergh (1975b) finds $H_0 = 94 \pm 7$ km s^{-1} Mpc^{-1} and very understandably states that past experience teaches that little trust has to be placed in the small formal errors obtained for H_0.

Epilog

The philosophy behind the use of many distance indicators to compensate possible variations in chemical composition and evolution is relied upon both by van den Bergh and de Vaucouleurs. Sandage and Tammann instead rely only on Cepheids as the sole primary precision indicator, and those are the reasons for the differences in moduli within the Local Group.

If Sandage and Tammann's linear extrapolation in the calibration of the H II diameters and luminosity classes were corrected, their value for H_0 would go up to 85 km s^{-1} Mpc^{-1} (Bottinelli and Gouguenheim 1976; Sérsic, 1978), the difference with the other estimations (100, 94) being only a consequence of the different sets of Local Group moduli.

V.3. Far Away and Long Ago

The exploration of the Universe in depth requires consideration of the effects that the redshift produces on the light we receive from the galaxies, The observation of distant galaxies implies a look-back time which is a sensible fraction of the age of the Universe. This causes seeing them at earlier evolutionary stages which we need to understand first in order to compare them with the homologous ones in our neighborhood.

To these ends, we need first to deal with the redshift and its cosmological nature.

V.3.1. THE REDSHIFT

Let Γ_0 be the world line of a galaxy which emits electromagnetic radiation of frequency ν_0 and Γ the world line of our observer who receives that radiation with frequency ν (Figure V.5). Let us now trace the respective null-geodesics corresponding to two nearby events A_0, B_0 on Γ_0 which intersect at two other events A, B in the future. If the space-time has a regular geometry, this will be a univocal construction and the ratio of proper times elapsed between A_0, B_0 and A, B namely ds_0/ds, will be wholly determined by the two world-lines Γ_0 and Γ and the event A_0 or, equivalently, by A.

Since null-geodesics are lines of a constant phase for radiation propagating along them, the phase difference between A_0, B_0 will be the same as that between A, B. The oscillation period for the radiation waves will be proportional to the elapsed times: for the source this is the proper time ds_0, and will be ds for the observer, and the general expression for the redshift will be in consequence,

$$1 + z = \frac{ds_0}{ds} = \frac{\nu}{\nu_0} \tag{V.11}$$

where ν_0 is the 'proper frequency'. For an atom moving with the source and emitting a particular spectral line, is the frequency we find in the tables of physical constants, it is the frequency the observer would measure in his own laboratory for the same kind of atom moving with him. The observed frequency ν is instead the frequency the observer meassures when he looks at the galaxy through a telescope.

Relationship (V.11) gives the total redshift, which involves the relative motion of the galaxy and the observer as well as the curvature of space-time. There is no general way to separate this effect into a Doppler part due to relative motion, and another of cosmologic nature due to the curvature.

Let us make a more specific application of the relationship (V.11) and consider a space-time satisfying a Robertson–Walker metric, as it is the case in the explosive

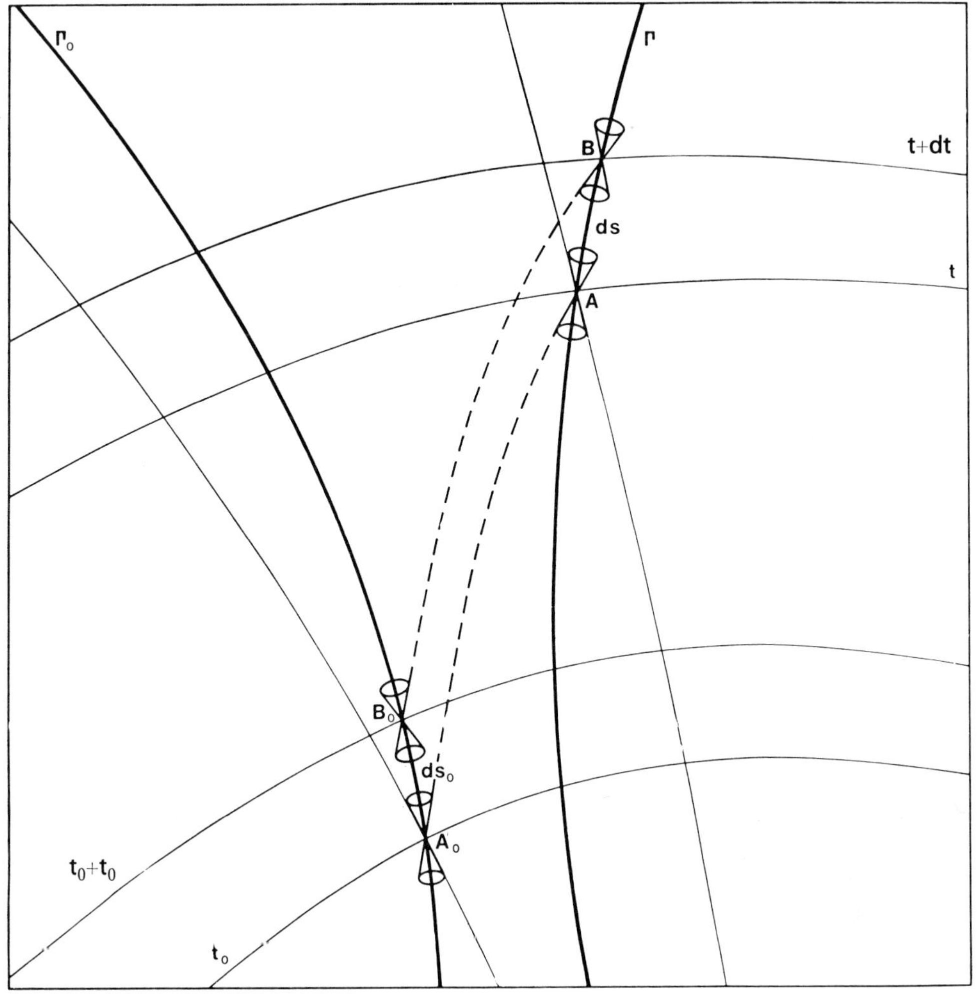

Fig. V.5. Space-time diagram to define the redshift (see text).

cosmologies (Section VI.1). We have for the arc-element ds

$$ds^2 = dt^2 - \left(\frac{R}{c}\right)^2 d\sigma^2, \tag{V.12}$$

where $R(t)$ is the curvature of space-time, $d\sigma$ the space arc-element, and c the speed of light. The equation of a null-geodesic ($ds = 0$) gives us

$$dt = \left(\frac{R}{c}\right) d\sigma_r \tag{V.13}$$

where dr is the component of $d\sigma$ along the light trajectory. After considering Figure

V.5 we may write for each geodesic A_0A and B_0B,

$$\int_{t_0}^{t} \frac{\mathrm{d}t}{R(t)} = \frac{1}{c} [\sigma_r(A) - \sigma_r(A_0)]$$

$$\int_{t'_0}^{t'} \frac{\mathrm{d}t}{R(t)} = \frac{1}{c} [\sigma_r(B) - \sigma_r(B_0)]$$

and noticing that $t' = t + \delta t$, $t'_0 = t_0 + \delta t_0$ and also $\sigma_r(B) = \sigma_r(A) + \delta\sigma_r$, $\sigma_r(B_0) = \sigma_r(A_0) + \delta\sigma_{r_0}$, where $\delta t \, \delta r_0$, $\delta\sigma_r$, $\delta\sigma_{r_0}$ are small quantities of the first order whereby we have taken advantage of the regular geometry postulated for the space-time.

We also have the relationship

$$\int_{t'_0}^{t'} \frac{\mathrm{d}t}{R} = \int_{t_0+\delta t}^{t_0} + \int_{t_0}^{t} + \int_{t}^{t+\delta t}$$

so that

$$\frac{1}{c} [\delta\sigma_r - \delta\sigma_{r_0}] = \frac{\delta t}{R(t)} - \frac{\delta t_0}{R(t_0)} + \text{higher order terms.}$$

After defining the radial velocity $c\beta_r$ of each event with respect to the comoving system defined by $R(t)$

$$\beta_r = \frac{R(t)}{c} \frac{\delta\sigma_r}{\delta t}$$

we obtain

$$\frac{\delta t_0}{\delta t} = \frac{R(t_0)}{R(t)} \frac{1 + \beta_r}{1 + \beta_{r_0}}.$$

Now the total velocity $c\beta$ is related to $\mathrm{d}s$ and $\mathrm{d}t$ through Equation (V.2)

$$\mathrm{d}s^2 = \mathrm{d}t^2 [1 - \beta^2] \quad \text{where} \quad \beta = \frac{R}{c} \frac{\mathrm{d}\sigma}{\mathrm{d}t}$$

and we obtain at last

$$1 + z = \frac{\mathrm{d}s_0}{\mathrm{d}s} = \frac{R(t_0)}{R(t)} \frac{1 + \beta_r}{1 + \beta_{r_0}} \left(\frac{1 - \beta_0^2}{1 - \beta^2} \right)^{1/2}$$

for the redshift (V.11). Introducing now the cosmological redshift

$$1 + z_c = \frac{R(t)}{R(t_0)} \tag{V.14}$$

due only to the expanding background of the Universe, and also the Doppler shift

z_D originated in the relative motion of the galaxy and the observer

$$z_D = \beta_r - \beta_{r_0}$$

we obtain at last the so called 'composition law' for redshifts

$$1 + z = (1 + z_c)(1 + z_D) \tag{V.15}$$

if we disregard second order terms in the velocities.

V.3.2. THE HUBBLE LAW

The apparent bolometric luminosity l of a galaxy at a distance σ_r from the observer is related to its intrinsic luminosity L by

$$L/l = 4\pi(R(t)\sigma_r)^2 (1 + z)^2 \tag{V.16}$$

in an expanding Universe: l varies inversely proportional to the surface area $4\pi(R\sigma_r)^2$ of the wave front reaching the observer at time t. However, two other additional effects must also be taken into consideration. Because the frequency of the radiation is modified in the ratio $1 + z$ (Equation (V.11)), the energy of the photons is reduced at this ratio (Hubble's 'energy effect') but, since the photons are spread over a distance increased by the ratio $R(t)/R(t_0) = 1 + z$ between times of emission and reception, the photons arrive at a constant velocity c although with a rate reduced by this factor ('number effect').

In order to express the metric distance σ_r in terms of the redshift (which is an observable), we develop $R(t)$ in power series of $\Delta t = t_0 - t$, around present time t, which is possible on account of the regular nature of the function $R(t)$. Equation (V.14) leads to

$$z = H_0 \Delta t + (1 + \tfrac{1}{2}q_0)H_0^2(\Delta t)^2 + \cdots \tag{V.17}$$

after introducing the Hubble constant H_0 and the deceleration parameter q_0 through

$$H_0 = R'/R, \qquad q_0 = -RR''/(R')^2$$

both computed at present time t. The metric distance σ_r can be obtained from Equation (V.13)

$$\sigma_r = c \int_t^{t_0} \frac{dt}{R} = \frac{cz}{H_0 R} - \tfrac{1}{2}(1 + q_0) \frac{cz^2}{H_0 R} + \cdots$$

after developing R^{-1} in series and using Equation (V.17) to replace Δt by the redshift z. The luminosity ratio given by Equation (V.16) can be written now

$$L/l = 4\pi \left(\frac{cz}{H_0} \right)^2 [1 + (1 - q_0)z + \cdots]$$

which leads to

$$m_{\text{BOL}} = 5 \log(cz) + 1.086(1 - q_0)z + M_{\text{BOL}} - 5 \log H_0 + \cdots \tag{V.18}$$

after using Pogson's law to introduce the bolometric magnitudes.

The Redshift Magnitude Diagram

Expression (V.18) gives Hubble's law in our cosmic neighborhood. The linear term in z expresses the departures from linearity in the redshift-magnitude plane which are used, at least in principle, to estimate the deceleration parameter q_0 and to find in this way the space curvature of the Universe (Section VI.6.2).

The application of Equation (V.18) however, is not straightforward because the observed magnitudes are not bolometric and suffer two major corrections:

(i) The aparture effect: Because the small distant galaxies may have their luminosities included within the measuring diagram whereas the outer parts of the nearby ones are not measured, all measurements must be corrected to some adopted standard diameter.

(ii) The K-effect: It is composed of two contributions. The redshift causes the fixed photometric window of the photometer to measure different regions of the spectrum in the redshifted galaxies. Moreover, the redshift reduces the number of photons passing through the filter in a factor $1 + z$, thus

$$K(k) = 2.5 \log(1 + z) + 2.5 \log \frac{\int_0^\infty E(\lambda_0) S_k(\lambda) \, d\lambda}{\int_0^\infty F(\lambda_0/1 + z) \, S_k(\lambda) \, d\lambda},$$

where $F(\lambda_0)$ is the intrinsic flux distribution of the galaxy at zero redshift and $S_k(\lambda)$ the filter transmission which defines the photometric window. The sign of $K(k)$ has to be defined so that the transformation from the observed magnitude $m(k)$ to the bolometric scale is

$$m_{\text{BOL}} = m(k) - K(k) + \text{const.}$$

It is clear that to make the corrections tractable, it is convenient to use a standard type of galaxy. The giant ellipticals (gE) dominate rich clusters of galaxies (Section IV.2.3) and the brightest of these objects has been used to trace the Hubble law (Figure V.6). Oke and Sandage (1968) have tabulated K-corrections for several photometric windows using $F(\lambda_0)$ measured in a number of nearby gE's galaxies (Table V.9).

Equation (V.18) now becomes

$$m_k - K(k) - A_k = 5 \log(cz) + 1.086(1 - q_0)z$$
$$+ M_k - 5 \log H_0 + 5, \tag{V.19}$$

where galactic absorption A_k has also been considered.

TABLE V.8

K-corrections*

z	K_a	K_V
0.0	$0\overset{m}{.}00$	$0\overset{m}{.}00$
0.05	0.23	0.10
0.10	0.47	0.19
0.15	0.71	0.27
0.20	0.94	0.42
0.25	1.18	0.59

* After Oke and Sandage (1968).

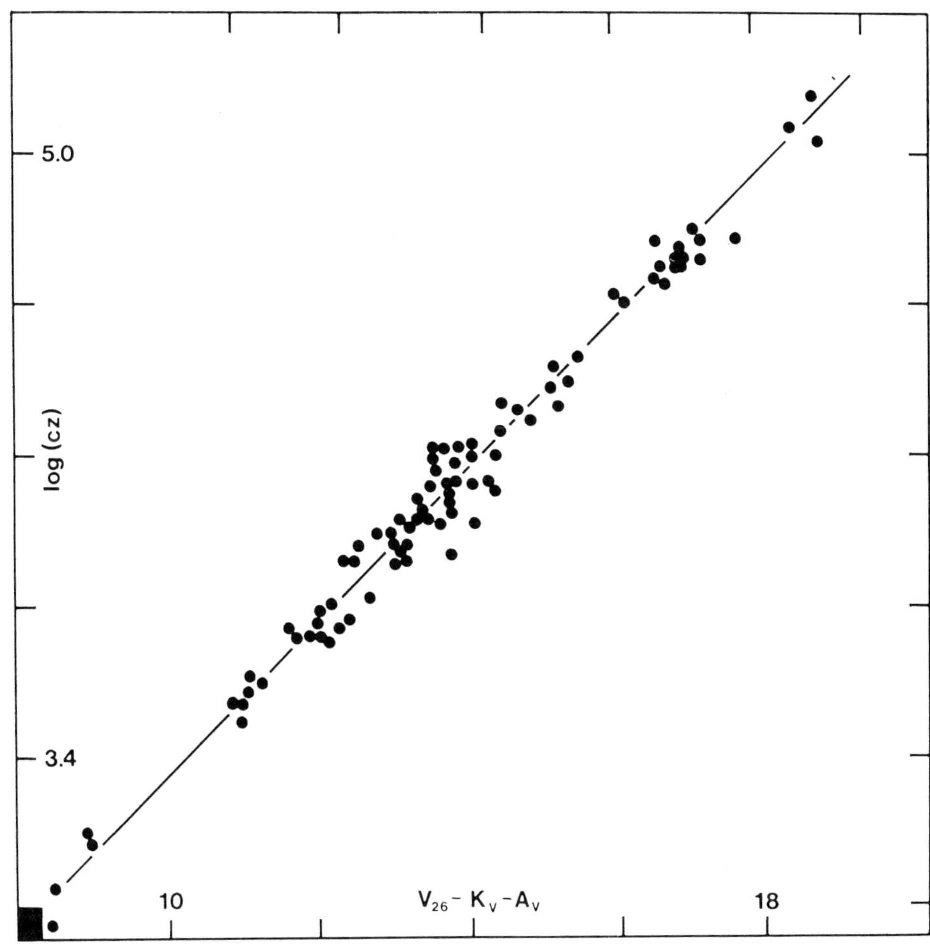

Fig. V.6. The Hubble law for large and distant clusters of galaxies.

The present redshift-magnitude diagram (Figure V.6) is based on 84 galaxy clusters collected from observations made by different sources: 33 by Peterson (1970b), 10 by Westerlund and Wall (1969) and 41 by Sandage. Most of the clusters were selected in a non-random fashion before Abell's catalogue was published. The measured redshifts range from $z = 0.037$ for the VC to $z = 0.461$ for the SRG 3C295, measured by Minkowski (1960). A fundamental limitation arises in obtaining spectral absorption lines beyond apparent magnitude 18: the surface brightness of the galaxies becomes less than that of the night-sky. Only through emission line measurements it becomes possible to measure high redshifts in galaxies (Spinrad, 1976).

The properties of the redshift-magnitude diagram have been discussed by Sandage (1972), who shows the observed slope to be 5 within experimental error. The least-square fit of a line with slope 5 to the points in Figure V.6 gives

$$m_V - K(V) - A_V = 5 \log(cz) - 6^m76 \pm 0^m32$$

The meaning of the observed scatter in the points defining Hubble's law can be attributed either to the redshifts or to the magnitudes. In the first case we find $\Delta \log(cz)$ is a constant at a given m, independently of z, the dispersion being $\sigma(\Delta z/z) = 0.064$. Because there is no physical explanation for that, Sandage is inclined to think that the second alternative is the valid one. In that case the source of the scatter is to be found in the absolute magnitudes of the first ranked cluster galaxies, whose mean value results

$$\langle M_V \rangle_1 = 21^m76 \pm 0^m32 - 5 \log(H/100).$$

The dispersion $\sigma(M_V) = 0^m32$ means that these objects are excellent distance indicators. An analysis of the residuals with respect to cluster type reveals that both $\langle M_V \rangle$ and $\sigma(M_V)$ are independent of the cluster richness, so that the Neyman–Scott effect (1957) is absent. An interpretation of these properties follows naturally from the merging process taking place in the central parts of rich clusters (Section IV.1.1).

The Deceleration Parameter

The first estimate of the deceleration paramter q_0 by Humason *et al.* (1956) was made on the basis of 17 cluster galaxies with $z \leqslant 0.2$. A formal value ($q_0' \approx 1$) can be obtained from the points of the Figure V.6, depending critically on the three last points in the Hubble diagram. There is some doubt, however, regarding the intrinsic luminosities of the farthest galaxies. The look-back time is so large as to be 30 to 50% of the age of the Universe and evolutionary effects on the intrinsic luminosity of the galaxy should be apparent. If $L = L(t)$ we should use

$$L(t_0) = L(t)\left[1 - E\frac{z}{H_0} + \cdots\right] \quad \text{where} \quad E = \left(\frac{d \ln L}{dt}\right)$$

in Equation (V.16) instead of L, to derive Equation (V.18). Consequently the formal value q_0' derived from the Hubble diagram is related to the true value q_0 through

$$q_0' = q_0 - \frac{E}{H_0}$$

Tinsley (1972, 1976) has computed the evolution of the luminosity and color of gE galaxies under the assumption that they resulted from a single burst of stellar formation and that the IMF for $m \leqslant 1.5$ solar units is a Salpeter-type power-law (Section II.3). She found that stellar evolution in E galaxies makes their luminosities grow fainter at a rate $d \ln L/d \ln t \approx -1$ which gives $E < 0$ and consequently $q_0 < q_0'$.

The dynamical evolution of bright galaxies in clusters also influences in q_0' probably in the same sense (Tinsley, 1976). All that can be said is that q_0 probably is close to zero but in no case as large as 1.

References

Ambartsumian, V.A.: 1963, in F.J. Kerr and A.W. Rodgers (eds.), 'The Galaxy and the Magellanic Clouds', *IAU Symp.* **20**, 122.
Arp, H.: 1956, *Astron. J.* **60**, 15.
Arp, H.: 1960, *Astron. J.* **60**, 404.

Arp, H.: 1976, *IAU Colloq.* **37** and CNRS, No. 263 p. 13.

Baade, W.: 1926, *Astr. Nachr.* **228**, 359.

Baade, W.: 1945, *Astrophys. J.* **102**, 309.

Baade, W.: 1952, *Trans. IAU* **8**, 397.

Baade, W. and Swope, H.: 1963, *Astron. J.* **68**, 435.

Bok, B.J. and Bok, P.: 1962, *Monthly. Notices. Roy. Astron. Soc.* **124**, 435.

Bottinelli, L. and Gouguenheim, L.: 1976, *Astron. Astrophys.* **51**, 275.

Bottinelli, L., Gouguenheim, L., Paturel, G., and de Vaucouleurs, G.: 1980, Preprint.

Buscombe, N. and de Vaucouleurs, G.: 1955, *Observatory* **75**, 170.

Carranza, G.J.: 1968, Thesis, Universidad Nacional, Cordoba.

Cohen, M.H., Moffet, A.T., Rommey, J., and Schilizzi, R.T.: 1976, *Nature* **259**, 17.

Corwin, H.G.: 1971, *Publ. Astron. Soc. Pacific* **83**, 320.

de Vaucouleurs, G.: 1958, *Astron. J.* **63**, 253.

de Vaucouleurs, G.: 1976, *IAU Coll.* **37** and CNRS No. 263. p. 71.

de Vaucouleurs, G.: 1977, *Nature* **266**, 126.

de Vaucouleurs, G.: 1978a, *Astrophys. J.* **223**, 351.

de Vaucouleurs, G.: 1978b, *Astrophys. J* **223**, 730.

de Vaucouleurs, G.: 1978c, *Astrophys. J.* **224**, 14.

de Vaucouleurs, G.: 1978d, *Astrophys. J.* **224**, 710.

de Vaucouleurs, G. and Bollinger, B.: 1979a, *Astrophys.* **227**, 380.

de Vaucouleurs, G. and Bollinger, B.: 1979b, *Astrophys. J.* **227**, 729.

de Vaucouleurs, G. and Bollinger, B.: 1979c, *Astrophys. J.* **233**, 433.

de Vaucouleurs, G. and *Malik*, G.M.: 1969, *Monthly Notices Roy Astron. Soc.* **142**, 387.

de Vaucouleurs, G., de Vaucouleurs, A., and Corwin, H.G.: 1976, *Second Reference Catalogue of Bright Galaxies*, Texas Univ. Press.

Gum, C.S. and de Vaucouleurs, G.: 1953, *Observatory* **73**, 152.

Hanson, R.B.: 1975, *Astron. J.* **80**, 379.

Hodge, P.: 1976, *R.G.O. Bull.* **182**, 170.

Holmberg, E.: 1950, Lund. Medd. No. 136.

Holmberg, G.: 1958, Lund. Medd. Ser. II, No. 136.

Hubble, E.: 1934, *Astrophys. J.* **79**, 8.

Hubble, E.: 1936, *Astrophys. J.* **84**, 158.

Hubble, E. and Sandage, A.R.: 1953, *Astrophys. J.* **118**, 353.

Humason, M.L., Mayall, N.U., and Sandage, A.R.: 1956, *Astron. J.* **61**, 97.

Humphreys, R.M.: 1978, *Astrophys. J. Suppl.* **38**, 309.

Humphreys, R.M. and Sandage, A.R.: 1980, *Astrophys. J. Suppl.* **236**, L1.

Kennicutt, R.C.: 1979a, *Astrophys. J.* **228**, 394.

Kennicutt, R.C.: 1979b, *Astrophys. J.* **228**, 696.

Kellermann, K.I. an Shaffer, D.B.: 1977, Coll. Int. CNRS No. 263, p. 346, Paris.

Kirschner, R.P. and Kwan, J.: 1974, *Astrophys. J.* **193**, 27.

Kowal, C.T.: 1968, *Astrophys. J.* **73**, 1021.

Kowal, C.T.: 1969, *Publ. Astron. Soc. Pacific.* **81**, 608.

Kukarkin, B.V.: 1974, *The Globular Star Clusters*, Moscow.

Kron. G.E. and Mayall, N.U.: 1960, *Astron. J.* **65**, 581.

Lundmark, K.: 1926, Uppsala Medd. I, 19b, No. 8.

Lynden-Bell, D.: 1977, *Nature* **270**, 396.

Maza, J. and van den Bergh, S.: 1976, *Astrophys. J.* **204**, 519.

McLaughlin, D.B.: 1945, *Publ. Astron. Soc Pacific.* **57**, 69.

Mineur, H.: 1938, *Ann. Astrophys.* **1**, 97.

Minkowski, R.: 1960, *Astrophys. J.* **132**, 908.

Minkowski, R.: 1966, *Astron. J.* **71**, 371.

Mould, J., Aaronson, M., and Huchra, J.: 1980, *Astrophys. J.* **238**, 458.

Oort, J.H.: 1938, *Bull. Astron. Inst. Neth.* **8**, 233.

Oke, J.B. and Sandage, A.R.: 1968, *Astrophys. J.* **154**, 21.

Peterson, B.A.: 1970a, *Astron. J.* **75**, 695.

Peterson, B.A.: 1970b, *Astrophys. J.* **159**, 333.
Pfau, W.: 1976, *Astron. Astrophys.* **60**, 113.
Rees, M.: 1967, *Monthly Notices. Roy. Astron. Soc.* **35**, 345.
Roberts, M.S.: 1972, in D.S. Evans (ed.), 'External Galaxies and QSOs', *IAU Symp.* **44**, 12.
Rossino, L.: 1964, *Ann. Astrophys.* **27**, 498.
Rubin, V.C., Ford, K.C., and Rubin, J.S.: 1973, *Astrophys. J.* **183**, L111.
Rubin, V.C., Ford, K.V., Thonnard, N., Roberts, M.S., and Graham, J.: 1976, *Astron. J.* **81**, 687.
Sandage, A.R.: 1958, *Astrophys. J.* **127**, 513.
Sandage, A.R.: 1961, in McVittie (ed.), 'Problems of Extragalactic Research', *IAU Symp.* **15**, 359.
Sandage, A.R.: 1964, *Observatory* **84**, 245.
Sandage, A.R.: 1968, *Observatory* **88**, 91.
Sandage, A.R.: 1972, *Quant. J. Roy. Astron. Soc.* **13**, 202.
Sandage, A.R. and Tammann, G.A.: 1968, *Astrophys. J.* **151**, 825.
Sandage, A.R. and Tammann, G.A.: 1974a, *Astrophys. J.* **190**, 525.
Sandage, A.R. and Tammann, G.A.: 1974b, *Astrophys. J.* **191**, 603.
Sandage, A.R. and Tammann, G.A.: 1974c, *Astrophys. J.* **194**, 223.
Sandage, A.R. and Tammann, G.A.: 1974d, *Astrophys. J.* **194**, 559.
Sandage, A.R. and Tammann, G.A.: 1975a, *Astrophys. J.* **196**, 313.
Sandage, A.R. and Tammann, G.A.: 1975b, *Astrophys. J.* **197**, 265.
Sandage, A.R. and Tammann, G.A.: 1975c, *Astrophys. J.* **202**, 563.
Sandage, A.R. and Tammann, G.A.: 1976, *Astrophys. J.* **210**, 7.
Scott, E.L.: 1957, *Astron. J.* **62**, 248.
Schmidt-Kaler, T.: 1957, *Z. Astrophys.* **41**, 182.
Schurmann, S.R., Arnett, W.D., and Falk, S.W.: 1979, *Astrophys. J.* **230**, 11.
Shane, C.D. and Wirtanen, C.A.: 1967, *Publ. Lick Obs.* **22**, 1.
Sérsic, J.L.: 1958, *Observatory* **78**, 24.
Sérsic, J.L.: 1959, *Observatory* **79**, 54.
Sérsic, J.L.: 1960, *Z. Astrophys.* **50**, 168.
Sérsic, J.L.: 1978, in A. Gutierrez and H. Moreno (eds.), Vol. III, Dept. Astronomia, Univ. Chile, p. 31.
Sérsic, J.L.: 1980, *Ciencia y Cultura* **32**(1), 13.
Seyfert, C.K.: 1940, *Astrophys. J.* **91**, 528.
Smith, M.: 1977, Colloq. No. 263 CNRS, p. 75, Paris.
Spinrad, H.: 1976, *Publ. Astron. Soc. Pacific* **88**, 565.
Stebbins, J.: 1950, *Monthly Notices Roy. Astron. Soc.* **110**, 416.
Tammann, G.A.: 1970, in W. Becker and G. Contopoulos (eds.), 'The Spiral Structure of our Galaxy', *IAU Symp.* **38**, 236.
Tammann, G.A.: 1977, *Ann. N.Y. Acad. Sci.* **302**, 61.
Tinsley, B.M.: 1972, *Astrophys. J.* **173**, L93.
Tinsley, B.M.: 1976, Colloq. CNRS No. 263, p. 223, Paris.
Tully, R.B. and Fisher, J.R.: 1977, *Astron. Astrophys.* **54**, 661.
Tully, R.B. and Fisher, J.R.: 1978, in M.S. Longair and J. Einasto (eds.), 'The Large Scale Structure of the Universe', *IAU Symp.* **79**, 214.
van den Bergh, S.: 1960, *J. Roy. Astron. Soc. Can.* **54**, 49.
van den Bergh, S.: 1967, *Astron. J.* **72**, 70.
van den Bergh, S.: 1970, *Nature* **225**, 503.
van den Bergh, S.: 1975a, *Stars and Stellar Systems*. in A.R. Sandage, M. Sandage, and J. Kristian (eds.), Chapter 12, Vol. IX, Chicago Univ. Press.
van den Bergh, S.: 1975b, Colloq. CNRS No. 263, p. 13.
van den Bergh, S.: 1975c, in A.R. Sandage, M. Sandage, and J. Kristian (eds.), *Galaxies and the Universe*, p. 509.
van den Bergh, S.: 1977, 'Redshift and Expansion of the Universe', *IAU Symp.* **37**, 73.
van den Bergh, S.: 1980, *Astrophys. J.* **235**, 1.
van Herk, G.: 1965, *Bull. Astron. Inst. Neth.* **18**, 71.
Vigier, P. and Pecker, J.: 1976, Colloq. CNRS No. 263, p. 477.
Visvanathan, N. and Sandage, A.R.: 1977, *Astrophys. J.* **216**, 214.

Wagoner, R.V.: 1979, *Comm. Astrophys.* **8**, 121.
Westerlund, B.E. and Wall, J.V.: 1969, *Astron. J.* **74**, 335.
Wray, J.: 1978, in R.M. West and J.L. Heudier (eds.), *Modern Technique in Astronomical Photography.*
Wray, J.W. and de Vaucouleurs, G.: 1980, *Astron. J.* **85**, 1.
Whitford, A.E.: 1958, *Astrophys. J.* **63**, 201.

CHAPTER VI

COSMOLOGY

The impulse of observational and theoretical research in the field of cosmology has grown very rapidly in the last decade and has led to very fruitful progress on a wide front. Observational results of optical and radio astronomy have shown unequivocally that cosmological theories when related to observations can, according to the classical scientific procedure, also be submitted in their turn to decisive test.

The two theories most seriously considered in recent years are Steady State Theory and Relativistic Cosmology. The bulk of observational evidence does not now favor the Steady State Theory in its original and simple form. Measurements of redshifts and apparent magnitudes which lead to Hubble's law, the radio sources counts, background-radiation and the He and D abundances are all observables favoring a Universe which was denser and hotter in the past and whose expansion is at present slowing down. Besides, most recent observations tend to demonstrate that the intergalactic space is extremely empty and its density is much inferior to that of the luminous matter.

Up to now it has not been possible to construct a model of the Universe that lacks a singular state at its origin that is based on the original equations of General Relativity. There is much discussion for and against the necessity of a singularity in the cosmological equations. After all, the singularity may be considered as a unique characteristic of the real world. The practical confirmation of the cosmologic character of quasars redshift would imply that these features play an essential role in the structure and evolution of the Universe.

This chapter, however, has not been written as an account of Cosmology itself: there are many other theories and models other than those described here, as well as several excellent books on the subject in the literature. We intend, however, to give a description of the scenario where the extragalactic objects form and evolve.

Since the observations of the evolutionary processes in galaxies suggest that most of them were formed a long time ago, we need a theoretical frame to reconstruct the conditions prevailing in the Universe at that time. Reciprocally, as we shall see later in this chapter, it is through observations in our cosmic neighborhood that some progress has been made in the selection of a 'best fit' model for the Universe.

VI.1. Basic Assumptions

The formulation of relativistic models of the Universe as a hot primeval explosion is actually founded only on working hypotheses which up to now have given good results. These scientific postulations concerning the general nature of the Universe are few and difficult to establish. Therefore some authors call them 'principles'. In this section we describe and discuss them, following on an outline given by Ehlers (1976).

(i) The geometry of space-time and gravitation is described by Einstein's General Relativity Theory (GR). The remaining laws of the Physics are locally applied.

The use of the GR theory is justified as long as no simpler one agreeing with the observations is known.

By the laws of physics we mean the structure and interpretation of the basic equations of physics as established in our cosmic neighborhood. Since one of the leading ideas in Cosmology is the belief that we do not occupy a privileged location in space-time, the local validity of the other physical laws is equivalent to a prohibition to use laws or postulates which can be proved only at cosmological scales e.g. the continuous creation of matter, as in the steady-state theory.

This point of view does not guarantee that the laws of physics are everywhere the same; therefore it is necessary to appeal to the observations before its adoption.

(ii) For all times and space in the Universe there is a mean state of motion whose observable properties are independent of direction.

We must disregard all irregularities such as planets, stars, galaxies and clusters of galaxies in order to obtain the general characteristics of the Universe. These properties are those in volumes with a linear size of the order of 100 Mpc.

Assumption (ii) is justified by the great simplicity in its application to theory and is supported empirically by the Hubble law (Section V.3.2) and the high degree of isotropy in the background radiation (Section VI.5).

From (i) and (ii) the existence of a *cosmic time t* is deduced which agrees with the proper time (atomic time) of the so-called *fundamental observers* which participate in the mean motion mentioned in (ii).

The points in space-time (events) with a fixed value of t are simultaneous for the fundamental observers in the sense of GR. These produce a constant curvature of the space k/R^2 where the index k is equal to $+1$, 0, or -1 and the factor of scale (also known as radius of curvature) $R(t)$ is always a positive function of time. The spatial homogeneity follows at once from the assumption of isotropy (ii). The distance between any two fundamental observers, or 'typical' galaxies, is then proportional to $R(t)$.

The case $k = 0$ is the critical one that divides the closed spherical ($k = +1$) models collapsing in the future from the open ($k = -1$) hyperbolic models expanding forever. From the field equations of GR and (i), (ii) it is also deduced that cosmic matter has a constant mass density $\rho(t)$ in space, and also an isotropic pressure p, which are related to the radius of curvature R by means of the so-called Friedman equation

$$(R')^2 = \frac{8\pi G}{3} R^2 - kc^2 + \tfrac{1}{2} \Lambda c^2 R^2 \tag{VI.1}$$

and also through the local energy balance equation

$$(\rho c^2 R^3)' + P(R^3)' = 0 \tag{VI.2}$$

where the symbol ($'$) means time derivative, G is the newtonian constant of gravitation, and c is the velocity of light. The 'cosmological constant' Λ has to be set equal to zero because in (i) we have assumed GR to be a self-consistent theory. McCrea (1971) has given arguments in favor of this point of view based on the fact that GR alone cannot assign any meaning to Λ.

(iii)   There is a time t_0 for which the state of mean motion postulated in (i) is an *expansion*. That is

$$R'_0 = R'(t_0) > 0 \quad \text{when} \quad t_0 = \text{now.} \tag{VI.3}$$

In spite of some doubts about the interpretation of the redshift (Section III.8), at least a great part of it is of cosmological origin and seems to be well founded empirically on the observation of the Hubble flow (Chapter V).

(iv)   There is a lapse in the evolution of the Universe which is dominated by matter, when the total pressure (of matter and radiation) is negligible compared with the total density of energy.

This assumption is based on the fact that all values deduced from the observations together with theoretical estimates give at most a nominal contribution to the local energy density coming from the peculiar motions of galaxies, photons, etc. as we have seen in Chapter IV.

Since $p \ll \rho c^2$ is valid in the present (t_0), in a recent past should have been

$$\rho c^2 + 3p \geqslant 0. \tag{VI.4}$$

From Equations (VI.1), (VI.2), and (VI.4) we deduce that $R'' \leqslant 0$. From Equation (VI.3) we also obtain that when we go back into the past from t_0, the value of $R(t)$ decreases monotonically when Equation (VI.4) is valid. Equation (VI.4) is true at least up to nuclear densities, for in the past matter should have been in a much denser and hotter state than at present. If we accept the validity of Equations (VI.1), (VI.2) and (VI.4), as is usually done, we deduce the existence of an initial singularity, i.e. $R = 0$ and $\rho \to \infty$ in a finite time. This result, which for many people is shocking, has been demonstrated very generally, and is known as the Penrose-Hawking (Hawking and Penrose, 1969) theorem on the singularity.

It is quite difficult to violate the condition $\rho c^2 + 3p \geqslant 0$. Several authors have tried to construct models of the Universe at early stages with unusual quantum conditions to see if either ρ and p, or both, could become negative. Parker was able to demonstrate that in a very improbable coherent situation (laser type) we can have $\rho c^2 + 3p < 0$ and thus avoid the singularity. The existence of an initial coherent state in which all particles occupy a definite state is obtained in a routine way in laboratories working with masers and lasers. However, this is a highly artificial situation which requires a special arrangement. If it is true that nature prefers simplicity, the configuration set up by Parker is highly improbable.

According to Canuto (1978), Parker's results are only an indication of how difficult it is to violate the condition $\rho c^2 + 3p \geqslant 0$. Datta, Kallman and this author (based on an idea of Zel'dovich) have attempted to compute p and ρ separately for an extreme equation of state which includes all mesons and the mesonic field. When the densities are a hundred to a thousand times higher than nuclear density, the nucleons begin to feel the effect of the heavier mesons and, depending on the value one adopts for the coupling constant of the mesonic field, negative pressures can be obtained. Nevertheless, the value of $\rho c^2 + 3p$ *remains positive* even under these conditions, since the pressure is at least a factor of ten smaller in absolute value than the density. Since this approach is the best we can do to obtain negative pressures by means of the realistic use of forces between particles, the validity of the condition $\rho c^2 + 3p \geqslant 0$ is considerably strengthened even in these very early stages. Consequently, at the level of the GR equations, the cosmological singularity and all its implications seem inevitable.

(v) In the course of evolution of the Universe a (hot) state existed in which the matter was in thermodynamic and chemical equilibrium at a temperature of some 10^{11} K $\simeq$ 10 MeV.

This assumption is, according to Ehlers, an excellent constitutive hypothesis to describe the remote past with high temperatures, because of its simplicity. It is the basis of the calculations of the abundances of He and D of cosmical origin (cf. Section VI.4). The state of equilibrium with temperature $T = 10^{11}$ K is called the 'primeval state' without, however, implying any doubt about the existence of previous states which cannot yet be described.

By means of the postulates (i) to (v) introduced above, the *cosmological model* is determined in principle, including even some parameters which describe the primeval state, such as the chemical potentials for the elementary particles prevailing at these extreme conditions.

These initial data are, in principle, sufficient to compute the evolution of the model in t through Equation (VI.1) and the equation which describe the reactions between particles (provided the cross sections are known).

VI.2. Explosive Cosmologies

As we have established in Section VI.1 we shall limit ourselves to models based on the GR theory with zero cosmological constant ($\Lambda = 0$). All these models of the Universe which are spatially homogeneous and isotropic have a physical singularity in a finite past time. They are called the Robertson–Walker explosive models (cf. Chapter V). To describe the global properties of these models we follow a discussion given by Ellis (1976) for a different purpose.

Let us consider the representation of the curved space-time which is obtained when two spatial dimensions are suppressed. The resulting diagram has then only one spatial dimension which describes the whole history of the Universe (Figure VI.1). The curvature of the space-time causes the spatial sections to be represented by curved lines, and at every one of their points the local direction of time is normal to the spatial section. The application of postulates (i) and (ii) to the diagram leads to homogeneous and isotropic models, such as the one illustrated. The surfaces of constant cosmic time are then surfaces of spatial homogeneity, that is, all physical properties are the same for each event on these surfaces. The fundamental observers are represented by world-lines flowing normally to the spatial sections (Hubble-flow). The separations between world-lines at any given cosmic time represent the distances

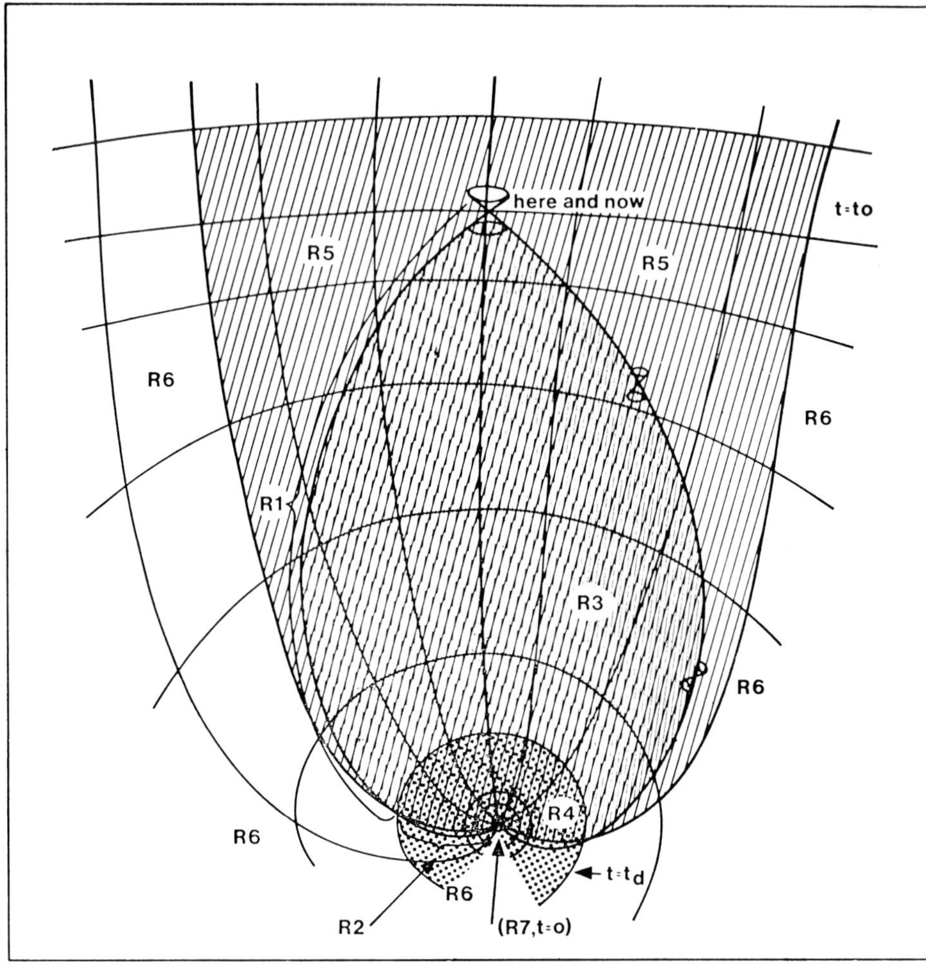

Fig. VI.1.   Ellis diagram (see Section VI.2).

between the objects associated to them. Any object (e.g. a galaxy) has a peculiar velocity with respect to the mean motion defined by the fundamental observers, and its world-line is not necessarily coincident with one of them. The difference is, however, small because the peculiar motions are much smaller than the speed of light.

The present cosmic time t_0 is represented by the surface $t = t_0$ and postulate (iii) dictates the behavior of the world-lines in its neighborhood: larger separations at later times ($t > t_0$) and closer together for times t earlier than t_0.

Hawking and Ellis (1968) have demonstrated that the energy density contained in the background radiation is enough to satisfy the Penrose–Hawking theorem, so that the world-lines converge in the past towards a compact region or singularity. Since that happens in a finite time in the past, it is useful to select that instant as the origin of cosmic time, i.e. $t = 0$. From a mathematical standpoint we are confronted with a singularity in the model ($R = 0$); from the physical point of view, the convergence of the world-lines towards the past leads to a condition of infinite energy density at $t = 0$. At and before the singular event ($t \leqslant 0$) the local physical laws are no longer applicable which prevents us from making predictions for earlier times. In this way our model is finite towards the past and t_0 may be called 'age of the (model) Universe'.

The regions of space-time which are or were accesible to observation lie on and inside our past light-cone. That cone represents parts of the Universe from which we can receive direct signals in the form of waves propagating with the speed of light (e.g. electromagnetic or gravitational radiation). It is also the frontier of that part of space-time from which we have not been able to receive any signal due to the finite speed of light. This way, all that could have influenced us in the past and is influencing us now lies inside or on the surface of our past light-cone.

The rising temperature of matter and radiation due to compression when we go towards the past, causes another important phenomenon: When radiation becomes energetic enough, it ionizes matter and the content of the Universe becomes a dense plasma opaque to electromagnetic radiation. This means that there is a cosmic time t_d ($0 < t_d < t_0$) when the adiabatic cooling of the expanding plasma lowers the common temperature of matter and radiation which become decoupled (cf. Section VI.5).

According to Ellis we may identify seven regions in this idealized model of the Universe (Figure VI.1), which have different observational status. Region R_1 is the part of our past light-cone since decoupling of the primeval plasma ($t_d < t < t_0$). To R_1 belongs the set of events from which we can receive information by electromagnetic means, particularly light or radio waves or other forms of radiation like gravitational waves. This is the maximum extension of the Universe we can really hope to see. Region R_2 is the part of our past light-cone previous to decoupling ($0 < t < t_d$). Although we cannot receive electromagnetic signals because the plasma absorbs or scatters the photons going through it, we would be able, in principle, to 'see' events that generate gravitational or neutrinos waves if we had a suitable means to detect them.

Region R_3 is the interior of our past light-cone since decoupling at t_d. It is a region we cannot observe by any means and which corresponds to the absolute past in the Minkowskian space-time. Nevertheless, we have enough information available, according to Ellis, either by direct observation of the light-cone or by the past history of our world-line through geological data as to have a general idea of the conditions prevailing in it.

In principle the same can be said of region R_4, i.e., the interior of our past light-cone prior to decoupling ($t < t_d$) but, in practice, we cannot reconstruct a precise image of what happens there because of the 'damping' produced by the plasma. It is well known that different initial conditions can lead to very similar final states; reciprocally the observation of the final state gives us little or no information about the initial conditions. The enormous entropy per nucleon produced in these early stages (cf. Section VI.5) is responsible for this 'forgetting' of the initial conditions.

The R_5 and R_6 regions are parts of the Universe with which we have not had any causal communication. The difference between these two regions is that the events contained in R_5 could possibly be observed in our future while we will never receive any signal from the events occurring in R_6. Through extrapolation from R_3 we may make predictions at some extent in the local future, but we do not have, on the other hand, any information on the matter in R_6 at any stage of its history, therefore we cannot say anything at all about it. Nevertheless, it is possible, in principle, to detect its presence by its effects on the structure of space, although it is not feasible to establish the detailed distribution of matter in R_6.

When we follow our history back in time, we obtain a shrinking light-cone that contains smaller and smaller amounts of matter causally connected with us. This shrinking horizon poses the problem of how we should expect spatial homogeneity in early stages of the Universe when there was no interaction mechanism to produce it, since the different regions were not causally connected. In fact, effective communication ends at time t_d, as all information from events inside R_2 is damped by the plasma pervading the Universe for $t < t_d$. When $t = 0$ we are confronted with the breakdown of the local physical laws at the singularity (R_7 in Figure VI.1). This phenomenon forbids any insight on processes there or before the singular event R_7. We then have to resign ourselves to think that the model of Universe begins at this event.

Although postulates (i) to (v) have allowed cosmologists to built models of the Universe which describe its global properties in reasonable agreement with those observed in R_1 and compatible with geologic evidence found in our world-line inside R_4, the doubt has been raised as to whether the predictions of the models in these far away regions like R_5 which are infinitely remote in time, are not 'artifacts' of the system of postulated (i) to (v). This argumentation (due to Ellis) points to the concrete predictions made by the models for these regions for which there is no possibility of observational validation. However, the importance of this model lies – according to Weinberg (1972) – "not in its truth, but in the common meeting ground for an enormous variety of observational data".

VI.3. Thermal Evolution of the Universe

Let us assume there is a black-body radiation pervading the Universe for $t_e > t_d$, with a spectrum

$$B_\nu = \frac{2h\nu^2}{c^2}\left[\exp\left(\frac{h\,\nu}{kT}\right) - 1\right]^{-1}.$$

If for convenience's sake we take $R(t_e) = 1$, the redshifted frequency for an observer at t_0 is $\nu_0 = \nu/R(t_0)$. As we assume t_e, $t_0 > t_d$, there is no interaction between matter and radiation, the number of protons will be conserved and the relation

$$\frac{B_{\nu_0} R^3(t_0)\,\mathrm{d}\nu_0}{h\,\nu_0} = \frac{B_\nu\,\mathrm{d}\nu}{h\nu}$$

will be applied. Hence,

$$B_{\nu_0} = \frac{\nu_0\,\mathrm{d}\nu}{R^3\nu\mathrm{d}\nu_0}\,B_\nu = \frac{2h\nu^3}{c^2}\left[\exp\left(\frac{h\nu_1 R}{kT}\right) - 1\right]^{-1}$$

and the *observed* spectrum behaves as a Planck function with temperature.

$$T_0 = T/R.$$

The planckian character of the radiation is conserved provided there is no interaction with matter, but the spectrum is shifted towards longer wavelengths by a factor R.

Introducing now the redshift $1 + z$ (Section V.1) we obtain

$$T/T_0 = R/R_0 = 1 + z$$

which shows how the temperature of black-body radiation is related to the redshift. Since $T_0 = 2.8$ K at the present time $t = t_0$ (cf. Section VI.4) for values of $1 + z$ of the order of $10^3 (R = 10^{-3})$, the value of T is of the order of 2800 K. Since a Planck distribution of this temperature has enough quanta with energies to photoionize H, matter and radiation become coupled (region R_4 in Figure VI.1) and the Compton effect redistributes the photons efficiently on all frequencies, so that the Planck distribution is maintained. Let us now see whether the temperature maintains the same dependence on the redshift $1 + z$.

With $R^{-1} = 1 + z$, Equation (VI.2) can be applied to the medium formed by matter with a particle density N and a radiation field with a density aT^4/c^2, so strongly coupled that they share the same equilibrium temperature T. Let $\rho c^2 = (\frac{3}{2})kNT + aT^4$ be the total energy density of the mixture and $p = kNT + (\frac{1}{3})aT^4$ the pressure. If we recall that the conservation of the number of barions requires $d(R^3N) = 0$, Equation (VI.2) leads to the following relation

$$\frac{d \ln T}{d \ln R^3} = - (\Gamma - 1).$$

This is an adiabatic process, where the adiabatic coefficient of the mixture is given by

$$\Gamma = \frac{4}{3} + \frac{1}{3 + B(aT^3/kN)} ,$$

and $4aT^3/3kN$ is the ratio of the specific heats between the two components of the mixture. If this ratio is very large, the Universe behaves like a gas of radiation ($\Gamma = \frac{4}{3}$) in adiabatic expansion, and then we have in good approximation $T = T_0(1 + z) = T_0R^{-1}$ even if matter and radiation were coupled. Since $N = N_0R^{-3} = N_0(1 + z)^3$, it is easy to see that $4aT^3/3kN = 4aT_0^3/3kN_0 \approx 10^8$ keeps its value until processes taking place at $T \approx 10^{10}$ K inject energy into the medium and break up the adiabatic condition.

The conservation of nucleons implies that the matter density ρ_M is proportional to R^{-3}. On the other hand, the radiation density $\rho_R = aT^4/c^2$, varies like R^{-4}. Thus $\rho_M = \rho_R$ at some time in the past, since $\rho_R \ll \rho_M$ at present. Given that $\rho_R = 4 \times 10^{-34}$ cm^{-3} and taking $\rho_M \approx 5 \times 10^{-31}$ g cm^{-3}, we find that $\rho_M = \rho_R$ when $R = 8 \times 10^{-4}$ that is to say for $1 + z = 1250$. At that time $T = 3.5 \times 10^3$ and $\rho_M = \rho_R = 10^{-21}$ g cm^{-3} (Figure VI.2).

Figure VI.2 describes the thermal history of the Universe. The decoupling of matter from radiation takes place at $t_d = 2 \times 10^5$ yr, if we assume a Universe with zero curvature ($k = 0$), while the transition from a radiation-dominated to a matter-dominated Universe occurred a short time before, in $t = 10^5$ yr ($T \approx 4500$ K). The temperature $T = 10^9$ K in which electron pairs are created is reached only 100 s after the singularity.

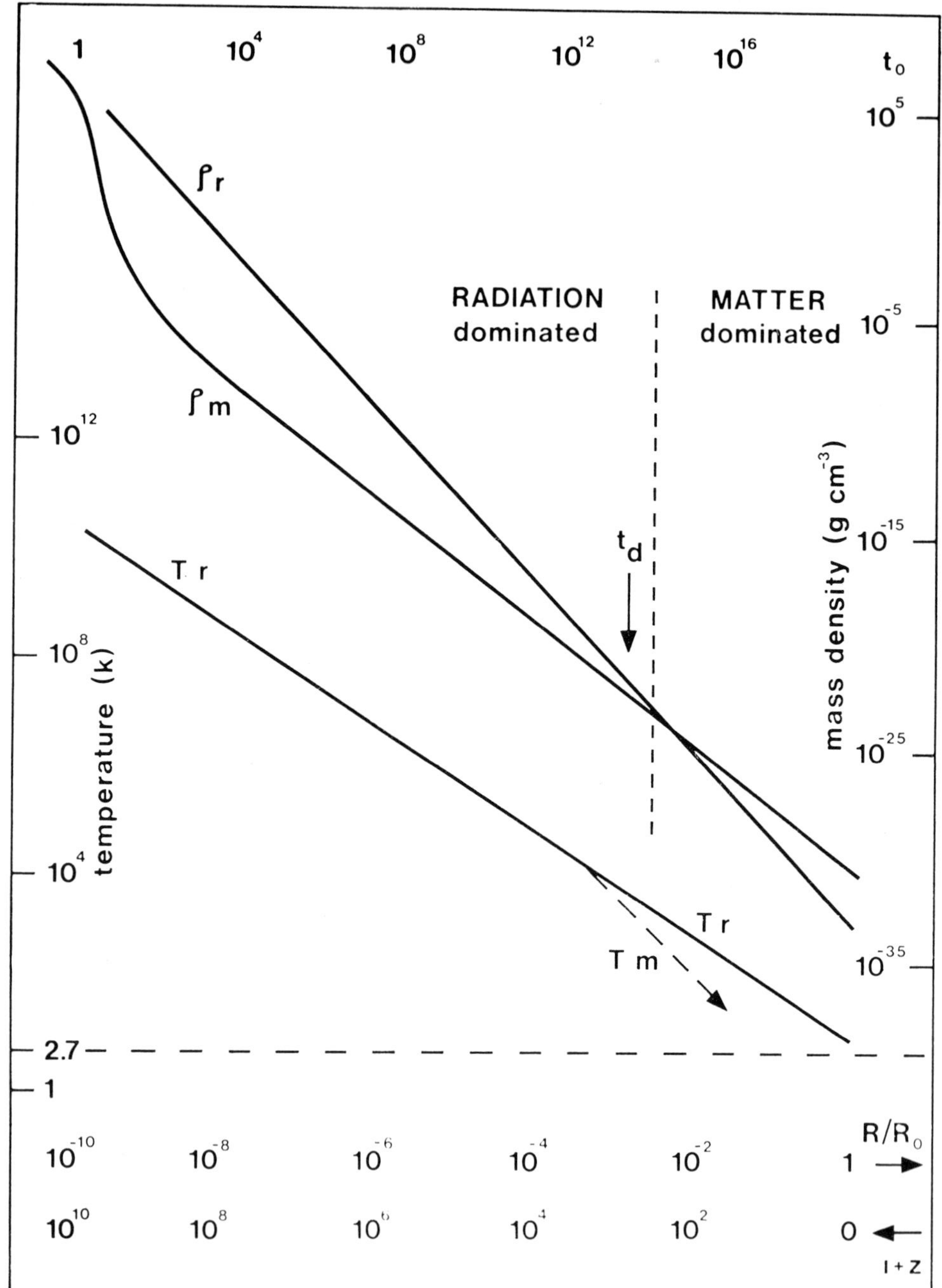

Fig. VI.2.   Thermal history of the Universe.

After decoupling, matter and radiation follow separate adiabatics. We have $T_M \sim \rho_M^{2/3} \sim R^{-2}$ for matter. If other processes had not intervened, the present value of T_M would be very low, in the order of 0.003 K but the formation of galaxies and related processes such as cosmic rays, ultraviolet emission, non-thermal radiation, etc. heated the gas again, probably to above 10^4 K (cf. Section VII.3).

VI.4. Primeval Nucleosynthesis

When $t = 100$ s and $T = 10^9$ K ($\approx 10^5$ eV), the photons of the light energy wing of the Planck distribution have energies larger than 1 MeV and can initiate nuclear reactions. At still earlier times, temperatures were higher and space was filled with e^+, e^- pairs. Still earlier the neutrinos took part in the thermal equilibrium through the reaction

$$\mu^{\pm} \leftrightarrow e^{\pm} + \nu_e + \nu_\mu.$$

At this stage the number of neutrinos, electrons and photons was nearly the same, about 10^8 times the number of barions. Under such conditions, reactions such as

$$n + \nu_e \leftrightarrow p + e^- \quad \text{and} \quad n + e^+ \leftrightarrow p + \nu_e$$

brought the abundance of neutrons to the equilibrium value. From these neutrons and the protons, stable nuclei were formed. Alpher and Herman (1948) thought they found a mechanism which would enable them to explain the synthesis of all elements of the periodic table in the first hundred seconds of expansion, a time scale which is also of the same order of the collision probability between the particles. When T decreases, thermal equilibrium requires that the ratio n/p drop. This fraction remains, however, somewhat above the instantaneous equilibrium value due to the fact that the neutron mean life is comparable to the age of the Universe (t) in the critical period. Therefore the abundance computations are very sensitive to the value of the mean-life of the neutrons. Nevertheless, nothing prevents the synthesis of the lighter elements. Gamow (1953) proposed that if there were free neutrons, D could be synthetized by means of the reaction

$$p + n \leftrightarrow D + \gamma.$$

The great advantage of this reaction is that it takes place only through strong interactions and is consequently faster than the reaction occurring nowadays in the interior of stars. The cross section is $\sigma = 10^{-29}$ cm^2 so that the reaction time is $\tau = (Nv\sigma)^{-1}$ and the abundance of D only becomes noticeable when T has dropped to values of the order of 10^9 K, when the number of protons capable of photo-dissociation is drastically reduced. The resulting D is then reprocessed and transformed into ^{3}He and T by means of reactions with protons and neutrons. Finally, these nuclei react once more to produce ^{4}He, the most accessible stable nucleus. The absence of stable nuclei with atomic number $A = 5$ to 8 proved to be an insurmountable difficulty for the theory.

In this way the Universe emerges from the era of primeval nuclear reactions with a composition of H, ^{4}He, and small quantities of D and ^{3}D. Traces of ^{7}Be and ^{7}Li are also formed. Figure VI.3 shows the result of computations of density, assuming that the contemporaneous density ρ_M can be written $\rho_M = \rho_0(T/T_0)^3$, where ρ_0 is the

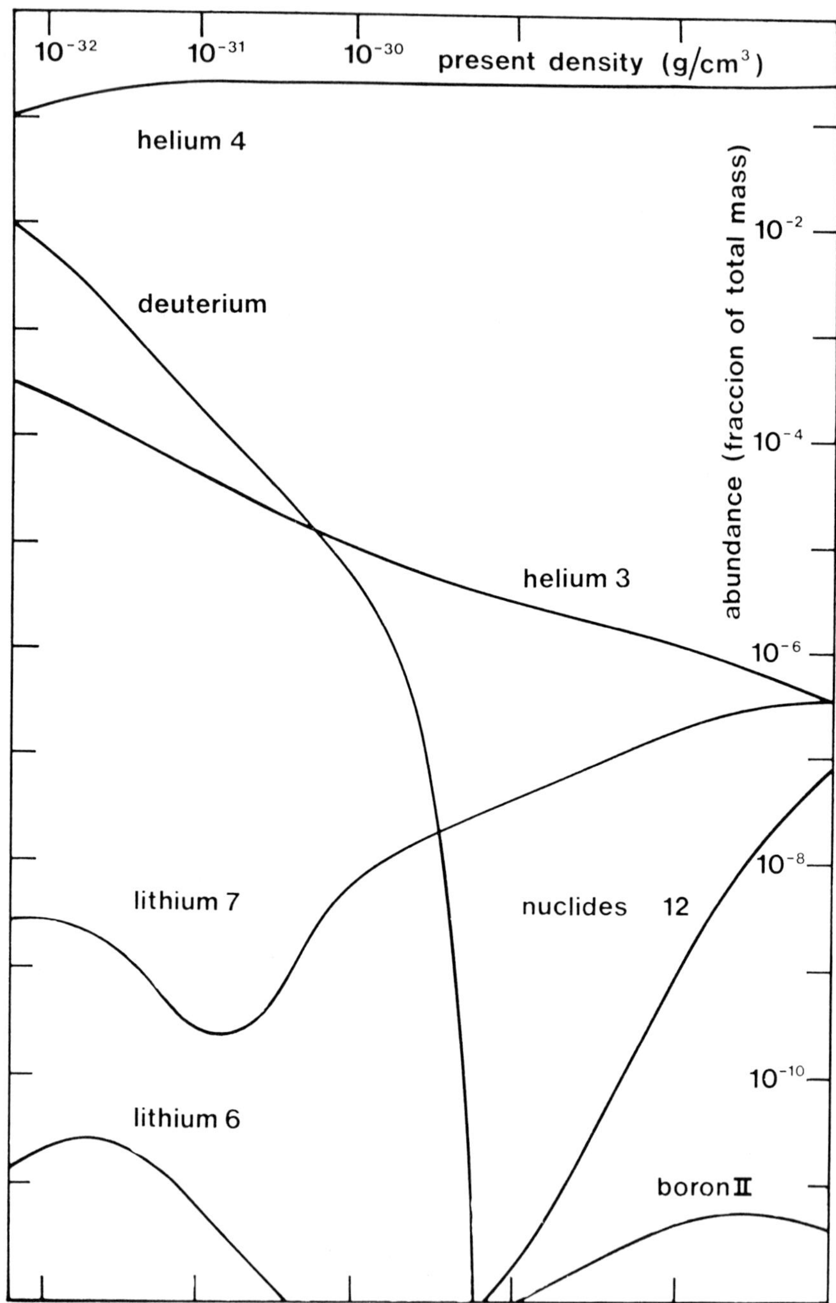

Fig. VI.3.   Element synthesis in the early Universe.

present density of matter. The abundances of the heavier elements increase with the parameter ρ_0 because at the time the reactions took place (when $T \approx 10^9$ K), their number increased with ρ_M.

Since D is an intermediate product, its abundance depends on the contemporary density (ρ_M). Because of this, the observational determination of the cosmic abundance of D is an empirical means to determine the actual density of matter in the Universe (Section VI.2). This is obviously based on the hypothesis that the observed abundance X(D) is of strictly cosmological origin. Otherwise one would have to ask where else in space D could be formed or destroyed. To this purpose Hoyle and Fowler (1973), and later Wagoner and Schram (1977) examined the nuclear reactions which would be able to produce or destroy D From the required energies and densities, they deduced possible scenarios where such processes could take place. Since the observed abundance is much higher than the one produced in ordinary stars, its origin should be sought only in violent events such as supernovae explosions or the origin of the Universe itself. A comparison with the abundances of the other light elements, however, seems to rule out shock waves in supernovae which leaves the primeval synthesis as the most probable origin of most of the existing D.

VI.5. The Background Radiation

At the early stages of evolution in models of the Universe described by Friedman Equation (VI.1), when $z \approx \Omega_0^{-1}$ (cf. Section VI.6), we have

$$H^2 = 8\pi G\rho/3, \qquad \rho = \rho_M + aT^4/c^2$$

but also

$$aT^4 \gg \rho_M c^2$$

so that with $H = R'/R = -T'/T$ we can set up a simple equation which allows computation of the age of the Universe t in the radiation dominated era as a function of the temperature T:

$$t = [(3/32)\pi Ga]^{1/2}(c/T^2).$$

Alpher and Herman (1948), and later Gamow (1953), estimated the nucleon density N on the hypothesis that the age of the Universe t at the time of primeval nucleosynthesis should not be shorter than the mean life τ of neutrons. We have already seen (Section VI.4) that $\tau = (N\sigma v)^{-1}$, where σ is the cross-section for neutron captures and v its velocity ($v \approx c$ at these temperatures). This condition can be written

$$N \approx (32\pi Ga/3\sigma^2)^{1/2} \, (T/c)^2.$$

But the present density of matter is $N_0 = N(T_0/T)^3$, so that we have

$$N_0/T_0^3 \approx (32\pi Ga/3\sigma^2)^{1/2}/c^2T^3.$$

Gamow as well as Alpher and Herman used this relationship to predict the radiation temperature T_0 at present in the Universe, adopting for T the value 10^9 K prevailing at the end of the primeval nucleosynthesis. With an estimated value of the present density $N_0 \approx 10^{-7}$ cm^{-3} they deduced values of T_0 between 10 and 5 K.

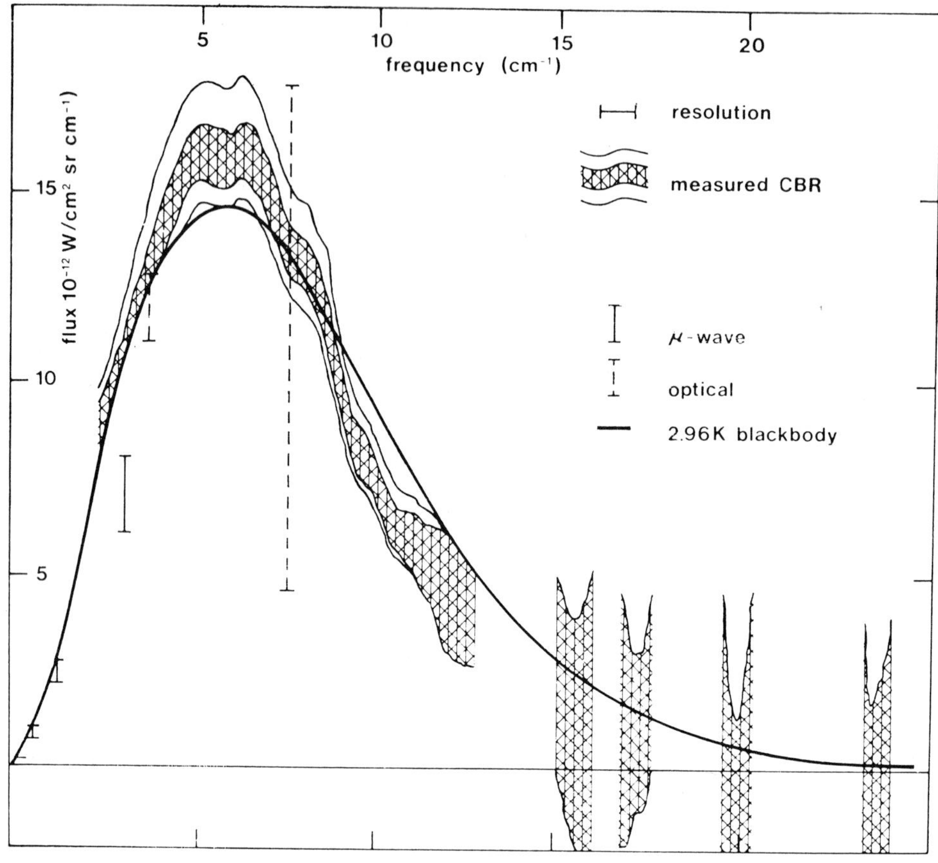

Fig. VI.4.   Spectrum of the background radiation.

Penzias and Wilson (1965) detected the existence of the radiation first predicted by Alpher, Herman, and Gamow. Its existence is taken as a proof of a very dense and hot Universe in the past. Up to now the interpretation of that radiation as a residue of a period of thermal equilibrium between matter and radiation is the simplest.

The planckian character of the microwave spectrum and an extreme isotropy is essential to validate the explosive model.

Most recent observations of the background radiation are due to Woody and Richards (1979), who carried them out with a radiometer taken in the gondola of a stratospheric ballon. Comparison of the observed spectrum (Figure VI.4) with a planckian curve corresponding to the same total intensity leads to a temperature $T_0 = 2.79$ K. Both the observed spectrum and the planckian curve are very similar in a long range of wavelengths, but the observations show some small departures. These could be explained, according to some authors, if the quanta of radiation had spin. Postulating the conservation of the angular momentum of the background radiation, the theory predicts a spectrum very similar to the one found.

The isotropy of the background radiation was shown to be within 0.1 % between any two directions of space. This conclusion is of great interest as it agrees with the predictions of the explosive models of Universe.

At any rate, it is also important to see whether small differences in brightness exist, since these should be expected as a consequence of the peculiar motion of the earth with respect to the primeval radiation background. The first clear detection of this effect was made by Smoot *et al.* (1977). Later Cheng *et al.* (1979) at Princeton confirmed these results, finding a sine-variation with amplitude in the order of 0.1 % of the observed flux. If an observer moves with respect to the radiation background, the measured temperature will be proportional to the redshift originated by the velocity of the observer (cf. Sections VI.3). When observations were corrected for the motion of the Earth around the Sun, it was found that it moves towards a point in the sky in $L = 287°$ and $B = 61°$ at a velocity of 332 ± 36 km s^{-1}. After applying corrections for the motion of the Sun in the Galaxy and of the Galaxy in the Local Group, the resulting velocity of the baricenter of the Local Group with respect to the primeval background is of the order of 540 km s^{-1} towards a point $L = 280°$ and $B = +30°$, near β Hydrae. This direction does not agree with the results of motion in relation to the galaxies of our cosmical neighborhood (Section IV.2.4) and the significance of this discrepancy is not yet clear.

VI.6. Observational Cosmology

As we have seen in the foregoing chapters, the empirical information about the Universe is obtained basically through electromagnetic radiation. This means that the validation of cosmological models depends almost only on the data originated on events located on our retarded light-cone (region R_1 in Ellis' diagram of Figure VI.1). We are thus unable to observe either the present state (at $t = t_0$) or the evolution of distant isolated object. Cosmological information presents itself as an ordered succession of different evolutive stages when we look back into our past light-cone. The selection of a cosmological model based on observations is thus possible if we know the process of formation and evolution of the structures that populate the Universe, particularly the galaxies and systems of galaxies.

VI.6.1. OBSERVABLE PARAMETERS

The observable cosmological magnitudes are the Hubble constant $H_0 = R'_0/R_0$, the deacceleration parameter $q_0 = R''_0 R_0/R'^2_0$, the present density of matter $\rho_0 = m_p N_0$ resulting from the density ρ_G in the form of galaxies and ρ_{IG}, the density of the intergalactic medium, the present temperature T_0 of the background radiation, the relative cosmical abundances of H, He, D, etc. in very old objects, and the age t_0 of the Universe (or at least a lower bound such as the age of the oldest astrophysical objects).

There are various relations between these parameters which should be pointed out for their future use in this chapter. If we divide Equation (VI.1) by R^2 we find for any time

$$H^2 = (\tfrac{8}{3})\pi G\rho - k/R^2. \tag{VI.5}$$

Notice that for a flat Universe ($k = 0$) we have $3H^2 = 8\pi G\rho$. It is convenient then to introduce the *density parameter* Ω through

$$\Omega = \frac{\rho}{\rho_c} = \frac{8\pi G\rho}{3H^2} \tag{VI.6}$$

which measures the ratio between the actual density ρ in terms of the critical density $\rho_c = 3H^2/8\pi G$ for which the Universe would have $k = 0$. Since $k = R^2(\Omega - 1)$, the knowledge of H_0 and ρ_0 provides information on the sign of curvature k in the present time t_0.

By means of the Hubble diagram (Chapter V) it is possible to know, in principle, the parameter q_0: from Equations (VI.1) and (VI.2) we deduce

$$R'' = -\frac{4\pi G}{3}\left(\rho + \frac{3p}{c^2}\right) R$$

and from the definition of q we find for the later times when $3p \ll \rho c^2$

$$q = \frac{4\pi G}{3H^2} = 2\,\Omega.$$

This relationship is true provided the cosmological constant Λ is zero, as we have already assumed Section (VI.1). If it were possible to make independent estimates of the present values of Ω_0 and q_0, we would be able to test the validity of the choice $\Lambda = 0$.

It is useful now to relate Ω and H with its present values Ω_0, H_0 through an observable quantity such as the redshift. As $R = R_0(1 + z)$ and $\rho = \rho_0(1 + z)^3$, Equation (VI.1) together with Equation (VI.6) gives

$$H^2 = H_0^2(1 + z)^2(1 + \Omega_0 z). \tag{VI.7}$$

Moreover, if we observe that $\Omega/\Omega_0 = (H_0/H)^2(1 + z)^3$, we also have

$$\Omega = \Omega_0\frac{1 + z}{1 + \Omega_0 z}. \tag{VI.8}$$

The Einstein – de Sitter model universe has $\Omega = \Omega_0 = 1$ for all its evolution. Notice however, that other models with $\Omega_0 < 1$ approach the Einstein – de Sitter condition for $z \gg \Omega_0^{-1}$ as can be seen in Equation (VI.8). This result is particularly relevant for the gravitational instability theory of galaxy formation (Section VII.3).

In the models of Universe described by Friedmann, Equation (VI.1), the age of the Universe t_0 is the cosmic time elapsed since the singularity $R = 0$ to the present $R = R_0$. This can be computed with

$$t_0 = \int_0^{R_0}\frac{dR}{HR} = \int_0^{\infty}\frac{dz}{H(1 + z)} = \frac{1}{H_0}\int_0^{\infty}\frac{dz}{(1 + z)^2(1 + 2q_0 z)^{1/2}}$$

after having used Equation (VI.7) and put $\Omega_0 = 2q_0$. From the preceding expresion it follows that the product $H_0 t_0$ is a monotonically decreasing function of q_0 (see Table VI.1).

If it were possible to make independent determinations of H_0, and t_0, the relation $H_0 t_0 = F(q_0)$ would allow us to estimate the value of q_0. Actually, it is only possible

TABLE VI.1

Values of $H_0 t_0$

q_0	$H_0 t_0$
0.00	1.00
1.00	0.90
0.10	0.85
0.2	0.78
0.5	0.67
0.8	0.60
1.0	0.57

to have a lower bound for t_0, so that this procedure provides only an upper limit for q_0.

VI.6.2. FITTING THE MODEL TO THE UNIVERSE

Gunn and Gott (1972) have taken advantage of the independence of Ω from H, to discuss all cosmological information from observables in terms of these quantities as parameters. This idea became apparent in the (H, Ω)-diagram illustrated in Figure VI.4. The limits for H and Ω by the consideration of different observable parameters are traced and a region is defined which provides the best compromise for the present values of H and Ω.

We shall discuss now the different observables and the limits defined in the (H, Ω)-diagram. The Hubble constant, as we have seen in Chapter V, is comprised within the bounds

$$55 \text{ km s}^{-1} \text{ Mpc}^{-1} < H_0 < 100 \text{ km s}^{-1} \text{ Mpc}^{-1}$$

which define a horizontal strip in Figure VI.5.

The density of visible matter in the form of galaxies supplies a lower bound for the total mass density ρ_M, since M/L-ratios are measured in isolated galaxies and there is no way to estimate the contribution of the intergalactic medium. With the figures of Chapter IV we have $\Omega > 0.01$. The dynamics of clusters and groups of galaxies leads, as we have also seen in Chapter IV, to M/L ratio larger than those found for isolated objects by a factor of about of five. If this is due to the presence of massive haloes, this factor can be applied to the previous estimate, from which we get the upper bound $\Omega \leqslant 0.05$.

In a low density medium ($\Omega \approx 0$) the gravitational effect of matter concentrations is very small. If, however, Ω is large, a small density perturbation $\delta\rho$ may counteract the expansion. In fact, let $\delta M = (4\pi/3)\, l^3\, \delta\rho$ be the excess of mass in an accumulation of matter of size l and distant r from a test particle and δv the perturbation it produces in the particles's velocity, then we have

$$v\delta v = -\, G\delta M/r.$$

Assuming now $v = Hr$ to be the Hubble velocity, we get

$$\delta v = -\, \frac{1}{2} Hr \left(\frac{\delta\rho}{\rho}\right)\left(\frac{l}{r}\right)^3 \Omega$$

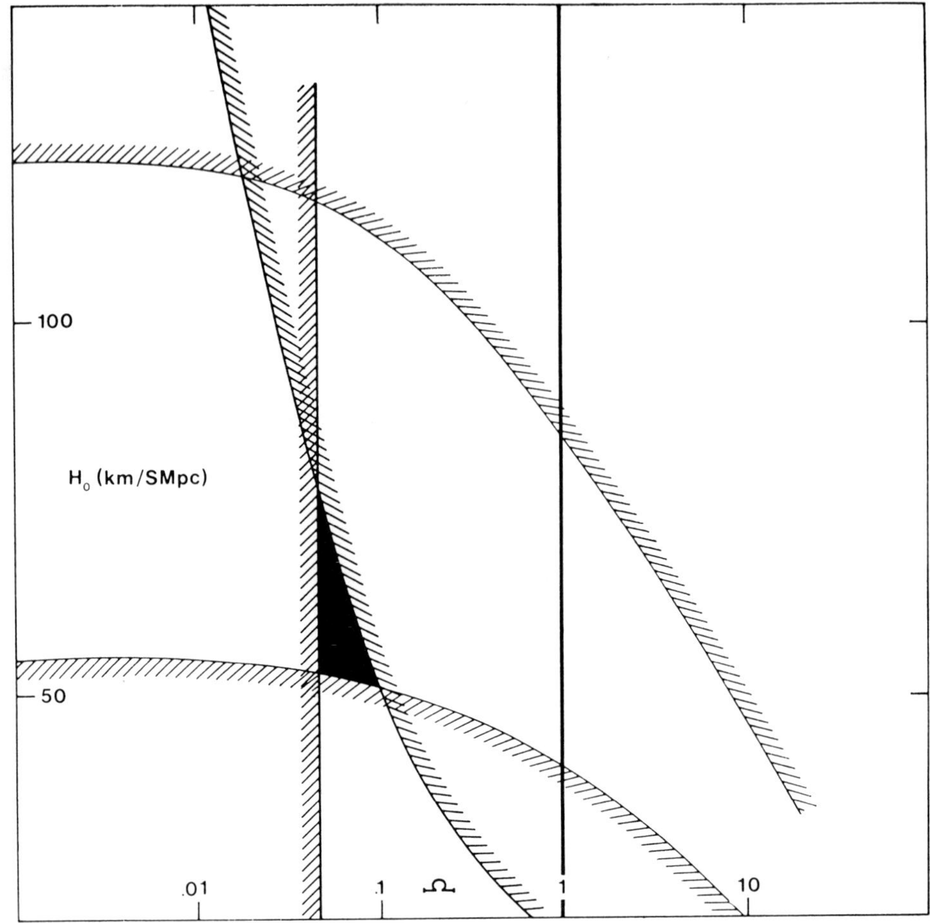

Fig. VI.5.   The (H, Ω)-diagram.

an expression which allows us to appreciate the critical role Ω plays in the velocity field when nearby concentrations of matter are present.

Let us consider the case of the local Supercluster of galaxies (cf. Chapter IV) with its center in the Virgo Cluster. The Local Group is at a distance $l = r$, that is at the border of the Supercluster which is, besides, a concentration of matter in the Metagalactic field whose density is estimated at $10^{-29.4}$ g cm^{-3}. The mean density of the Supercluster on the other hand, is $10^{-28.5}$ g cm^{-3}, so that the density contrast is $\delta\rho/\rho \approx 8$. If we use the value found by Tammann $et\ al.$ (1979) for the peculiar velocity of the Virgo Cluster, $\delta v \approx 100$ km s^{-1}, the resulting estimate for Ω is 0.025. Other tests of this kind have been carried out with the perturbations of the Hubble flow at scales of 20 Mpc. Again from the data of Sandage $et\ al.$ (1972), $\Omega < 0.1$ was obtained.

The limits for Ω are

$$0.1 > \Omega > 0.025$$

which define a vertical strip in the (H, Ω) plane (Figure VI.5). This reduces the region of interest now to a rectangle.

Let us consider the 'age of the Universe' t_0, which is an observable depending on Ω and H. It is estimated that the oldest stars of population II must have been formed when the Galaxy still was collapsing. The timescale for this collapse is of the order of 10^9 yr, while the theory of stellar evolution gives an age range for the globular cluster going from 11 to 18 times 10^9 yr according to their He content of $Y = 0.2$ to 0.3. With these bounds for t_0, and those already used for H_0, together with the relationship $H_0 t_0 = F(\Omega_0)$ we can trace two limiting curves in the (H, Ω) plane. At least the lower one seems to be useful for the definition of a narrower critical area for H_0, Ω_0.

The observations we have discussed up until now provide certain direct evidence for the evolutionary history of the Universe after decoupling $(t > t_d)$, but they do not involve any evidence as to earlier times. These must then be looked for in the abundances of the elements resulting from the primeval nucleosynthesis. We have already seen (Section VI.4) that in the first few minutes after the singular state, at temperatures of the order of 10^9 K, reactions were produced that synthesized light nuclei, most of which were destroyed again. The only survivors with abundances by mass larger than 10^{-5} are ^{4}He, D, and ^{3}He.

From the knowledge of the present background radiation temperature T_0 and the abundance of D derived from the satellite UV observations of $L_{\beta, \gamma, \delta}$ and ε in β Centauri, $D/H = (1.4 \pm 0.2) \times 10^{-6}$, we deduce from Wagoner computations (Figure VI.3) a value of $\rho_0 = 5 \times 10^{-31}$ g cm^{-3} for the present density of matter in the Universe. This figure leads to write $\Omega_0 \leqslant 0.90 \, (60/H_0)^2$. According to Gunn, it is possible that the destruction of D in the interior of stars has reduced its abundance to one half of the original value, so that $\Omega_0 \geqslant 0.45 \, (60/N_0)^2$ and these limits, in spite of their uncertainty, circumscribe a very reduced area in the (H, Ω) diagram.

In this way the range of possible Friedmann models is reduced now to those with local hyperbolic spaces $(\Omega \approx 0.02, q_0 \approx 0.01)$ in which the expansion keeps on forever, without further contraction. It is noteworthy to see that the most definite information comes from the local observations (age t_0, D abundance, local dynamics) and not from the observation of the far away galaxies.

VI.6.3. Lifting the Restriction $\Lambda = 0$

An analysis similar to the foregoing one was carried out by Gunn and Tinsley (1975), to test the possibility of a cosmological constant $\Lambda \neq 0$. According to the results of their discussion, the most plausible models have a closed locally spherical space, the present density is given by $\Omega \geqslant 0.01$ and a Hubble constant $H_0 = 80$ km s^{-1} Mpc^{-1}. These models also expand irreversibly and their density is too large to allow for the synthesis of D at the primeval synthesis. The nature of the origin of D becomes critical in this way for the definition of the type of Universe model which fits the observations. However, the cosmologic origin of D seems well established (Section VI.4) and consequently the suitability of the case $\Lambda = 0$ lifts principistic difficulties with a self-consistent GR cosmology.

References

Alpher, R. A. and Herman, R.: 1948, *Nature* **162**, 774.

Canuto, V.M.: 1978, 'New Trends in Cosmology', *Nuovo Cimento* **2**, 1.

Cheng, E.S., Saulson, P.R., Wilkinson D.T., and Corey, B.E.: 1979, **230**, L139.

Ehlers, J.: 1976, *Mitt. Astron. Gesellsch.* **38**, S41.

Ellis, G.R.: 1976, *Quart. J. Roy. Astron. Soc.* **16**, 245.

Gamow, G.: 1953, *Dan. Mat. Phys. Medd.* **27**, 1.

Gunn, J.E. and Gott, R.: 1972, *Astrophys. J.* **176**, 1.

Gunn, J.E. and Tinsley, B.M.: 1975, *Nature* **257**, 454.

Hawking, S.W. and Ellis, G.F.R.: 1968, *Astrophys. J.* **152**, 25.

Hawking, S.W. and Penrose, R.: 1969, *Proc. Roy. Soc.* **A314**, 529.

Hoyle, F. and Fowler, W.: 1973, *Nature* **241**, 384.

McCrea, W.; 1971, *Quart. J. Roy Astron. Soc.* **12**, 140.

Penzias, A. and Wilson, R.: 1965, *Astrophys. J. Letters* **142**. 419.

Sandage, A.R., Tammann, G.A., and Hardy, E.: 1972, *Astrophys. J.* **172**, 253.

Schram, D.N. and Wagoner, R.F.: 1977, *Ann. Rev. Nucl. Sci.* **27**, 27.

Smoot, F.G., Gorenstein, M.V., and Muller, R.A.: 1977, *Phys. Rev. Letters* **39** (14), 898.

Tammann, G.A., Yahil, A., and Sandage, A.R.: 1979, *Astrophys. J.* **234**, 775.

Weinberg, S.: 1972, *Gravitation and Cosmology*, Wiley and Sons, N.Y.

Woody, D.P. and Richards, P.L.: 1979, *Phys. Rev. Letters* **42**, 925.

Bibliography

Berry, M.: 1976, *Principles of Cosmology and Gravitation*, Cambridge.

Ehlers, J.: 1967, *Relativity Theory and Astrophysics*, Am. Math. Soc., Vol. 8, Lectures of Appl. Math., Providence.

Gunn, J.E., Longair, M.S., and Rees, M.J.: 1978, *Observational Cosmology*, SAAS-FEE.

Jakobsen, H.P., Kon, M., and Segal, I.E.: 1979, *Phys. Rev. Letters* **42** (26), 1788.

North, J.D.: 1965, *The Measure of the Universe*, Oxford.

Peebles, P.J.E.: 1971, *Physical Cosmology*, Princeton.

Sciama, D.W.: 1971, *Modern Cosmology*, Cambridge.

CHAPTER VII

GRAVITATIONAL INSTABILITY
AND GALAXY FORMATION

Two lines of thought regarding the origin of structures in the expanding Universe have been worked out up to a certain detail.

One proposes the fragmentation of large elementary structures resulting from instabilities at or before recombination time into smaller and denser substructures. As a consequence, star formation occurs at relatively later times and dissipation plays an important role in the process.

According to the clustering hypothesis, however, small objects are first formed and then, through gravitational interactions, collect themselves into larger and larger systems. In this way stars and star clusters are formed shortly after recombination, so that clustering becomes a rather dissipationless process.

Both hypotheses probably play their role in the actual Universe. The same system may fragment but also go through clustering at the same or a later time.

Gravitational instability can be invoked to describe either the successive fragmentations or the condensation of stars in the clustering hypothesis, and Jeans' length is a key concept for it.

The concept of Jeans' wavelength was introduced in 1902 when Jeans calculated the modification of the propagation of sound waves in the presence of gravitation. The pressure gradients are overcome by gravitation for perturbations with wavelength λ large enough in a gaseous medium. There is a critical value λ_J above which the gas becomes unstable against gravitation and is fragmented in self-gravitating clouds.

Let us consider a perturbation in a gaseous cloud with a mean density ρ. The gravitational scale of time is $\tau_G = (G\rho)^{-1/2}$ and the adiabatic velocity of sound $c_s = (dP/d\rho)^{1/2}$, where P is the pressure. We may now form a length $\lambda_J = c_s\tau_G = (\pi/G\rho)^{1/2}(dP/d\rho)^{1/2}$ which measures the distance travelled by the pressure waves in a time τ_G. From the physical point of view the cloud can withstand the gravitational forces, provided it has time to generate the pressure gradient necessary to counteract them. This happens only if the perturbation wavelength is $\lambda < \lambda_G$, when the process reduces itself to the propagation of sound waves. If, on the contrary, we have $\lambda > \lambda_J$ the cloud is unable to avoid the gravitational collapse, the density perturbations grow exponentially (at least in the linear regime) and the system fragments into clouds with sizes of the order of λ_J.

VII.1. Theory of Jeans' Wavelength and Mass

Let us consider an extended homogeneous gaseous medium with fluctuations in density ($\delta\rho$), pressure (δp), temperature (δT), and gravitational potential ($\delta\phi$). We have the following equations

$$\rho \frac{\delta v}{\delta t} = - \text{ grad } \delta p - \rho \text{ grad } \delta\phi,$$

$$\frac{\delta}{\delta t} \delta\rho = - \rho \text{ div } \delta v; \quad \frac{\delta p}{p} = \frac{\delta\rho}{\rho} + \frac{\delta T}{T}, \qquad\qquad (\text{VII.1})$$

$$\nabla^2\delta\phi = 4\pi G\delta\rho,$$

between the respective fluctuations, provided the velocities v were small, so that their squares and products can be neglected, as well as those of their derivatives.

Let us look for a solution of Equation (VII.1) which represents the propagation of plane waves along the z-axis. We have

$$\rho \frac{\delta v_z}{\delta t} = - \frac{\delta}{\delta z} \delta p - \rho \frac{\delta}{\delta z} \delta\phi,$$

$$\frac{\delta}{\delta t} = - \rho \frac{\delta v_z}{\delta z}, \qquad\qquad (\text{VII.2})$$

$$\frac{\delta^2}{\delta z^2} \delta\phi = 4\pi G\delta\phi.$$

If we neglect the thermal conduction of the medium, we will be in an adiabatic situation, and then

$$\delta p = c_s^2 \delta, \qquad\qquad (\text{VII.3})$$

where $c_s^2 = (dp/d\rho)$ is the adiabatic velocity of sound. To solve system (VII.2) we write

$$\frac{\delta}{\delta t} = iw \quad \text{and} \quad \frac{\delta}{\delta z} = - ik,$$

where w is the frequency and k the wave number of the perturbation. Taking the fore-going expressions to Equation (VII.2) and keeping in mind Equation (VII.3), we are led to a homogeneous linear system of equations in the perturbations $\delta\rho$, $\delta\phi$, and v_z which can be written as

$$\begin{pmatrix} i\rho w & i\rho k - ikc_s^2 \\ i\rho k & 0 & iw) \\ 0 & k^2 & 4\pi G \end{pmatrix} \begin{pmatrix} v_z \cdot \\ \delta\phi \\ \delta\rho \end{pmatrix} = 0.$$

In order to obtain a non-trivial solution, the determinant of the system has to be equated to zero, which leads to the following equation

$$\frac{w}{k} = k^2 c_s^2 - 4\pi G\rho \qquad\qquad (\text{VII.4})$$

since ρ as well as k are different from zero by hypothesis.

To have stable solutions (i.e. waves), w/k has to be real. If w/k were imaginary, the amplitude of the waves would grow indefinitely, leading to an unstable condition.

Therefore we have

stability for $\quad k^2 c_s^2 > 4\pi G\rho,$

and

instability for $\quad k^2 c_s^2 \leqslant 4\pi G\rho,$

which constitutes the criterion given by Jeans.

To the critical wave number k_J for which $w = 0$, we associate Jeans' wavelength

$$\lambda_J = 2\pi/k_J = (\pi c_s^2/G\rho)^{1/2} \tag{VII.5}$$

as discussed before. With it Jeans' criterion for stability assumes the form $\lambda < \lambda_J$ for perturbations with wavelength λ.

VII.1.1. THE JEANS' MASS

With the Jeans' wavelength λ_J and the density ρ of the medium, we may compute a "Jeans' mass"

$$M_J(T, \rho) = \pi^{3/2} C_s^3 \rho^{-1/2} G^{-3/2}$$

whose significance is that of a lower bound for the masses able to collapse and fragment in case of gravitational instability. If M is the mass associated to a given perturbation, it will be unstable and fragment if $M \geqslant 2M_J$. The new temperatures T' and densities ρ' prevailing in the fragments define a new Jeans' mass M'_J. If the mass M'' of a fragment again satisfies the relation $M'' > 2M'_J$, the instability will continue once again. From this we take that the condition for a continuated process of fragmentation leading to a hierarchy of systems is that the Jeans' mass $M_J(T, \rho)$ be a decreasing function of the density.

In the case of a medium which follows an equation of state like $p \propto \rho^\gamma$, the speed of sound is $c_s^2 = \mathrm{d}p/\mathrm{d}\rho \propto \gamma \rho^{\gamma-1}$ and $M_J \propto \rho^{1/2(3\gamma-4)}$. The importance of the heat losses by the fragments in the continuation of the process is now evident. When $\gamma < \frac{4}{3}$ the Jeans' mass decreases with increasing density: that is the case of a cooled or an isothermal collapse. When the process is adiabatic ($\gamma = \frac{5}{3}$) as is the case of a monoatomic gas, M_J increases with ρ and fragmentation does not proceed.

VII.2. Gravitational Instability in an Expanding Universe

The manifest non-uniformity of the present-day Universe does not render Friedmann models inapplicable. They give a valid overall description provided the typical fluctuations $\delta\rho$ of the mean density over a length scale d satisfy $G\delta\rho d^2 \ll c^2$ for all d. This ensures that the fractional departures from the Robertson–Walker metric are everywhere smaller than one and that the peculiar velocities induced by the inhomogeneities are much smaller than c. For both galaxies and clusters of galaxies this relation is fulfilled by a margin of 10^4 and the isotropy of the background radiation strongly suggests that it holds on all larger scales up to the Hubble radius.

The study of the time-evolution of the density perturbation of expanding a gaseous medium presents insuperable difficulties when posed in its widest generality. That happens both within the frame of the Newtonian Theory and the General Theory of Relativity. Only through the limitation of the problem to small disturbances of a medium otherwise homogeneous and isotropic, some interesting progress can be made. This is due to the fact that the parameters describing the behavior of the medium in its undisturbed state are then functions of cosmic time, independent of spatial coordinates, which enables the linearized equations also to be separable.

The consideration of perturbations with linear dimensions smaller than the horizon ct also having peculiar velocities v small compared with that of light poses the problem within the newtonian approximation as was shown by Bonnor (1957). The solution of the same problem, but dealing with the equations of GR theory, is due to Lifshitz (1946).

Let us consider now an inertial frame, i.e. at rest with respect to the surrounding matter. Let $\rho_1(t, r)$ be its density such as observed from our reference system. In absence of perturbations, $\rho_1(t, r)$ is reduced to the uniform function of time $\rho(t)$. The density contrast $s(t, r)$ is thus defined with

$$\rho_1(t, r) = \rho(t)[1 + s(t, r)]$$

which is assumed to be a small quantity of first order.

The velocity field, such as observed from the origin of the reference system, takes the form

$$v(t, r) = H \cdot r + u(t, r)$$

since the medium is expanding. Here $H = R'/R$ is the expansion parameter and $u(t, r)$ the peculiar velocity of the perturbations. It will also be considered a first order small quantity, so that both the squares as well as the products of u and s will be neglected.

Let p now be the pressure and ϕ the gravitational potential. Then the linearized equations of motion, continuity and Poisson's are, respectively,

$$\frac{\delta u}{\delta t} + H(u + r \operatorname{grad} u) = - \operatorname{grad} \delta\phi - knT \operatorname{grad} S - \gamma u,$$

$$\frac{\delta s}{\delta t} + H \cdot r \operatorname{grad} s + \operatorname{div} u = 0, \qquad\qquad \text{(VII.6)}$$

$$\nabla^2 \delta\phi = 4\pi G\rho \cdot s.$$

We have assumed in Equation (VII.6) that the medium is an isothermal ideal gas with $P = knT$. The term $-\gamma u$ describes a process controlled by the viscosity coefficient γ whose meaning and range of validity will be discussed later.

A comparison of Equation (VII.6) with Equation (VII.2) shows at once the relevance of the expansion flow for the system of equations.

TABLE VII.1

Transformation to the co-moving system

Inertial system	Co-moving system
r	Rr_e
$\dfrac{\partial}{\partial t} + H \operatorname{grad}$	$\dfrac{\partial}{\partial t_c}$
$R \operatorname{div}$	div_c
$R \operatorname{grad}$	grad_c
$R^2 \nabla^2$	∇^2

Let us transform now the coordinates through

$$r_c = r/R(t)$$

so that we shall write Equation (VII.6) in a reference system co-moving with the perturbation. By applying the equivalences given in Table VII.1 we arrive at once at the new system of equations

$$\frac{\delta u}{\delta t_c} + \left(\frac{R'}{R} + \gamma\right) u = -\frac{1}{R} \operatorname{grad} \delta\phi - \frac{1}{R} knt \cdot \operatorname{grad} s,$$

$$\frac{\delta s}{\delta t_c} + \frac{1}{R} \operatorname{div}_c u = 0, \qquad\qquad (\text{VII.7})$$

$$\nabla_c^2 \delta\phi = 4\pi G\rho s R^{-2}.$$

Taking now the divergence of the first Equation (VII.7) and using repeatedly the second to eliminate $\operatorname{div}_c u$, we obtain

$$\frac{\delta^2 u}{\delta t^2} + \left(2\frac{R'}{R} + \gamma\right)\frac{\delta s}{\delta t} = 4\pi G\rho s + knT \frac{\nabla_c^2 s}{R^2} \qquad\qquad (\text{VII.8})$$

an equation in which we have disregarded the subscript referring to the co-moving system and shall use it from now on.

Equation (VII.8) is separable and can be Fourier-analyzed through

$$S(r, t) = \sum_k S_k(t) \cdot e^{ikr}$$

that is, with a sum of spatial waves of length $\lambda = 2\pi R/k$ and time-dependent amplitudes $S_k(t)$. The differential equation for S_k will then be

$$S_k'' + \left(2\frac{R'}{R} + \gamma\right) S_k' = 4\pi G\rho(t)\left[1 - \left(\frac{\lambda_J}{\lambda}\right)^2\right] S_k, \qquad\qquad (\text{VII.9})$$

where we have introduced the Jeans' wavelength Equation (VII.5)).

In order to discuss the solutions of Equation (VII.9) it is convenient to consider the cases separately according to whether the coefficient γ is important or not.

The intense radiation field in early times of the Universe $(t < t_d)$ interacts with the electrons through the Thompson scattering causing a viscous drag force which opposes the independent motion of matter. The negative acceleration $-\gamma u$ appearing in the equation of motion in Equation (VII.6) and subsequent systems is then effective before the decoupling time.

If σ_T is the Thompson cross section for the electrons, T the common temperature of matter and radiation, the coefficient γ of radiative viscosity can be written

$$\gamma = \frac{4}{3}\frac{\sigma_T a T^4}{m_p c} \simeq \left(\frac{1+z}{10^4}\right)^4 S_k^{-1}$$

since $T = 2.8\,(1 + z)\,K$. This expression tells us that for $z < 10^3$ the effect is negligible but, on the other hand, it is easy to see that for $z > 10^3$, $\gamma \gg R'/R$ and therefore Equation (VII.9) can be written simply

$$S_k' = \frac{4\pi G\rho}{\gamma} S_k \qquad (z > 10^3)$$

for unstable gravitational modes, i.e. those with $\lambda > \lambda_J$ as we shall see at once.

Since for very large values of z the actual universe approaches the Einstein-de Sitter model $\Omega(z) \to 1$, (cf. Section VI.6), and $R \sim t^{2/3} \sim (1 + z)^{-1}$ so that the foregoing equation for S_k admits the solution

$$S_k \propto \exp\left(\frac{150}{1 + z}\right)^{5/2}. \tag{VII.10}$$

Thus the density contrast of a perturbation experiences a rather slow increase during the era prior to decoupling, due to the efficiency of the radiative damping. So e.g. for $z = 10^3$ we have only $S_k = 1.008$ and the perturbations, if they exist, do not increase significantly but stay 'frozen' until recombination time t_d.

Let us consider now the circumstances when the perturbations increase after recombination. Since then $z < 10^3$, we may set $\gamma = 0$ and Equation (VII.9) reduces to

$$S_k'' + 2\frac{R'}{R}\, S_k' = 4\pi G\rho(t)\left[1 - \left(\frac{\lambda_J}{\lambda}\right)^2\right] S_k. \tag{VII.11}$$

In order to emphasize the role played by λ_J in this equation we change the independent variable t for θ so that the term in S_k' is eliminated. With $\theta' = (4\pi G\rho_*)^{1/2}\,(R_*/R)^2$, where R_* and ρ_* refer to the instant when $\theta = t$, we obtain

$$\frac{d^2 S_k}{d\theta^2} = \frac{R}{R_*}\left[1 - \left(\frac{\lambda_J}{\lambda}\right)^2\right] S_k \tag{VII.12}$$

after recalling that $\rho = \rho_*(R_*/R)^3$. Equation (VII.12) shows how a component S_k of the spectrum of the density perturbations varies as a function of θ. When $\lambda > \lambda_J$ the second term is positive and the larger S_k gets, the more its growth is accelerated: then, there will be gravitational instability. Otherwise, if $\lambda < \lambda_J$ we are confronted with oscillations and the mode is stable. Equation (VII.12) leads to Jeans' criterion (Section VII.1) for an expanding uniform medium. If the elastic reaction of the medium can be neglected, i.e. for perturbations with $\lambda \gg \lambda_J$, Lipshitz has shown that Equation (VII.11) becomes

$$S_k'' + 2\frac{R'}{R}\, S_k' = 4\pi G\rho S_k \qquad (\lambda \gg \lambda_J)$$

and admits algebraic solutions. Through the new independent variable $\tau = t/t_0$ and assuming an expansion law $R/R_0 = \tau^h$, we obtain

$$\frac{d^2 S_k}{d\sigma^2} + \frac{2h}{\tau}\frac{dS_k}{d\tau} = \frac{3}{2}\,\Omega\, h^2 S_k \tau^{-3h}$$

after introducing the density parameter Ω. A solution can be tried in the form $S_k = \tau^b$, and the equation to determine the exponent b is

$$b^2 + \tfrac{1}{3}b - \tfrac{2}{3}\Omega = 0$$

which results from the necessity of making $h = \tfrac{2}{3}$ to obtain solutions of the adopted form. This gives an expansion law $R/R_0 = (t/t_0)^{2/3}$ identical to that of the Einstein–de Sitter model ($\Omega = 1$) or very similar to the expansion law at early times ($z \gg \Omega_0^{-1}$) for models with $\Omega_0 < 1$ (cf. Section VI.6.1). The roots of the characteristic equation (with $\Omega = 1$) are $b_1 = \frac{2}{3}$ and $b_2 = -1$, so that the general solution takes the form

$$S_k = At^{2/3} + Bt^{-1} \qquad (\Omega = 1). \tag{VII.13}$$

We are obviously interested in the way the density contrast increases with time, i.e. we set $B = 0$. Although the growth rate of the contrast is slower than that of Jeans' classical solution, it is argued that when it reaches the value $S_k = 1$, the non-linear effects take over and accelerates the process.

VII.2.1. THE ERA OF GRAVITATIONAL INSTABILITY

The results of Section VII.3 allow us to set bounds for the redshift range in which gravitational instability could have been efficient to produce structures in the Universe.

Since the radiative damping freezes the growth of perturbations for any time before decoupling (t_d or z_z) and also because Lipshiftz's algebraic solutions are possible only for $\Omega = 1$, Equation (VI.8) requires $z \gg \Omega_0^{-1}$ in order to have $\Omega \approx 1$, so that the limits

$$\Omega_0^{-1} \ll z < z_d \tag{VII.14}$$

define the range of z for gravitational instability to be operative in an expanding Universe. The expression (VII.14) shows how model dependent are the processes which are supposed to give structure to the otherwise homogeneous and isotropic Universe.

VII.3. Protogalaxies

Because of the strong coupling between matter and radiation prior to decoupling ($z > z_d$), the Universe behaves as a single fluid whose pressure is dominated by radiation (Section VI.3), $p = \frac{1}{3}\rho_R c^2$, so that the speed of sound is

$$c_s = \left(\frac{\mathrm{d}p}{\mathrm{d}\rho}\right)^{1/2}_{\mathrm{Ad}} = \frac{c}{\sqrt{3}}$$

for redshifts $z > 1250$, when both matter and radiation densities are equal. The associated Jeans' mass (Section VII.1.1) is then

$$\mathfrak{M}_J(z \gg z_d) = 1.5 \times 10^{19} \left(\frac{35\,000}{1+z}\right)^2 \text{ solar masses}$$

and assumes a rather constant value $\mathfrak{M}_J \approx 10^{19}$ solar masses for $35\,000 \gg z > z_d$. The decoupling of matter and radiation causes a sharp decrease in $\mathfrak{M}_J$, because the speed of sound is now given by the elastic properties of a perfect gas, so that we have

$$c_s = \left(\frac{\mathrm{d}p}{\mathrm{d}\rho}\right)^{1/2}_{\mathrm{Ad}} = \left(\frac{kT}{m_H}\right)^{1/2}.$$

The mass scale of objects which can then collapse depends upon the exact temperature of the medium at that time, and since we may assume the matter temperature T to be close to the radiation temperature $T_R = T_{OR}(1 + z)$ and $\rho_M = \rho_0(1 + z)^3$ at that epoch (cf. Section VI.3) we obtain

$$\mathfrak{M}_J(z_d) = \left(\frac{\pi k}{m_H}\right)^{3/2} \left(\frac{T^3_{\mathrm{OR}}}{\rho_0}\right)^{1/2} (1 + z_d)^{-3/2} = 1.3 \times 10^5 \text{ solar masses}$$

a value comparable to the mass of a globular cluster. The range of Jeans' masses at or near decoupling covers, in consequence, the range of masses of galaxies; and all fluctuations with mass greater than $\mathfrak{M}_J(z_d)$ can collapse after recombination.

A detailed discussion of the behavior before and around recombination has been carried out by several authors (see for example the review by Jones, 1976). Because of the diffusion of radiation through matter when density declines with expansion and Compton scattering becomes less effective, a density perturbation with a mass of the order of a typical galaxy first oscillates like pressure waves and then rapidly dampens as decoupling approaches. If these perturbations survive until recombination, they will grow again because of the decrease of the Jeans mass at z_d.

In this way we become confronted with a pressure free density enhancement in a Friedmann Universe. The expansion of such a region lags behind and stops at a critical time t_p with associated redshift z_p. The value of z_p depends on the density contrast $s = \delta\rho/\rho$ of the fluctuation at recombination time t_d, but the density at t_p has a definite value with respect to the critical background density ρ_c of the Universe at that time, as we shall see in the next section.

VII.3.1. Non-linear growth of a Density Perturbation

Newtonian mechanics is quite sufficient to discuss the dynamics of the expanding Universe (McCrea and Milne, 1934) in a region where both GR and Newtonian mechanics are equally valid, as we have seen in Section (VII.2).

Let us consider a spherical region of space containing a mass M. The equation of motion for the radius R of the spherical surface enclosing that mass is

$$R'' = -\frac{GM}{R^2}$$

which, after one integration, may be written

$$\left(\frac{dp}{d\theta}\right)^2 = 1 - \frac{1}{P} \tag{VII.15}$$

with the introduction of the new variables P, θ through

$$P = R/R_p, \qquad t = (8\pi G \rho_p/3)^{-1/2}\,\theta$$

and imposing the condition $4\pi R_p^3 \rho_p = 3M$ at $R'_T = 0$. R_p means the maximum radius of the perturbation before further collapse, and ρ_p its mean density at the same time.

The parametric solution of Equation (VII.15) is

$$P(\eta) = 1 - \cos\eta, \qquad \theta(\eta) = \eta - \sin\eta \tag{VII.16}$$

so that the maximum of P (or R) is attained when $\theta_M = \pi/2$.

Now we assume the region we are considering is a disturbance in an otherwise homogeneous Universe. Since we are interested in the evolution of a density perturbation originated in the gravitational instability process, our scenario has to be placed in a epoch where $\Omega \approx 1$ according to the results in Section VII.2. For small P Equation (VII.15) may be written

$$P\left(\frac{dp}{d\theta}\right)^2 = 1$$

in good approximation. But this is also the equation for the scale factor Q of an Einstein–de Sitter universe ($\Omega = 1$), so that both solutions become coincident for small values of P. The Einstein–de Sitter solution is

$$Q(\eta) = (\tfrac{3}{2}\eta)^{2/3} \tag{VII.17}$$

since now $\eta = \theta$. Because the ratio of densities between the perturbation and the background at the moment of maximum expansion of the former is $[Q(\pi/2)/P(\pi/2)]^3$ we get

$$\rho_p = \left(\frac{3\pi}{4}\right)^2 \rho_c = \left(\frac{3\pi}{4}\right)^2 \frac{3H^2}{8\pi G}$$

after recalling the meaning of the critical density ρ_c (cf. Sections VI.5 and VI.6.1). With Equation (VI.7) the foregoing relation for ρ_p can be written

$$\rho_p = \left(\frac{3\pi}{4}\right)^2 \frac{3H_0^2}{8\pi G} (1 + z)^2 (1 + \Omega_0 z) \tag{VII.18}$$

in terms of the present values of the Hubble and density parameters H_0, Ω_0.

Detailed computations of the nonlinear growth of density perturbations in an expanding Universe are due to Peebles (1967), Field (1975), and most recently to Gott (1980).

The shift from the Einstein–de Sitter solution (VII.17) to the perturbation solution (VII.16) poses a very fundamental problem. Since the density contrast $s = \delta\rho/\rho$ develops only as an algebraic power of θ (Section VII.2), its value at recombination has to be small but hardly infinitesimal. For earlier times the contrast is kept frozen because of damping and consequently we are forced to explain an early Universe with small but finite density perturbations.

Two lines of thought have been advanced in this respect regarding the origin of the density fluctuations at recombination: Either the inhomogeneities are primordial, which presupposes that structure, however rudimentary, originates with the Universe, or they are spontaneous.

In this latter case the inhomogeneities are produced through mechanisms originated in the properties of matter at different temperatures and densities. Thermal and also gravitational instabilities for $t > t_d$ would then produce spontaneous density variations.

Harrison (1973) has criticized both points of view and asks whether they really explain anything fundamental, since the growth of the density contrast requires a definite 'seed' to start with.

VII.4. Galaxy Formation Through Dissipative Collapse

We may conceive protogalaxies as resulting from the collapse and fragmentation of clouds of primeval composition formed in the gravitational instability era of the Universe (Section VII.2.1). Consider such a cloud and assume it initially neutral. As the collapse proceeds, it would heat up adiabatically. If the cloud were initially inhomogeneous (a plausible hypothesis because its large dimensions lead to a large Reynolds number and the regime becomes turbulent), the density contrast of regions with scales larger than the Jeans' length will be enhanced and sub-condensations will

appear. In order to keep these fragments permanently separated, gravitational instability does not suffice. Hoyle (1953) was the first to point out the important role of
the cooling mechanisms in these processes because the fragments radiating away
enough gravitational energy become appreciably bound.

Since the time scale for cooling (through radiative recombinations) is $\tau_c \propto \rho^{-1}$
whilst that for collapse is $\tau_G \propto \rho^{-1/2}$ there will be regions inside the cloud where the
cooling-rate will overtake the collapse rate. At some time we shall have $\tau_c \leqslant \tau_G$ and
these regions become detached fragments able to survive: the protogalaxies. They are
now individually capable of going through the same stages of cooling and collapse,
repeating the cycle many times up to the moment when the fragments are dense
enough to become opaque, ceasing to radiate its entropy production. An adiabatic
condition is set up again, internal temperature rises with collapse of the last fragments
until nuclear processes are turned on and provide the source of entropy with a very
long time-scale which keeps the just born stars in metastable thermodynamical
equilibrium. The condition $\tau_G = \tau_c$ is consequently a critical one to define the properties of progalaxies and also galaxies.

This point of view has been worked out by Rees and Ostriker (1977) and also by
Silk (1977).

Two principal cooling mechanisms are considered: Compton scattering caused by
the cooler background radiation on the electrons in the cloud, and radiative recombinations in a medium of primaeval composition.

The critical density at which cooling and compression heating due to collapse are
equalized is obtained from Equation (VII. 24)

$$J_q \delta t = -(p + \Pi)\delta V$$

of next section. Here J_q is the energy flux of radiation escaping from the volume element δV in the time interval δt. P and Π, respectively, are the pressures of thermal
and dynamic origin. We shall disregard mass motions (cf. Section VII.24) and take
$\delta V = (16\pi G\rho)^{-1/2} \delta t$ for a pressure supported spherically symmetric collapse (Penston,
1969).

The Compton contribution to cooling is (Silk, 1977)

$$(J_q)_c = AT_R^4(T - T_R)n_e \approx AT_{OR}^4 Tn_e(1 + z)^4 \qquad (T \gg T_R),$$

where n_e is the number density of free electrons, T the gas temperature, $T_R = T_{OR}(1 + z)$
the background radiation temperature, and

$$A = 4\sigma Tac^3 k/n_e.$$

The Compton critical density now becomes

$$\rho_c = 10^{-33.65}(1 + z)^8 \text{ g cm}^{-3}$$

for a fully ionized gas. With $H_0 = 100$ km s^{-1} Mpc^{-1}, the density of a protogalaxy,
when collapse starts, is

$$\rho_p(z) = 10^{-28}(1 + z)^2 (1 + \Omega_0 z) \text{ g cm}^{-3}$$

after Equation (VII.18) and the redshift for this object when $\rho_p = \rho_c$ sets a lower
limit for the redshift of galaxy formation. An upper bound results from the observa-

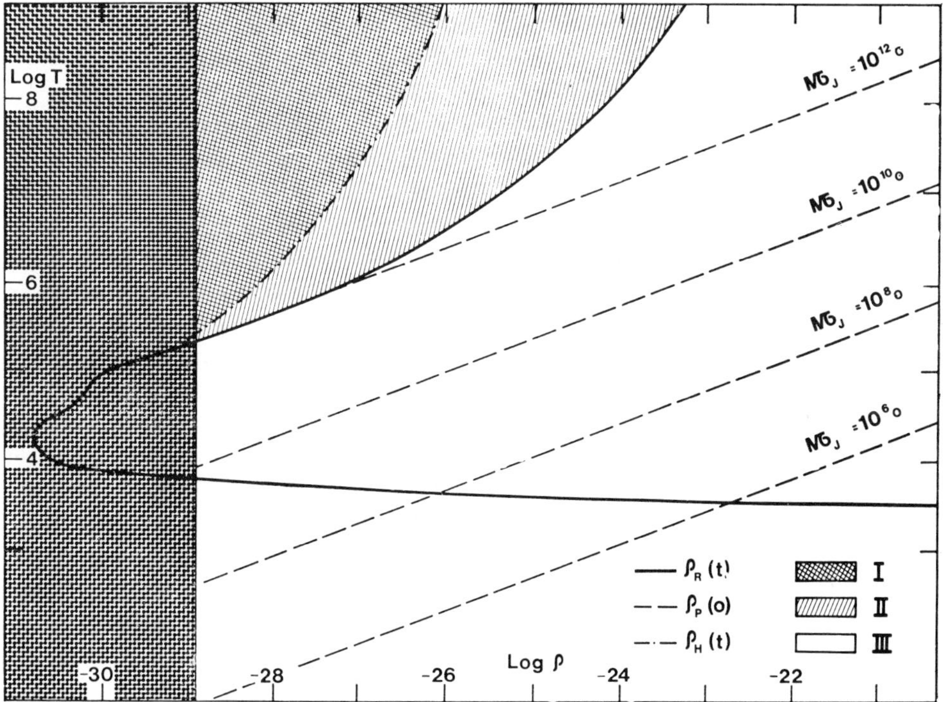

Fig. VII.1.   Temperature-density diagram (see text).

tion that $\rho_p \leqslant \rho_G/L^3$, where $L = \frac{2}{3}$ ln (Number of stars) is the compression factor relating the sizes of the galaxy and its protogalaxy (Section VII.4.4.). With $L \approx 10$ for the average galaxy, we get

$$10 \lesssim z_{GF} < 24 \left(\frac{\rho_G}{10^{-23}}\right)^{1/2} \tag{VII.19}$$

for the redshift of the era of galaxy formation through dissipative collapse.

The consideration of radiative recombinations in a transparent gas of primeval composition leads to the cooling flux

$$(J_q)_R = L(T)\rho^2$$

where $L(T)$ is a function of the temperature and the composition of the protogalaxy. The critical density at which Equation (VII.24) is satisfied

$$\rho_R(T) = 16\pi G \left(\frac{kT}{\mu_L(T)}\right)^2$$

is now temperature-dependent, and defines a locus in the (T, ρ)-plane (Figure VII.1). The protogalaxy will lose energy through radiation and consequently will describe a track in the (T, ρ)-plane. Rees and Ostriker (1977) have discussed the properties of protogalaxies evolving in it. To follow their discussion, let us introduce some other loci in the (T, ρ)-plane, besides $\rho_R(T)$. The vertical line $\rho_p(0)$ is the limiting density for disturbances of the primeval medium that even today ($z = 0$) have not yet started to collapse as protogalaxies. Hence, none is to be found for $\rho \leqslant \rho_p(0)$.

The line $\rho_H(T) = (\frac{1}{6}\rho_R\rho_c)^{1/2}$ defines the locus where the protogalactic cooling time equals the Hubble time in an Einstein–de Sitter universe. Several lines for representative values of constant Jeans' mass were also traced in the (T, ρ)-plane.

From Rees and Ostriker's discussion, three regions relevant to protogalaxy evolution can be distinguished in the (T, ρ) plane: Region I, where the cooling time is longer than the age of the Universe. If any protogalaxy in it achieves virial equilibrium, it will remain there forever, without change. Region II: the cooling time is shorter than the Hubble time, but longer than the dynamical scale of time. The protogalaxy evolves along a track $T \propto \rho^{1/3}$ in quasi-static collapse until the locus $\rho_R(T)$ is encountered and enters region III. Since cooling is more effective than compressional heating here, the temperature drops to 10^4 K, from where isothermal collapse proceeds in a free-fall time, unimpeded by gas pressure.

The gross properties of $L(T)$ can be derived if we use a simplified expression for the cooling rate of a pure hydrogen cloud (Silk, 1976), namely

$$\rho^2 L_*(T) = A_{ff}T^{1/2}\left(1 + \frac{T_*}{T}\right)\rho^2 \text{ erg cm}^{-3}\text{ s}^{-1},$$

where

$$A_{ff} = \frac{2^{9/2}\,\pi^{1/2}\,e^4\alpha k^{1/2}}{3^{3/2}\,m_e^{3/2}\,m_p^2\,c^2} \quad \text{and} \quad T_* = \alpha^2 mc^2/k = 10^{5.5} \text{ K}.$$

Here A_{ff}, $A_{bf} = A_{ff}T_*\phi(T)$ are the free-free and bound-free emission coefficients and $\phi_1(T)$ a slowly varying function which we set equal to one.

Silk (1976) now expresses the Jeans' mass (Section VII.1) through the use of $\rho_R(T)$ and obtains

$$\mathfrak{M}_J = \left(\frac{k}{\mu}\right)^{1/2}\frac{\pi^2}{(24)^{1/2}\,G^2}\,A_{ff}T\left(1 + \frac{T_*}{T}\right)\left(\frac{\rho_R}{\rho}\right)^{1/2}$$

for a protogalaxy with temperature T and density ρ which loses energy through the cooling function $L_*(T)$. For $T \ll T_*$ a characteristic mass $\mathfrak{M}_*$ is obtained through elimination of (ρ_R/ρ),

$$\mathfrak{M}_* = \tfrac{1}{6}\pi\,\mathfrak{M}_J = \frac{2^{5/2}\pi^{7/8}}{27}\left(\frac{\hbar c}{Gm_p}\right)^2\alpha^5\left(\frac{m_p}{m_c}\right)^{1/2}m_p$$

after setting $\mu = (\frac{1}{2})m_p$. Also a characteristic radius can be derived doing the same with the Jeans' length (Section VII.1),

$$R_J = \pi\frac{(\mu/k)^{/21}}{(24)^{1/2}\,G}\left(\frac{\rho}{\rho_R}\right)^{1/2}A_{ff}\left(1 + \frac{T_*}{T}\right)$$

which, for $T \gg T_*$ reduces to

$$R_* = \tfrac{1}{2}R_J = (2\,\sqrt{\pi}/9)(\mu/m_p)^{1/2}\left(\frac{\hbar c}{Gm_p^2}\right)(m_p/m_e)^{1/2}\,(h/m_e c).$$

The expressions for $\mathfrak{M}_*$, R_* were originally given by Ostriker (1975). When values for $L(T)$, which include the cooling due to He and other refinements are used, the characteristic mass and radius become

$$\mathfrak{M}^* \approx 10^{12} \text{ solar masses}; \qquad R_* \approx 75 \text{ kpc}. \qquad\qquad \text{(VII.20)}$$

Protogalaxies with $\mathfrak{M} \leqslant \mathfrak{M}_*$ have $\tau_c < \tau_G$ for all densities larger than $\rho_p(o)$, while those with $\mathfrak{M} \gg \mathfrak{M}_*$ have $\tau_c < \tau_G$ provided their radii were $R < R_*$. If $R > R_*$. collapse will heat up the gas to its virial equilibrium temperature and then will begin to cool quasi-statically (region II) until the critical radius R_* is reached (region III). Here it cools quickly to 10^4 K and free-fall collapse begins. At this point star formation proceeds through isothermal stationary state leading to a hierarchy of fragmentations (Section VII.4.1), to end up with a dissipationless collapse (Section VII.5).

Protogalaxies with $\mathfrak{M} < \mathfrak{M}_*$ would go sooner through the star formation episode, while those more massive than $\mathfrak{M}_*$, although with low density, would spend a fairly large 'waiting time' under quasi-static contraction. In some cases that time is longer than the Hubble time. In this way $\mathfrak{M}_*$ assumes a significant role to define the upper end of the mass function for galaxies.

VII.4.1. STATIONARY STATE OUT OF EQUILIBRIUM

The usual equilibrium configuration of a 'laboratory' gas is one of uniform temperature and density as a consequence of the short-range forces in it. A self-gravitating system however, is controlled by long range forces and cannot remain uniform, but tends instead to collapse. There is no maximum for the entropy for such a system, at least in principle, but internal forces may prevent, or at least delay, the ultimate collapse (Islam, 1977). In this way the entropy of a self-gravitating system achieves a local maximum. The entropy production comes from the collapse itself, the fluctuations, and the growth of structures. The outcome of the process is a more ordinate configuration whose entropy production has been partially radiated away and partially used up to increase the internal energy of the system.

If the end products of the process is a great number of small-cross-section objects like stars, the internal heat goes to dynamical motions, which prevents the system from further collapse (actually, it delays the collapse of the system because of the increased relaxation time).

Fragmentation and growth of structures become in this way an efficient mechanism to convert a large scale collapse into a hierarchy of local small scale ones.

Let us consider now a protogalaxy with mass $\mathfrak{M}$, volume V, mean density $\rho_{PG} = \mathfrak{M}/V$ and temperature T. The balance equation for the entropy S of the system is

$$T \, \delta S + J_q \, \delta t = -\delta W, \qquad\qquad \text{(VII.21)}$$

where J_q is the heat flux exchanged with the surrounding medium (in our case through radiation) and W the gravitational energy.

Assume now the gas is perfect, with pressure P, molecular weight μ and internal energy $(U/\mu) = 3PV/2$. In the presence of inhomogeneities, the change δW of the gravitational energy is composed of two contributions: $\delta_i W$ due to the collapse of the inhomogeneities and $\delta_T W$, originated in the tidal interactions between them. For a homogeneous cloud we have $\delta_T W = 0$, while in general is $\delta W = \delta_i W + \delta_T W$.

To take into consideration these facts, let us introduce the fluxes J_i, J_T through

$$W_i J_i \delta t = \delta_i W - P\,\delta V \quad \text{and} \quad W_T J_T \delta t = \delta_T W - \Pi\,\delta V, \qquad \text{(VII.22)}$$

where Π is the pressure of the mass motions associated with the mutual displacement of the inhomogeneities, while the coefficients W_i, W_T will be defined in later sections.

When we take these fluxes to balance Equation (VII.22), namely

$$T\,\delta S + J_q\,\delta t = -\,W_i J_i\,\delta t - W_T J_T\,\delta t - (P + \Pi)\,\delta V$$

a stationary state may be defined with the conditions

$$J_i = J_T = 0 \quad \text{and hence} \quad \delta S = 0.$$

We have in consequence the relations

$$\delta_i W = P\,\delta V \qquad\qquad\qquad\qquad\qquad\qquad \text{(VII.22)}$$

$$\delta_T W = \Pi\,\delta V \qquad\qquad\qquad\qquad\qquad\qquad \text{(VII.23)}$$

$$J_q\,\delta t = -(P + \Pi)\,\delta V \qquad\qquad\qquad\qquad \text{(VII.24)}$$

to describe the properties of the protogalactic cloud in such a condition.

The entropy production is different from zero, so the maximum of S is a local one. This situation conserves the net value of the entropy ($\delta S = 0$) while the excess produced is radiated away as it is seen in Equation (VII.24). This possibility requires the system to be open or, in other words, the gas of the protogalaxy has to be transparent for the radiation to go unimpeded through it.

VII.4.2. HIERARCHY OF FRAGMENTATIONS

Let λ be the compression factor which describes the change in linear scale of a cloud during collapse. It is easy to see that $\delta \ln V = -3 \ln \lambda$ and $P\,\delta V = -(2U/\mu) \ln \lambda$ for an elementary contraction process.

Since contraction heats up the gas in the adiabatic regime, a change is to be expected in the molecular weight originated in the ionization. Equation (VII.24) together with the balance for the internal energy in the cloud,

$$\delta(U/\mu) + P\,\delta V = -J_q\,\delta t$$

gives

$$\delta(T/\mu) = -T\left(\frac{3\Pi V}{2U}\right) 2 \ln \lambda \qquad\qquad\qquad \text{(VII.25)}$$

an equation for the temperature, which reduces to a constant value when the energy of mass motions can be neglected with respect to the internal energy of the gas. We shall see in the next section that this is the case, so that the stationary state is also a very approximate isothermal one.

Because $W_i = W(\delta \ln W/\delta \lambda)_0$, where the subscript means to keep all other parameters constant, the variation of the gravitational energy originated in the collapse is

$$\delta_i W = W(\lambda - 1)$$

for the same elementary process. $W_i = W$ as a consequence of the homogeneous

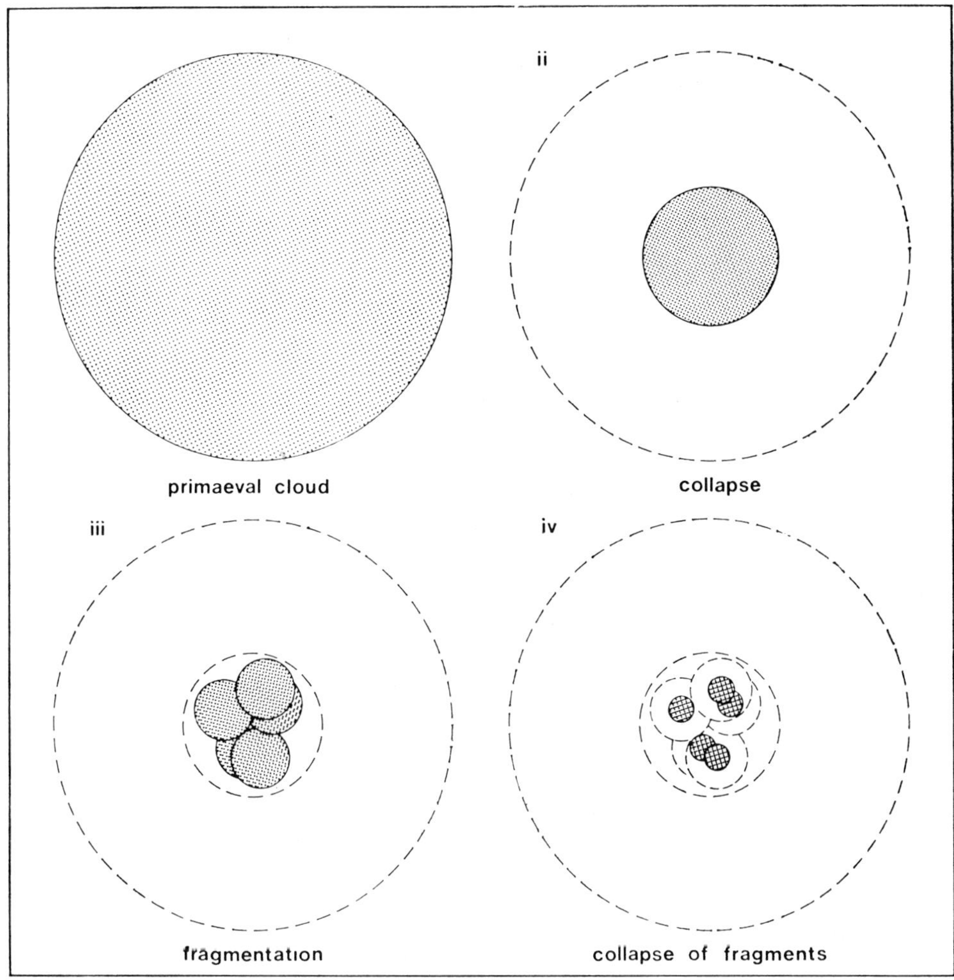

Fig. VII.2. Hoyle's hierarchical fragmentation. A protogalaxy (i), collapses reducing its radius a factor λ (ii), and fragments in $\lambda^{3/2}$ clouds (iii), which repeat the cycle (iv), many times until opacity break the isothermal condition.

character of W with respect to λ. From Equation (VII.22) and the isothermal nature of the process, we get

$$\mu = \left| \frac{2U}{W} \right| \frac{\ln \lambda}{\lambda - 1} \qquad \text{(VII.26)}$$

a relation first considered by Hoyle (1953) to estimate the maximum value of λ allowed for the contraction of an isothermal cloud.

Equation (VII.26) defines Hoyle's hierarchy of fragmentations (Figure VII.2). Starting from marginal instability ($2U + W \approx 0$) the protogalactic cloud collapses by a factor λ, its density ρ_{PG} going up to $\rho_{PG}\lambda^3$. At this stage fragmentation proceeds in N clouds with mass $\mathfrak{M}/N$ and radius R/N, again in marginal instability, so that Equation (VII.26) now applies to the fragments. Since the density of the collapsed cloud has to be equal to the density of the fragments prior to their own collapses, it follows at once that $N^2 = \lambda^3$.

If we introduce the step s in the hierarchy as an independent variable, the number of fragments $N(s)$, its density $\rho(s)$ and the time-scale for collapse are given by

$$N(s) = (\lambda^{3/2})^s; \qquad \rho(s) = \lambda^3 \rho_{PG} \quad \text{and} \quad \delta t(s) = \delta t(o) 3 \ln \lambda / N(s)$$

as follows from $\delta \ln \rho = 3 \ln \lambda$ and $\delta t(o) = (G\rho_{PG})^{-1/2}$. We shall use these relations in the following.

The realization of the hierarchy of fragmentations as a consequence of a stationary state with entropy production places this process within a whole class of physical phenomena. Saslaw (1968, 1969) has called attention to a close similitude between a phase transition in which a vapor condenses to a liquid and the condensation of discrete objects from a self-gravitating gas.

The hierarchy of fragmentations is also an example of the so-called 'dissipative structures' by Glandsorff and Prigogine (1971), which have been recognized to exist only far from equilibrium. These disorder-order processes only happen in open systems because the system has to get rid of its excess entropy in order to keep its degree of organization. This is done through dissipative structures which essentially consist in stationary flow patterns whose classical example is that of Bénard cells in a liquid heated from below. The particular self-gravitating nature of the astronomical systems imposes an irreversible trend towards more compact structures through a change of scale, although keeping the structural pattern.

VII.4.3. THE END OF THE HIERARCHY.

The change of the Jeans' mass $\mathfrak{M}_J(s)$ from one step to the next is

$$\delta \ln \mathfrak{M}_J = (3\gamma - 4)\tfrac{3}{2} \ln \lambda = -\tfrac{3}{2} \ln \lambda$$

because $\gamma = 1$ in the isothermal regime characteristic of the process. Since the proto-galaxy satisfies initially the instability condition $\mathfrak{M} > \mathfrak{M}_J$, this condition is fulfilled in successive steps provided the stationary state is maintained.

When the fragments become opaque enough to reabsorb its own radiation, the isothermality is broken, γ is larger than $\tfrac{4}{3}$ and the Jeans' mass would increase. This marks the final step (s_f) of the process, as well as the end of the stationary state defined in Section VII.4.1.

At the last step of the hierarchy the typical mass of a fragment is close to $\mathfrak{M}_J(s_f) \approx \mathfrak{M}_f^*$, which means the necessary condition for fragmentation to continue is that opacity should not impede a fragment from radiating its gravitational energy in a collapse time (Rees, 1976).

Absorption in a fragment becomes important when the absorption in the radiation field becomes comparable with the energy radiated by the gas, which implies the build up of a black-body distribution. In that condition the surface brightness of the fragment approaches that of a black-body and we may write for it

$$\delta_i W = 4\pi a c T_f^4 \, \delta t(s_f).$$

Since $\delta_i W = -G(\mathfrak{M}_f^2/R_f)(\lambda - 1)$ and $\delta t(s_f) = (16\pi G\rho_f)^{-1/2}$, we obtain after remembering (Section VII.1.1) the definition of the Jeans' mass associated to T_f and ρ_f,

$$\mathfrak{M}_J(s_f) = (4\sqrt{\pi}/3acG)^{1/2} (\pi k/m_p\mu)^{9/4} (\mathfrak{M}_f/\mathfrak{M}_J(s_f)^{1/2} T^{1/4}(\lambda - 1)^{1/2}.$$

Rees (1976) introduces the Chandrasekhar (1937) mass

$$\mathfrak{M}_{CH} = (\hbar c/Gm_p^2)^{3/2} mp$$

after remembering that $a = \pi k^4/15c^3\hbar^3$. With further handling we get

$$\mathfrak{M}_f \approx \mathfrak{M}_J(s_f) \approx \mathfrak{M}_{CH} \frac{\sqrt{5}}{2}\pi^{3/2}\left(\frac{kT_f}{m_pc^2}\right)^{1/4}(\lambda - 1)^{1/2}\mu^{-9/4} \qquad \text{(VII.21)}$$

an expression for the mass of the final fragments which will become the stars of the galaxy just formed. When numbers are introduced into Equation (VII.27), we obtain

$$\mathfrak{M}_f \approx 1.4\left(\frac{2}{\mu}\right)^{9/4}\left(\frac{T}{5000\ \text{K}}\right)^{1/4} \text{solar masses,}$$

with a value $\lambda = 3.5$.

It is worth noticing that $\mathfrak{M}_f$ is insensitive to the final temperature because of the power $\frac{1}{4}$, and also to composition, because μ cannot change very much for reasonable compositions. On the other hand, observes Rees (1976), the appearance of $\mathfrak{M}_{CH}$ in Equation (VII.27) explains why the final mass is a stellar one and not for example, a stellar cluster.

The efficiency of the fragmentation process is not complete however. A mass spectrum is to be expected at each step, and these fragments with masses smaller than the Jeans' mass at step s, $\mathfrak{M}(s)$, will stop fragmenting, and remain as inter-fragment gas. The stationary nature of the process requires, moreover, the mass spectrum to be a function independent of s, namely $F(\mu)$, where $\mu = \mathfrak{M}'/\mathfrak{M}_s$, $\mathfrak{M}'$ being the mass of a given fragment and $\mathfrak{M}_s$ the average mass of the fragments at step s.

The cut-off in the hierarchy will not be sharp because some of the fragments which are more massive than $\mathfrak{M}_f \approx \mathfrak{M}_f^*$, will survive as stars before the last step is attained (Fowler and Hoyle, 1967).

Let be $P(s)$ the probability a fragment in step s has for dropping out of the hierarchy, that is to say, not to fragment any more. We have

$$\mathfrak{M}_sN(s) = \mathfrak{M} - \sum_{k=1}^{s-1}\mathfrak{M}_kN(k)P(k),$$

where $\mathfrak{M}$ is the protogalactic mass. Notice the summation in the right side does not distinguish between fragments with masses smaller than $\mathfrak{M}_J(s)$ or stars more massive than $\mathfrak{M}_f^*$. In consequence the left side gives the mass which remains in the hierarchy at step s. Let $\eta(s)$ be defined by $\mathfrak{M}\eta(s) = \mathfrak{M}_sN(s)$, we have

$$\eta(s + 1) = \eta(s)[1 - P(s)]$$

with $P(o) = 0$, an equation whose solution is

$$\eta(s) = \prod_{k=1}^{s}[1 - P(k)] \qquad \text{(VII.28)}$$

The knowledge of $P(s)$ will provide values of $\eta(s)$. While fragmentation proceeds unimpeded, let us say for $s < s'$, the probability $P(s)$ will be a constant, namely

$$P(s) = h \tag{VII.29}$$

because $F(\mu)$ remains the same through the successive steps. After s' is reached, we have to consider star formation and the relevant Jeans' mass is now $\mathfrak{M}_s^*$, which is independent of s. The probability $P(s)$ now the average fragment has to become a stable star depends on the opacity of the fragment. Since the opacity is proportional to the density and to a typical length such as the radius R of the fragment, we have $P(s) \propto \rho R \propto \lambda^{2s}$ for $s \geqslant s'$. If we assume that all fragments at step s_f become stars, we readily get

$$P(s) = \lambda^{2(s-s_f)} \tag{VII.30}$$

The transition step s' where the two regimes meet is fixed noticing that $\mathfrak{M}_f(s') = \mathfrak{M}_{s'}^*$ is the maximum mass for a stable star ($\mathfrak{M}_{s'}^* \approx 60$ to 100 solar masses). With Equations (VII.29) and (VII.30) Equation (VII.28) gives

$$\eta(s') = (1 - h)^{s'} \quad \text{and} \quad \eta(s_f) = 0$$

which means the hierarchic fragmentation wastes a fraction $1 - \eta$ of the protogalactic mass as a consequence of the spread in the masses of the fragments, but uses all of the remaining fraction of the total mass $\eta(s')$ to efficiently convert it into stars.

The cumulative mass function $\xi(\mathfrak{M}_s)$ for stars is

$$\xi(\mathfrak{M}_s) = \sum_{s'}^{s} P(s) \qquad s \geqslant s'$$

or

$$\xi(\mathfrak{M}) \propto \left(\frac{\mathfrak{M}^*}{\mathfrak{M}} \right)^{4/3} - 1$$

after using Equation (VII.30), introducing the upper mass limit $\mathfrak{M}^*$ and dropping the index s as irrelevant. The exponent β in a power law approximation $\xi(\mathfrak{M}) \propto \mathfrak{M}^\beta$ gives

$$\beta = \frac{d \ln \xi(\mathfrak{M})}{d \ln \mathfrak{M}} = -\frac{4}{3} \left[1 + \left(\frac{\mathfrak{M}}{\mathfrak{M}^*} \right)^{4/3} \right]$$

and approaches the value $\frac{4}{3}$ for $\mathfrak{M} \ll \mathfrak{M}^*$ as in the case for Salpeter's empirical initial mass function (IMF) (Section II.3.2).

VII.4.4. Global Dynamics

The variation of the gravitational energy due to the changing mutual interactions between the increasing number of fragments is

$$\delta_T W = W \frac{\delta \ln N}{N - 1}$$

because $W_T = W(\delta \ln W / \delta \ln N)_0 = W/(N - 1)$ provided $N > 1$. From Equations (VII.3) and (VII.5) we obtain

$$\frac{3\Pi V}{2U} = \frac{3}{2(N-1)} \tag{VII.31}$$

an expression for the energy of mass motions in the cloud. The further fragmentation advances in the protogalaxy, the less gravitational energy is available for mass motions, their influence being noticeable only at the early steps of the hierarchy.

Equation (VII.31) justifies the assumption of the isothermal character of the process, since $(3\Pi V/2U)$ becomes very small after a few steps and the dissipation of energy reduces to a point where the temperature stabilizes at a constant value given by Equation (VII.25).

Because most of the hierarchical structure is subsequently disrupted by tidal interactions, the stored gravitational energy reverts to the protogalaxy as a whole. From Equations (VII.2), (VII.23) and (VII.27) we find for the total gravitational energy

$$W(s) = W - (2U/\mu)[\tfrac{2}{3} \ln N(s) + K(s)]$$

after summation over the first s steps. $K(s)$ is a slowly increasing function of s which approaches the asymptotic value $K(s_f) \approx 0.2$ after a few steps, for $\lambda = 3.5$.

Now the virial $2(U/\mu) + 3\Pi V + W(s)$ can be computed for the protogalaxy as a whole for any step of the hierarchy. When star formation is completed, at $N_f \approx (\lambda^{3/2})^{s_f}$, the system is in gross dynamical disequilibrium because

$$\text{VIRIAL} \approx -\tfrac{2}{3}(2U/\mu) \ln N_f = -(2U/\mu)S_f \ln \lambda.$$

Simultaneously the stars just formed (Section VII.4.4) go through nondissipative collapse and violent relaxation takes place (Section VII.5). Because of the rapid shortening of the time-scales for sucessive steps, the fragmentation process becomes equivalent to a sudden transformation of a dissipative transparent gas cloud into a system of point masses (stars) out of equilibrium which relaxes towards an equilibrium configuration of lower gravitational energy W_f given by

$$W_f \approx W - \tfrac{2}{3}(2U/\mu) \ln N_f = W - (2U/\mu)S_f \ln \lambda.$$

The compression factor L between the protogalaxy, when the hierarchy starts, and the galaxy, when violent relaxation has already decayed, is

$$L = (W_f/W) \approx (\tfrac{2}{3}\mu) \ln N_f = (s_f/\mu) \ln \lambda = s_f(\lambda - 1)$$

and the ratio between their respective mean densities results $\rho_G/\rho_{PG} = L^3$.

VII.4.5. THE INFLUENCE OF ROTATION

The origin of the angular momentum in galaxies was considered by Hoyle (1951) and later by Peebles (1969), within the frame of the gravitational instability theory. Also Zeldovich and Sunyaev (1969) and Doroshkevitch (1973) have proposed models of the same subject.

The leading idea is that the torque causing the spin-up is due to tidal forces acting on the distortions produced in the protogalaxy by masses passing by. The tidally induced motions possess a net angular momentum whose magnitude was estimated to be of the order of

$$|H| \approx 3\,(G\,\mathfrak{M}\mathfrak{M}_0/a^3)\theta R^2\,\Delta t,$$

where $\mathfrak{M}, \mathfrak{M}_0$ are the masses of the protogalaxy and the disturbing system respectively,

a its separation, θ a geometric factor which describes the orientation and shape of the protogalaxy, R the radius of gyration and Δt the time of the interaction.

Although Harrison (1971) has criticized Peebles' calculations, numerical simulations by Eftathiou and Jones (1980) seem to give support to the model.

Let us consider here how the results of the preceding sections are modified in the presence of a conserved angular momentum in the protogalaxy. If T_R is the kinetic energy of rotational motions, the marginal instability condition for the protogalaxy becomes

$$2U + 2T_R + W \approx 0$$

and consequently we have

$$\left| \frac{2U}{W} \right| = 1 - \left| \frac{2T_R}{W} \right| = 1 - 2t_i,$$

where t_i is the absolute value of the ratio between the initial values of the kinetic energy T_R of rotational motions and the gravitational energy W. Since the fragmentation process is a local one, at least in a first approximation, Equation (VII.22) is still valid in the presence of rotation, and Equation (VII.26) becomes

$$\frac{\ln \lambda}{\lambda - 1} = \frac{\mu}{1 - 2t_i}. \tag{VII.32}$$

The role of rotation is now apparent: it reduces the value of the compression factor λ. More steps in the hierarchy would be necessary to achieve the same fraction of the total mass to be converted into stars in a galaxy with rotation than in another without rotation. However, because s' is fixed by the ratio $\mathfrak{M}/\mathfrak{M}_s^*$, for two galaxies with the same total mass $\mathfrak{M}$, the gas content $1 - \eta$ is larger in that with larger t_i. In this way the model describes the presence of gas in galaxies with rotation, such as spirals and irregulars.

The global properties of galaxies depend on whether the angular momentum is acquired before or after stellar formation proceeds, at least as extreme possibilities.

If stellar formation is completed in a system with $t_i = 0$, λ has its maximum value, stellar formation is very efficient and any small rotational energy acquired in the meantime through interactions with neighboring galaxies will lead to a dissipationless collapse with rotation as discussed (Section II.3.1). The galaxies resulting from such a process are ellipticals and their gas content is low.

If the protogalaxy already has a small but finite rotational energy T_R prior to star formation, the collapse will be dissipative, the value of λ given by Equation (VII.32) is smaller and star formation becomes less efficient.

We may represent the final configuration of the system after collapse with McLaurin spheroids as we did in Chapter II. Although this procedure is only an approximation, it is enough to put in evidence the general properties of the process.

Before star formation begins, the protogalaxy is assumed to be in virial equilibrium, with rotational energy T_R. At that epoch we have the virial condition $2E = W$, while the final configuration of the relaxed galaxy after collapse has $2E_f = W_f$. Since angular momentum conservation is assumed, the kinetic energy becomes $T_R L^2$, L being the overall compression factor introduced before. With the McLaurin spheroid representation, we have

$$W_f = WLF(e) \quad \text{and} \quad 2T_R L^2 = -WLG(e)$$

where $F(e)$ and $G(e)$ are functions of the spheroid excentricity e. When $F(e)$, $G(e)$ are computed, it is found that F remains very close to one and G increases from zero to about 0.5, between $e = 0$ and $e^* \approx 0.98$. For $e > e^*$ both F and G go to zero at $e = 1$.

Since the total energy of the system has decreased through dissipation, $E_f - E = \frac{1}{2}(W_f - W) \approx -\frac{1}{2}(2U/\mu)s_f \ln \lambda$, and we have

$$L \approx 1 + (\lambda - 1)s_f; \qquad G(e) \approx t_i[1 + (\lambda - 1)s_f]$$

after using Equation (VII.32). These expressions reduce to the spherically symmetric case for $t_i = 0$. Since s_f is about 15 ± 3 for the range of masses observed among galaxies, t_i has to be of the order of $[2(\lambda - 1)s_f]^{-1} = 0.01$ to have stable final configurations.

Dissipative collapse with non-zero initial rotation leads, in consequence, to very flattened rotating systems with a sensible fraction of their mass still in gaseous form. The flattening of the system can be typically described by that of the ellipsoid which maximizes T_R, so that the true axial ratio $q_0 = c/a = \sqrt{1 - e_*^2} \approx 0.2$ as is the case of spiral and SO galaxies (Section I.3).

The extreme situation is reached when $t_i = (\frac{1}{2})(1 - \mu)$ in which case Equation (VII.32) gives $\lambda = 1$ and the hierarchy of fragmentations is aborted: the protogalaxy remains forever as a gaseous spheroid.

VII.5. Dissipationless Collapse

If a protogalaxy has converted all its gas into stars before its collapse is completed, a negligible amount of binding energy has been released during collapse, which becomes dissipationless. The maximum compression stage will be reached then when the age of the Universe is twice the time t_p when the protogalaxy had its most extended radius R_p (cf. Section VII.3.1). The characteristic formation time $t_c = 2t_p = (8\pi G\rho_p/3)^{-1/2}$ is related to the redshift in an Einstein–de Sitter universe through $t_c = (\frac{2}{3}H_0)(1 + z_c)^{-3/2}$. At the point of maximum expansion (t_p) the kinetic energy is small and the total energy is $E = W_p = -3GM^2/5R_p$. Since the collapse is dissipationless (Gott, 1977), the galaxy will reach virial equilibrium $2T + W = 0$ with the same energy $E = W_0 = (\frac{1}{2})W$ at an effective final radius $R_f \approx \frac{1}{2}R_p$. The mean density will be $\bar{\rho} \approx 8\rho_p$ and the redshift at which the galaxy completes a dissipationless collapse becomes

$$1 + z_c \approx 27\left(\frac{\bar{\rho}}{10^{-24}}\right)^{1/3}$$

when $\bar{\rho}$ is expressed in g cm^{-3} and $H_0 = 100$ km s^{-1} Mpc^{-1}.

The importance of violent relaxation for galaxy structure was first realized by Lynden–Bell (1967). A bound stellar system which is out of dynamical equilibrium goes through damped oscillations until it settles into an equilibrium configuration. A given star of mass m moves along a path without conservation of the mechanical energy mh, and we have

$$\frac{dh}{dt} = \frac{\delta\phi}{\delta t}.$$

Although the total energy is conserved, strong phase shifts take place between the star orbits. In this way the large scale oscillations go through Landau damping and the galaxy settles down in a time scale which is approximately $\frac{3}{2}$ times the collapse time t_p, as follows from numerical experiments done by several authors (Gott, 1977).

The numerical simulations made by this latter author (Gott, 1973, 1975) show that the density distribution produced by the initial collapse is highly concentrated and closely resembles the one observed in elliptical galaxies.

A rotating stellar system going through nondissipative collapse was discussed already in Section (II.3.1) and provided evidence in favor of elliptical galaxies with all the observed flattenings formed through this mechanism.

References

Bonnor, W.: 1957, *Monthly Notices Roy. Astron. Soc.* **117**, 104.

Chandrasekhar, S.: 1937, *Nature* **139** (3522), 757.

Doroshkevitch, A.G.: 1973, *Astrophys. Letters.* **14**, 11.

Eftathiou, G. and Jones, B.T.: 1980, *Comm. Astrophys.* **8**, 169.

Field, G.: 1975, *Stars and Stellar Systems* IX, Univ. of Chicago Press, p. 359.

Fowler, W.A. and Hoyle, F.: 1967, *Roy. Obs. Bull.* No. 67.

Glansdorff, P. and Prigogine, I.: 1971, *Structure, Stabilité et Fluctuations*, Masson, Paris.

Gott, R.J.: 1973, *Astrophys. J.* **186**, 482.

Gott, R.J.: 1975, *Astrophys. J.* **201**, 296.

Gott, R.J.: 1977, *Ann. Rev. Astron. Astrophys.* **15**, 235.

Gott, R.J.: 1980, Les Houches Lectures. Preprint.

Harrison, E.R.: 1971, *Monthly Notices Roy. Astron. Soc.* **154**, 167.

Harrison, E.R.: 1973, in E. Schatzman (ed.), *Cargese Lectures*, Vol. 6.

Hoyle, F.: 1951, in J.M. Burgers and H.C. van de Hulst (eds.), *Problems of Cosmical Aerodynamics*, International Union of Theoretical and Applied Mechanics and IAU, p. 195.

Hoyle, F.: 1953, *Astrophys. J.* **118**, 513.

Islam, J.N.: 1977, *Quart. J. Roy. Astron. Soc.* **18**, 3.

Jones, B.J.J.: 1976, *Rev. Mod. Phys.* **48**, 107.

Lifshitz, E.M.: 1946, *J. Phys. URSS* **10**, 116.

Lynden-Bell, D.: 1967, *Monthly Notices Roy Astron. Soc.* **136**, 101.

McCrea, W.H. and Milne, E.A.: 1934, *Q. J. Math.* **5**, 73.

Ostriker, J.P.: 1974, Quoted by Rees *et al.* (1977).

Peebles, P.J.E.: 1967, *Astrophys. J.* **147**, 859.

Peebles, P.J.E.: 1969, *Astrophy. J.* **155**, 393.

Penston, M.V.: 1969, *Monthly Notices Roy. Astron. Soc.* **144**, 425.

Rees, M.: 1976, *Monthly Notices Roy. Astron.Soc.* **176**, 483.

Rees, M. and Ostriker, J.P.: 1977, *Monthly Notices Roy. Astron. Soc,* **179**, 541.

Saslaw, W.C.: 1968 *Monthly Notices Roy. Astron. Soc.* **141**, 1.

Saslaw, W.C.: 1969, *Monthly Notices Roy. Astron. Soc.* **143**, 437.

Silk, J.: 1976, *Astrophys. J.* **211**, 638.

Zeldovich, Ya.B. and Sunyaev, R.A.: 1979, *Astrophy. Space Sci.* **4**, 301.

Bibliography

Hayli, A. (ed.): 1975, 'Dynamics of Stellar Systems', *IAU Symp.* **69**.

Peebles, P.J.: 1971, *Physical Cosmology*, Princeton Univ. Press.

Sandage, A.R., Sandage, M., and Kristian, J. (ed.): 1969, *Galaxies and the Universe*, Vol. IX of 'Stars and Stellar Systems', Chicago Univ. Press.

Shakeshaff, J. (ed.): 1974, 'Formation and Dynamics of Galaxies', *IAU Symp.* **58**.

Setti, G. (ed.): 1975, *Structure and Evolution of Galaxies*, D. Reidel Publ. Co., Dordrecht, Holland.

CHAPTER VIII

NOTES AND COMMENTS

We include in this chapter those subjects which are complementary to the main text. Some of the references are found at the end of the chapters which are complemented.

VIII.1. Catalogs and Atlas of Galaxies (cf. Chapter I)

Both for historical and practical reasons we give a brief description of the most used catalogs and atlas of galaxies.

Messier's Catalogue (M): Published first in the Connaissance des Temps for 1783 and 1784. Contains star clusters, gaseous nebulae and galaxies in 103 entries, all of them of diffuse appearance when observed with a small eyeglass.

Examples:

M31 = The Andromeda Galaxy. M83 = The Galaxy in Hydra.

M33 = The Galaxy in Triangulum. M87 = The elliptical galaxy in Virgo.

M81 = The spiral galaxy in Ursa Major.

New General Catalogue (NGC) and the Index Catalogues (IC): Published in 1888 J. L. E. Dreyer sponsored by the Royal Astronomical Society. The aim of this work was to revise, correct and enlarge the General Catalogue by Sir J. Herschel (1864). Since its publication, Dreyer's Catalogue has been a universal reference for nebulae, star clusters and bright galaxies with the abreviation NGC in astronomical literature. Herschel's General Catalogue contributed 5079 entries to the total of 7840 in the NGC. Whereas the remaining were taken from about fifty other additional sources. It was thought to contain all nebulae and clusters known at that time. There exist, understandably, an ample variation in quality between entries on account of the diversity of sources.

Two Index Catalogues IC_1 and IC_2 were added in 1895 and 1908.

The coordinates (RA and NPD) are given for the 1860 equinox, plus a synthetic description and references to the sources.

Examples:

M31 = NGC 224   M83 = NGC 5236

M33 = NGC 598   M87 = NGC 4486

M81 = NGC 3031.

The Shapley–Ames (SA) Catalogue: Published in Harvard Annals No. 2 (1932). It contains 1249 galaxies brighter than the photographic magnitude 12.9 and gives apparent magnitude, apparent dimensions and morphological types based on homogeneous criteria. de Vaucouleurs has studied the systematic errors of the magnitudes and dimensions in this catalogue.

Photometry of 300 Galaxies (1958): Published by E. Holmberg, contains P magnitudes and $P - V$ colors as well as isophotal dimensions for 300 boreal galaxies.

Palomar Sky Survey: A joint project between the National Geographical Society and the Palomar Observatory (Minkowski and Abell, 1963). Covers the sky north of declination $-27°$ with 935 pairs of photographs in the blue and red ranges. The 1.2 m Schmidt Camera at Palomar Mountain was used and the plate scale is 1 mm = $67''4$. The blue exposures were made on 103aO emulsions and the red ones on 103aE plus a red filter very similar to the Wratten 29.

Morphological Catalogue of Galaxies (1962–1964): By V. A. Vorontsov-Velyaminov, contains a detailed description of peculiar and interacting galaxies based on the PSS prints.

Reference Catalogue of Bright Galaxies (I and II): The first edition (I) by G. and A. de Vaucouleurs (1964) revised and considerably extended the SA Catalogue to contain 2599 galaxies, many of them fainter than the 13th photographic magnitude.

A second edition (with the collaboration of H. Corwin) appeared in 1976, which enlarged and revised the former. It gives 2×10^5 information elements regarding 4364 galaxies, mostly brighter than 16th magnitude. A special effort was made to include the major peculiar objects, blue objects, radio galaxies, compact, dwarf galaxies, interacting systems, etc.

Catalogue of Galaxies and Clusters of Galaxies (1961–1968): Six volumes edited by F. Zwicky and his associates. It contains, in correspondence to each PSS print, the contours of galaxy clusters and the coordinates and apparent magnitudes of individual galaxies up to $m_v = 14.5$. The catalog covers the same area of the Sky as the PSS.

The Revised New General Catalogue of Non Stellar Astronomical Objects (RNGC): Edited by the University of Arizona and prepared by J. Sulentic and W. G. Tifft (1973). It is essentially a revision of the NGC giving positions, magnitudes and references for 5000 NGC objects plus fifty corrections to the original Dreyer's catalogue. Each object in the Catalogue was checked in the Palomar Sky Survey, thus a modern description is available, including faint structures unnoticed by earlier visual observers.

Uppsala General Catalogue of Galaxies: Prepared by P. Nilson (1973). It contains 12921 galaxies north of Declination $-2°5$. It is based on the Palomar Sky Survey photographs and is assumed to be complete up to a limiting diameter of 1 arc min on the blue prints of the Sky Survey It includes also objects up to apparent magnitude 14.5 although it contains some with smaller apparent diameters taken from Zwicky's catalogue of galaxies and clusters.

Atlas of Peculiar Galaxies (1966): A collection of photographs of peculiar galaxies obtained by H. Arp mostly with the 5 m Palomar telescope.

Hubble Atlas of Galaxies (1961): A. R. Sandage defines in this very important atlas the modified Hubble system of classification for galaxies.

Reynolds Survey (1956): A homogeneous work on galaxies south of declination $-35°$

published by de Vaucouleurs (1956) and based on material obtained in Mt Stromlo with the Reynolds (0.75 m) reflector. It gives coordinates of many non-catalogued objects morphological types and approximate total magnitudes, besides an interesting study of the space distribution of southern galaxies.

VIII.2. Composite Spectra (Cf. Section II.1.2)

The spectroscopic information on the stellar content of galaxies can only be obtained from the composite spectrum of their main bodies (Section I.2). Let $f(M, C)$ be the spectral energy distribution emerging from a star with absolute magnitude M and color C (in a definite photometric system on which f_λ is univocally defined). Let also $N(M, C)\,dM\,dC$ be the number of stars in the cell $dM\,dC$ centered at M, C in the Magnitude–Color diagram of the observed region. The resulting integrated spectrum is then

$$F_\lambda = \iint f_\lambda(M, C) N(M, C)\,dM\,dC$$

the integration being extended over all values of M and C. The consideration of the velocity distribution of the individual stars requires the use of the convolved function $f_\lambda N$ with a function of λ corresponding to the line of sight velocity distribution, which is generally unknown.

Quantitative analysis of Composite Spectra

The shape of the absorption lines in a composite spectrum contains in principle, information on the luminosity and spectral type distribution of the stellar population (Whipple, 1935). The measurement of line profiles, however, is difficult to make from the low dispersion spectra used in extragalactic astronomy. The velocity dispersion of the individual stars distorts the profiles and some times becomes a dominant effect (Section II.2.1). On the other hand, the line intensities can be measured rather precisely in low dispersion spectra and are independent of the velocity dispersion.

The intensity of high excitation lines decreases rapidly toward late spectral types, the reverse happens with lines originated on law excitation states. Thus, the total and relative intensities of these two groups of lines in a composite spectrum are very sensitive to the frequency distribution of the elementary spectra assumed for the model galaxy. From the integrated color index, it is possible to obtain a defined solution for the frequencies of spectral types and luminosities.

The outlined approach is based in the following fundamental relations: Let $N(M, S)$ be the number of stars with absolute magnitude M and spectral type S in a cell $dM\,dS$ centered on M, S. The integrated luminosity from all the stars with spectral type S in the range dS will be

$$L(S) = \int_{-\infty}^{+\infty} N(S, M)\,\mathrm{Dex}(-0.4M)\,dM$$

and the total luminosity of all spectral types

$$L = \sum_k L(S_k) = \sum_k \int_{-\infty}^{+\infty} N(S_k, M)\,\mathrm{Dex}(-0.4M)\,dM.$$

The relative contribution of the spectral type S_k to the integrated surface brightness of the region, in a given wavelength λ, is given then by

$$k_\lambda(S_k') = L_\lambda(S_k)/L_\lambda.$$

This fraction varies rapidly with λ as a consequence of the different relative contributions of the stars of different spectral types. We will give next the results in terms of the visual luminosity and $k_\lambda \equiv k_v$ will then be referred to the isophotal wavelength $\lambda = 5450$ Å.

Let also $EW(s)$ be the mean equivalent HW of an absorption line in stars of spectral type S; the corresponding HW in the same line of the composite spectrum now becomes

$$\langle EW\rangle_\lambda = \sum k_\lambda(S_i)EW_\lambda(S_i)/\sum k_\lambda(S_i).$$

Finally, if $C(S)$ is the mean color index for stars of spectral type S, the composite color of the system becomes

$$\langle C_{\lambda'}\rangle = \sum k_{\lambda'}(S_i)C_{\lambda'}(S_i)/\sum k_{\lambda'}(S_i),$$

where λ' is a wavelength suitably selected between λ and λ_v.

The problem reduces now to find a plausible series of values k_λ for the six more important spectral types, which reproduce within the precision of the measurements, the observed equivalent widths and color index.

The data of equivalent widths for the absorption lines in stars can be obtained from the literature. The solution can be started with a first approximation assuming a spectral distribution similar to that in the solar neighborhood (Allen, 1955), from where through trial and error the equations can be satisfied.

G. and A. de Vaucouleurs (1959) have noticed that this type of analysis should give sufficiently detailed information on the distribution of spectral types of giant stars and the brighter part of the MS but cannot provide significant data on the dwarf population of the lower MS which only scarsely contributes to the luminosity of the system in the visual range. Through spectral work in the red and IR plus multicolor photometry beyond 9000 Å that information can be obtained.

VIII.3. The HI Spectrum (Cf. Section II.1.3)

Let us consider an electromagnetic radiation field in an emitting and absorbing gaseous medium such as neutral H. If $k(\nu)$ and $j(\nu)$, respectively, are the absorption and emission coefficients per unit volume, the energy transfer in the medium is governed by the well-known Equation

$$\frac{dI(\nu)}{dS} = j(\nu) - k(\nu)\,I(\nu)$$

whose solution is written

$$I(\nu) = \int_0^\infty j(\nu)\exp\left[-\int_0^s k(\nu)\,ds'\right]ds. \qquad \text{(VIII.1)}$$

Admitting that the wave frequencies are low enough to be in the range of radio frequencies, we can express the black body radiation by the Rayleigh-Jeans law, putting

$$I(\nu) = 2kT\nu^2/c^2$$

where k is Boltzmann constant, T the absolute temperature and c the speed of light. We can thus express $I(\nu)$ and $j(\nu)$ in terms of the temperature in the form

$$T(\nu) = I(\nu)\nu^2/2k\nu^2 \quad \text{and} \quad j(\nu) = j(\nu)c^2/2k\nu^2$$

so that Equation (VIII.1) can be written

$$T(\nu) = \int_0^\infty j(\nu) \exp\left[-\int_0^s k(\nu)\,\mathrm{d}s'\right]\mathrm{d}s \tag{VIII.2}$$

and $T(\nu)$ is known as the specific brightness temperature.

Line Formation

If we consider a specific transition in the medium such as 1–2 in the frequency ν_0, the spectrum can be separated into: (i) a continuum coming from non-specified transitions; and (ii) the 'line' coming from the transition 1–2.

Then let

$$J(\nu) = J_q(\nu_0) + J_b, \qquad k(\nu) = k_a(\nu_0) + K_b,$$

where $J_a(\nu_0)$ and $k_a(\nu_0)$ belong to the continuum, which is assumed to be constant in the neighborhood of the line. Then Equation (VIII.2) gives for the brightness temperature of the line.

$$T_{ab}(\nu) = \int_0^\infty J_a(\nu) + J_b \exp\left[-\int_0^s \left[k_a(\nu) + k_b\right]\mathrm{d}s'\right]\mathrm{d}s$$

while for the continuum,

$$T_b(\nu) = \int_0^\infty J_b \exp\left[-\int_0^s k_b\,\mathrm{d}s'\right]\mathrm{d}s.$$

The difference between both temperatures is called brightness temperature excess ΔT, which is written

$$\Delta T = T_{ab} - T_b.$$

A positive ΔT implies emission, a negative absorption. It can be demonstrated very easily that ΔT in an emission line does not exceed

$$(\Delta T)_{\mathrm{Max}} = \int_0^\infty J_a(\nu)\,\mathrm{d}s.$$

Similarly, the negative brightness temperature excess does not go beyond the value

$$(-\Delta T)_{\mathrm{Max}} = T_b(1 - e^{-\tau_a}) \leqslant T_b\tau_a,$$

where

$$\tau_a(\nu) = \int_0^\infty k_a(\nu)\,\mathrm{d}s$$

is the optical thickness of the medium.

Finally we shall demonstrats that in a uniform medium of constant k and J with optical thickness l, the brightness temperature excess is given by the approximate expression

$$\Delta T = T_k(1 - e^{-\tau}); \qquad \tau = \int_0^l k \, ds$$

provided that $\tau_b \ll 1$ and $T_k \gg T_b$ (Wild, 1952).

In fact, if $\tau_b \ll 1$ we obtain

$$T_b = \int_0^l J_b \, ds = J_b l \quad \text{and} \quad T_{ab} = \frac{J_a + J_b}{k_a} (1 - e^{-\tau_0})$$

hence the brightness temperature excess

$$\Delta T = \frac{J_a + J_b}{k_a} (1 - e^{-\tau_b}) - T_b.$$

Because of Kirchoffs law

$$J_a = k_a T_k,$$

we have at last

$$\Delta T = T_k\left[1 - e^{-\tau_a} + \frac{T_b}{T_k}\left(\frac{1 - e^{-\tau}}{\tau} - 1\right)\right],$$

an expression which reduces to Equation (3) when $T_k \gg T_b$, as we had wanted to demonstrate.

Excitation of the 21-cm Line

The upper hyperfine level (parallel spins) has a statistical weight of 3, the lower level one of 1. The galactic magnetic fields are too weak to produce a Zeeman effect. The probability of spontaneous transition from the upper to the lower level is 2.8×10^{-50} (Wild, 1952; Oort, 1954), corresponding to a mean life in an excited stage of 1.1×10^6 yr. The quantum numbers are 1 for the upper level F and 0 for the lower level F. Transitions from $F:0$ to $F:1$ and vise versa frequently occurs as result of collisions. Pourcel and Dicke have shown that of each 8 collisions 3 produce a change of spin (with distances smaller than 5 Å), which means that any transition from $F:0$ to $F:1$ or $F:1$ to $F:0$ can occur. In more distant collisions atoms behave momentarily like an H_2 molecule and then dissociate into H I atoms.

The frequency of $F:0$ to $F:1$ is found as follows: Let us suppose we have one atom per cm^{-3} with relative velocities of the order of 1 km s^{-1} and that the collision distance is $d = 5$ Å. The time between collisions is

$$t \approx (\pi n \, d^2 v)^{-1} \approx 1.3 \times 10^9 \, s \approx 40 \, yr$$

with 3 out of every 8 collisions effective in the production of spin changes, the intervals between them are of the order of 100 yr. The process is, therefore, more effective than the spontaneous transitions.

The basic expression for relative populations of the $F: 0$ and $F: 1$ states for LTE is

$$\frac{n''}{n'} = 3e^{-h\nu/kT},$$

where n', n'' are the populations at levels $F: 1$ and $F: 0$, respectively, and $V = 1420$ MHz. In this formula we normally use T_s, the spin temperature, for T but due to the great efficiency of the collision processes, we replace it by T_k the collision temperature, which according to Spitzer and Savedoff would be from 50 to 100 K.

The observations in 21 cm suggest, however, 110 K as this seems to be the black-body temperature registered in low latitudes for a saturated line which fills the whole antenna pattern.

Optical Depth and Absorption

For a source which fills the beam of the radio telescope, the observed brightness temperature is given by

$$\Delta T_\nu = T_e(1 - e^{-\tau_\nu})$$

where

$$\tau_\nu = \int_0^l k_\nu \, ds$$

k_ν being the mass-absorption coefficient at the frequency ν. The brightness tempera-ture exceeds at no point the brightness temperature T_e for an infinite optical depth. As a consequence, a saturated 21 cm line would have a flattened profile (Wild, 1952; Heeschen, 1953). The limiting cases are

$$\text{for } \tau_\nu \text{ small} \quad \Delta T \approx \tau_\nu T_e$$
$$\text{for } \tau_\nu \text{ large} \quad \Delta T \approx T_e.$$

For a LTE condition, $T_e = T_k = T_s$ and, because spontaneous emission contributes little or nothing, we have

$$k_\nu = n'\alpha_\nu\left(1 - \frac{B_{21}}{B_{12}}\frac{n''}{n'}\right),$$

where α_ν is the atomic absorption coefficient. Now we have

$$\frac{B_{21}}{B_{12}} = \frac{w_1}{w_2} \quad \text{and} \quad \frac{n''}{n'} = \frac{w_2}{w_1} e^{-h\nu/kT_e}$$

and thus

$$k_\nu = n_1\alpha_\nu(1 - e^{-h\nu/kT_e})$$

which for $h/kT_e \ll 1$ reduces to $k_\nu = (h\nu/kT_e)n'\alpha_\nu$. It is easy to derive the value of α_ν (see Wild, 1952; or Heeschen, 1953) resulting then

$$\tau_\nu = 2.58 \times 10^{-15} \int_0^l \frac{n}{T_e} f(\nu) \, ds,$$

where $n = 4n'$ is the total number of H I cm^{-3}, and $f(\nu)$ is a normalized frequency function depending on the distribution of turbulence velocities along the line of sight.

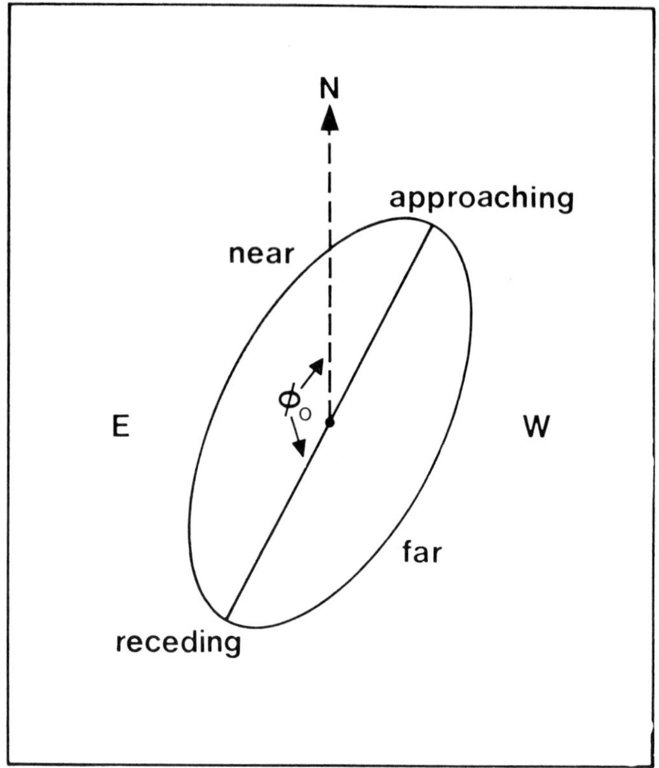

Fig. VIII.1. Relating the coordinates in the plane of the sky with those in the plane of symmetry of the galaxy. (*Adapted from van der Kruit et al.* (1976).)

VIII.4. Determination of Rotation Curves (Cf. Section II.2.1)

Let R, θ be a set of polar coordinates where R is the distance on the plane of symmetry from the center of rotation of the galaxy, and θ the longitude measured from the line of nodes. Let also be δ, θ' another set of polar coordinates on the plane of the sky, with the same origins (Figure VIII.1). Then, if $V(R)$ is the circular velocity at distance R and V_G the systemic radial velocity (that is the radial velocity of the center of rotation), the radial velocity due to rotation is

$$V_R(\delta, \theta') = V_G + V(R) \cos \theta \sin i, \tag{VIII.3}$$

where i is the inclination of the plane of the galaxy (Section II.2.1). Warner *et al.* (1973) assume now a set of values for V_G, i, the coordinates of the origin, and the position angle of the line of nodes, PA. Through the transformations

$$\mathrm{tg}\,\theta = \mathrm{tg}(\theta' - PA) \sin i$$

and

$$R^2 = \delta^2[\cos^2(\theta' - PA) + \sec^2 i \sin^2(\theta - PA)]$$

the coordinates R, θ on the symmetry plane are now computed and carried to Equation

(3) to have $V_R(\delta, \theta')$ on each observed point. If a large set of measurements is available, a least-square solution can be made for those points falling inside a set of discrete rings $R - \Delta R$, $R + \Delta R$ covering the whole galaxy. A simpler procedure is to take averages of $V(R)$ duly weighted along the same rings. Warner *et al.* (1973) used a weighting function $W(R, \theta)$ proportional to the signal-to-noise ratio of its observations at the considered point and also the the modulus $|\cos \theta|$ of the radial component of $V(R)$. They then wrote $V(R) = \sec i \sum W(R, \theta)[(V_R - V_G)/\cos \theta]/ \sum W(R, \theta)$, the summation being carried out on points inside the ring $R + \Delta R$, $R - \Delta R$.

VIII.5. Rotation Period of Central Regions in Galaxies (Cf. Section II.2.1)

In a edge-on, or a very inclined galaxy, one end of the projected images has a radial component of the rotational velocity towards the observer, while the other has the component in the opposite sense. If the slit of the spectrograph is placed along the major axis of the projected image, the spectral lines will be displaced by $\Delta\lambda/\lambda = (V_c/c) \sin i$ relatively to the wavelength corresponding to the center. In the central region the angular velocity is nearly constant, that is $V_c/R \approx cte$. In the focal plane of the spectrograph the linear distance to the center is $y \propto r$ and $\Delta\lambda \propto V_R$, so that $\Delta\lambda = ky$.

The spectral displacement $\Delta\lambda$ is proportional to the distance y to the nucleus in the plate. This means that the spectral line is straight and inclined on the plate. This effect is not observed when the slit is placed along the minor axis of the projected image (unless there were expanding motions out of the nucleus). If the spectral lines used were in absorption the range in y is limited. This situation improves considerably when emission lines from the interstellar medium in the galaxy are used (generally $H\alpha$, [O II] λ 3727). The linearity of the curve of rotation in the central regions of the galaxies provides a simple means for computing rotation periods for that region of galaxies.

Let D be the distance of a galaxy. The apparent radius is given by $r = R/D$, so that we have

$$k = \frac{c}{\lambda} \frac{D}{R} \Delta\lambda$$

and the period of rotation results

$$P = 2\pi R/V_c = 2\pi D/k = 2\pi\lambda R/c \, \Delta\lambda$$

for a given spectral line and spectrograph. The slope of the spectral line in the spectrum will be $S = \Delta\lambda/kr$ and it is easy to see that

$$P = (2\pi\lambda/kc)(D/S) = A(D/S),$$

where A depends only an the line and the spectrograph. If V_c is given in km s^{-1}, r in arc sec and D in Mpc, we have

$$P = 2.97 \times 10^6 \, D(r/V_c) \text{ yr.}$$

Notice P is proportional to D, the scale of distances.

VIII.6. Note for Section V.1.3

In fact, the bivariate correlation found by de Vaucouleurs can be written

$$M^0_{SAS} = -13.9 + 0.67(M^0_T + 18.6) + 2.5[(B - V)^0_T - 0.45] \qquad \text{(VIII.4)}$$

using $L_c = 3$ as zero point. Now, for $L_c \geqslant 3$ the absolute magnitude dependence in Equation (VIII.4) can be eliminated with Equation (VIII.7) of Section V.1.2, and we obtain

$$M^0_{SAS} + \langle Q_3 \rangle_{H\,II} = 19.6 + 2.5[(B - V)^0_T - 0.45] \quad L \geqslant 3. \tag{5}$$

For $L_c \leqslant 3$ a linear relationship

$$(B - V)^0_T \approx 0.45 - 0.11(L_c - 3) \qquad L \leqslant 3 \tag{6}$$

can be deduced from Figure II.4 and the approximate relationship $t = 2L_c$ (de Vaucouleurs, 1977). In a similar fashion we have

$$M^0_T = -18.6 - 1.00(L_c - 3) \qquad L \leqslant 3 \tag{7}$$

from Table V.4. Through elimination of L_c between Equations (VIII.6) and (VIII.7) and from Equation (VIII.7), Section V.1.2. it follows that

$$M^0_{SAS} + \langle Q_3 \rangle_{H\,II} = 19.6 - 0.57[(B - V)^0_T - 0.45] \qquad L \leqslant 3. \tag{8}$$

Equations (5) and (8) together constitute Equation (9), Section V.1.3.

References

Allen, C.W.: 1955, *Astrophysical Quantities*, London.

Arp, H.C.: 1966, *Atlas of Peculiar Galaxies*, Calif. Inst. of Technology.

de Vaucouleurs, G.: 1956, *Comm. Obs.* **31**, No. 13.

de Vaucouleurs, G.: 1977, in B.M. Tinsley and R.B. Larson (eds.), *The Evolution of Galaxies and Stellar Populations*, Yale Univ. Press, p. 43.

de Vaucouleurs, G. and de Vaucouleurs, A.: 1959, *Publ. Astron. Soc. Pacific* **71**, 83.

de Vaucouleurs, G. and de Vaucouleurs, A.: 1964, *Reference Catalogue of Bright Galaxies*, Univ. of Texas Press, Austin.

de Vaucouleurs, G., de Vaucouleurs, A., and Corwin, H.G.: 1976, *Second Reference Catalogue of Bright Galaxies*, Univ. of Texas Press, Austin.

Dreyer, J.L.: 1888, *A New General Catalogue of Nebular and Clusters of Stars*. Mem. Roy. Astron Soc Vo.. 49, Part I.

Heeschen, D.: 1953, *Astron. J.* **58**, 40.

Herschel, J.: 1864, *A New General Catalogue of Nebulae and Clusters of Stars*, Phill. Trans., Part I.

Homberg, E.: 1958, *Fotometría de 300 Galaxies*, Lund. Medd. No. 136.

Minkowski, R.: and Abell, F.O.: 1963, *The National Geographical Society Palomar Observatory Sky Survey. Strand: Basic Astronomical Data (Stars and Stellar System)*, Vol. III, Univ. of Chicago Press.

Nilson, P.: 1973, *Uppsala General Catalogue of Galaxies*.

Sandage, A.R.: 1961, *The Hubble Atlas of Galaxies*, Carnegie Inst. of Washington.

Shapley, H. and Ames, A.: 1932, *Ann. Harv. Coll. Obs.* **88**, 2.

Sérsic, J.L.: 1968, *Atlas de Galaxias Australes*, Univeridad Nacional Córdoba.

Sulentic, J.W. and Tifft, W.G.: 1973, *The Revised New General Catalogue of Nonstellar Astronomical Objects*, Univ. of Arizona Press, Tucson.

Vorontsov-Velyaminov, B.A. and Archipova, V.P.: 1964, *Morphological Catalogues of Galaxies*, Univ. of Moscow, Moscow.
Whipple, F.L.: 1935, Harvard Obs. Circ. No. 404.
Wild, R.: 1952, *Astrophys. J.* **115**, 206.
Zwicky, F. *et al.*: 1961–1968, *Catalogues of Galaxies and of Clusters of Galaxies*, 6 vols, Calif. Inst. of Tech., Pasadena.
Warner, P.J., Wright, M.C.H., and Baldwin, J.E.: 1973, *Monthly Notices Roy. Astron. Soc.* **163**, 163.

GEOPHYSICS AND ASTROPHYSICS MONOGRAPHS

AN INTERNATIONAL SERIES OF FUNDAMENTAL TEXTBOOKS

Editor:

BILLY M. McCORMAC (Lockheed Palo Alto Research Laboratory)

Editorial Board:

R. GRANT ATHAY (High Altitude Observatory, Boulder)
W. S. BROECKER (Lamont-Doherty Geological Observatory, New York)
P. J. COLEMAN, Jr. (University of California, Los Angeles)
G. T. CSANADY (Woods Hole Oceanographic Institution, Mass.)
D. M. HUNTEN (University of Arizona, Tucson)
C. DE JAGER (the Astronomical Institute at Utrecht, Utrecht)
J. KLECZEK (Czechoslovak Academy of Sciences, Ondřejov)
R. LÜST (President Max-Planck-Gesellschaft zur Förderung der Wissenschaften, München)
R. E. MUNN (University of Toronto, Toronto)
Z. ŠVESTKA (The Astronomical Institute at Utrecht, Utrecht)
G. WEILL (Institute d'Astrophysique, Paris)